(*continued on back*)

# APPLIED LINEAR REGRESSION

# Applied Linear Regression

## SECOND EDITION

---

**SANFORD WEISBERG**

*University of Minnesota*
*St. Paul, Minnesota*

**JOHN WILEY & SONS**

New York • Chichester • Brisbane • Toronto • Singapore

*Library of Congress Cataloging in Publication Data :*

Weisberg, Sanford, 1947–
   Applied linear regression.
   (Wiley series in probability and mathematical
statistics. Applied probability and statistics,
ISSN 0271-6356)
   Bibliography: p.
   Includes indexes.
   1. Regression analysis.   I. Title.   II.   Series.
QA278.2.W44    1985      519.5′36      85-5356
ISBN 0-471-87957-6

Printed in the United States of America

10 9 8 7 6 5

*To my parents and to Carol*

# PREFACE

In the five years since the writing of the first edition of this book, research in the area of linear regression has proceeded at a furious pace. Many new procedures have been suggested that should modify the way practitioners carry out regression analysis. This revision is designed to include many of these new ideas, as reflected by the 60 or so post-1980 literature citations given in the references.

Most of the text has been rewritten, to clarify existing discussions and introduce new ideas. Several new homework problems, worked examples, and figures have been added. A major addition is a new concluding chapter that provides an introduction to nonlinear, logistic, and generalized linear regression models. However, I have attempted to retain the brevity of the first edition by balancing the introduction of new topics with the deletion of older, less important material.

As with the first edition, I am grateful to many of my colleagues and friends for encouragement on this project. Michael Lavine read the final manuscript and corrected many errors that might otherwise have remained. Stephen Stigler has been especially helpful with many comments on the first edition that led to changes in the second edition. I am also indebted to Dennis Cook for our continuing dialogue on regression modeling.

SANFORD WEISBERG

*St. Paul, Minnesota*
*April 1985*

# PREFACE TO THE FIRST EDITION

Linear regression analysis consists of a collection of techniques used to explore relationships between variables. It is interesting both theoretically because of the elegance of the underlying theory, and from an applied point of view, because of the wide variety of uses of regression that have appeared, and continue to appear every day. In this book, regression methods, used to fit models for a dependent variable as a function of one or more independent variables, are discussed for the reader who wants to learn to apply them to data. The central themes are building models, assessing fit and reliability, and drawing conclusions. If used as a textbook, it is intended as a second or third course in statistics. The only definite prerequisites are familiarity with the ideas of significance tests, $p$-values, confidence intervals, random variables, estimation of parameters, and also with the normal distribution, and distributions derived from it, such as Student's $t$, and the $F$, and $\chi^2$. Of course, additional knowledge of statistical methods or linear algebra will be of value.

The book is divided into 12 chapters. Chapters 1 and 2 provide fairly standard results for least squares estimation in simple and multiple regression, respectively. The third chapter is called "Drawing Conclusions" and is about interpreting the results of the methods from the first two chapters. Also, a discussion of the effects of independent variables that are imperfectly measured is given. Chapter 4 presents additional results on least squares estimation. Chapters 5 and 6 cover methods for studying the lack of fit of a model, checking for failures of assumptions, and assessing the reliability of a fitted model. In Chapter 5, theoretical results for the necessary statistics are given, since these will be unfamiliar to many readers, while Chapter 6 covers graphical and other procedures based on these statistics, as well as possible remedies for the problems they uncover. In Chapter 7, the topics covered are relevant to problems of model building, including dummy variables, polynomial regression, and principal components. Then, Chapter 8 provides methods for selecting a model based on a subset of variables.

In Chapter 9, special considerations when regression methods are to be used to make predictions are discussed. In each of these chapters, the methods discussed are illustrated by examples using real data.

The next two chapters are shorter than the earlier ones. Chapter 10 gives guidelines for analysis of partially observed or incomplete data. Finally, in Chapter 11, alternatives to least squares estimates are discussed.

Several of the chapters have associated appendixes that have been collected at the end of the text, but are numbered to correspond to the chapters. For example: Appendix 1A.2 is the second appendix for Chapter 1. The chapters are ordered for a semester or quarter course on linear regression, and Chapters 1 to 8 make up a rigorous one-quarter course.

Homework problems are provided for each of the first nine chapters. The theoretical problems are intended only for students with the necessary statistical background. Problems that require analysis of data are intended for everyone. Some of these have been left vague in their requirements, so that they can be varied according to the interests of the students. Most of the problems use real data and can be approached in many ways.

*Computers.*   The growth of the use of regression methods can be traced directly to wider availability of computers. While this book is not intended as a manual for any specific computer program, it is oriented for the reader who expects to use computers to apply the techniques learned. High quality software for regression calculations is available, and references to the necessary sources are in the text, in the homework problems, and in the appendixes.

*Acknowledgments.*   I am grateful to the many people who have commented on early drafts of the book, supplied examples, or through discussion have clarified my own thoughts on the topics covered. Included in this group are Christopher Bingham, Morton Brown, Cathy Campbell, Dennis Cook, Stephen Fienberg, James Frane, Seymour Geisser, John Hartigan, David Hinkley, Alan Izenman, Soren Johansen, Kenneth Koehler, David Lane, Kinley Larntz, John Rice, Donald Rubin, Wei-Chung Shih, G. W. Stewart, Douglas Tiffany, Carol Weisberg, Howard Weisberg, and an anonymous reader. Also, I wish to thank the production staff at the University of Minnesota, Naomi Miner, Sue Hangge, Therese Therrien, and especially Marianne O'Brien, whose expert assistance made completion of this work a reality.

During the writing of this book, I have benefited from partial support from a grant from the U.S. National Institute of General Medical Sciences. Additional support for computations has been provided by the University Computer Center, University of Minnesota.

Sanford Weisberg

*St. Paul, Minnesota*
*February 1980*

# CONTENTS

# APPLIED LINEAR REGRESSION

# 1

# SIMPLE LINEAR
# REGRESSION

Regression is used to study relationships between measurable variables. Linear regression is used for a special class of relationships, namely, those that can be described by straight lines, or by generalizations of straight lines to many dimensions. These techniques are applied in almost every field of study, including social sciences, physical and biological sciences, business and technology, and the humanities. As illustrated by the examples in this book, the reasons for fitting linear regression models are as varied as are the applications, but the most common reasons are description of a relationship and prediction of future values.

Generally, regression analysis consists of many steps. To study a relationship between a number of variables, data are collected on each of a number of units or cases on these variables. In the regression models studied here, one variable takes on the special role of a response variable, while all of the others are viewed as predictors of the response. It is often convenient, and sometimes accurate, to view the predictor variables as having values set by the data collector, while the value of the response is a function of those variables. A hypothesized model specifies, except for a number of unknown parameters, the behavior of the response for given values of the predictors. The model generally will also specify some of the characteristics of the failure to provide exact fit through hypothesized error terms. Then, the data are used to obtain estimates of the unknown parameters. The method of estimation studied in most of this book is *least squares*, although there are in fact many estimation procedures. The analysis to this point is called *aggregate analysis*, since the main purpose is to combine the data into aggregates and

**1**

summarize the fit of a model to the data. The next, and equally important, phase of a regression analysis is called *case analysis*, in which the data are used to examine the suitability and usefulness of the fitted model for the relationship studied. The results of case analysis may lead to modification of the original prescription for a fitted model, and cycling back to the aggregate analysis after modifying the data or assumptions.

The topic of this chapter is simple regression, in which there is a single response and a single predictor. Of interest will be the specification of an appropriate model, discussion of assumptions, the least squares estimates, and testing and confidence interval procedures.

### *Example 1.1    Forbes' data*

---

In the 1840s and 1850s a Scottish physicist, James D. Forbes, wanted to be able to estimate altitude above sea level from measurement of the boiling point of water. He knew that altitude could be determined from atmospheric pressure, measured with a barometer, with lower pressures corresponding to higher altitudes. In the experiments discussed here, he studied the relationship between pressure and boiling point. His interest in this problem was motivated by the difficulty in transporting the fragile barometers of the 1840s. Measuring the boiling point would give travelers a quick way of estimating altitudes.

Forbes collected data in the Alps and in Scotland. After choosing a location, he assembled his apparatus, and measured pressure and boiling point. Pressure measurements were recorded in inches of mercury, adjusted for the difference between the ambient air temperature when he took the measurements and a standard temperature. Boiling point was measured in degrees Fahrenheit. The data for $n = 17$ locales are reproduced from an 1857 paper in Table 1.1 (Forbes, 1857). On reviewing the data, there are several questions of potential interest. How are pressure and boiling point related? Is the relationship strong or weak? Can we predict pressure from temperature, and if so, how well?

Forbes' theory suggested that over the range of observed values the graph of boiling point versus the *logarithm* of pressure yields a straight line. Following Forbes, we take logs to the base 10, although the base of the logarithms is irrelevant for the statistical analysis. Since the logs of the pressures do not vary much, with the smallest being 1.318 and

**Table 1.1   Forbes' data giving boiling point (°F) and barometric pressure (inches of mercury) for 17 locations in the Alps and in Scotland**

| Case Number | Boiling Point (°F) | Pressure (in. Hg) | Log(Pressure) | 100 × Log(Pressure) |
|---|---|---|---|---|
| 1 | 194.5 | 20.79 | 1.3179 | 131.79 |
| 2 | 194.3 | 20.79 | 1.3179 | 131.79 |
| 3 | 197.9 | 22.40 | 1.3502 | 135.02 |
| 4 | 198.4 | 22.67 | 1.3555 | 135.55 |
| 5 | 199.4 | 23.15 | 1.3646 | 136.46 |
| 6 | 199.9 | 23.35 | 1.3683 | 136.83 |
| 7 | 200.9 | 23.89 | 1.3782 | 137.82 |
| 8 | 201.1 | 23.99 | 1.3800 | 138.00 |
| 9 | 201.4 | 24.02 | 1.3806 | 138.06 |
| 10 | 201.3 | 24.01 | 1.3805 | 138.05 |
| 11 | 203.6 | 25.14 | 1.4004 | 140.04 |
| 12 | 204.6 | 26.57 | 1.4244 | 142.44 |
| 13 | 209.5 | 28.49 | 1.4547 | 145.47 |
| 14 | 208.6 | 27.76 | 1.4434 | 144.34 |
| 15 | 210.7 | 29.04 | 1.4630 | 146.30 |
| 16 | 211.9 | 29.88 | 1.4754 | 147.54 |
| 17 | 212.2 | 30.06 | 1.4780 | 147.80 |

the largest being 1.478, we shall multiply all the values of log(pressure) by 100, as given in column 5 of Table 1.1. This will avoid studying very small numbers, without changing the major features of the analysis.

A useful way to begin a regression analysis is by drawing a graph of one variable versus the other. This graph, called a _scatter plot,_ can serve both to suggest a relationship, and to demonstrate possible inadequacies of it. Scatter plots can be drawn by hand on ordinary graph paper. The x or horizontal axis is usually reserved for the variable that is to be the predictor or independent variable. In Forbes' data this is the boiling point. The y or vertical axis is usually for the quantity to be modeled or predicted, often called the response or the dependent variable. In the example, the values for the y axis are 100 × log(pressure). For each of the n pairs (x, y) of values in the data, a point is plotted on the graph. Most computer programs for regression analysis will produce this plot.

The overall impression of the scatter plot for Forbes' data in Figure 1.1 is that the points generally, but not exactly, fall on a straight line;

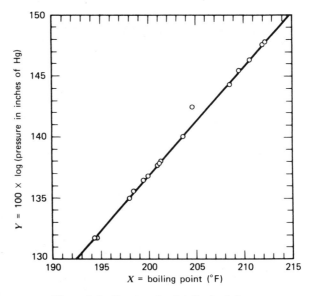

**Figure 1.1**   Scatter plot for Forbes' data.

the line drawn on Figure 1.1 will be discussed later. This suggests that the relationship between the two variables may be described at least as a first approximation by specifying an equation for a straight line.

As we progress through this chapter, the methods studied will be applied to these data.

## 1.1   Building a simple regression model

In simple regression, the relationship between two quantities, say $X$ and $Y$, is studied. First, we hope that the relationship can be described by a straight line. For this to be reasonable, we may need to transform the scales of the quantities $X$ and/or $Y$, as was done in Forbes' data, where pressure was transformed to log(pressure). In this chapter, observed values of the quantities $X$ and $Y$ are denoted by subscripted lowercase letters: $(x_i, y_i)$ are the observations on $X$ and $Y$ for the $i$th case in the study. The major features of the simple regression model are given here. A more formal approach is given in Appendix 1A.1.

**Equation of a straight line.**   A straight line relating two quantities $Y$ and $X$ can be described by the equation

$$Y = \beta_0 + \beta_1 X \tag{1.1}$$

In equation (1.1), $\beta_0$ is the _intercept_, the value of $Y$ when $X$ equals zero. The slope, $\beta_1$, is the rate of change in $Y$ for a unit change in $X$; see Figure 1.2. The numbers $\beta_0$ and $\beta_1$ are called _parameters_ and, since they range over all possible values, they give all possible straight lines. In most applications of statistical modeling, parameters are unknown and they must be estimated using data.

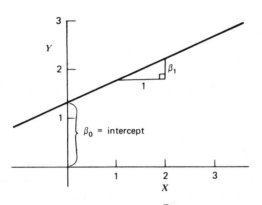

**Figure 1.2**   A straight line.

**Errors.**   Real data will almost never fall exactly on a straight line. The differences between the values of the response obtained and the values given by the model (for simple regression, the observed values of $Y$ minus $(\beta_0 + \beta_1 X)$) are called statistical _errors_. This term should not be confused with its synonym in common usage, "mistake." Statistical errors are devices that account for the failure of a model to provide an exact fit. These errors can have both fixed and random components. A fixed component of a statistical error will arise if the proposed model, here a straight line, is not exactly correct. For example, suppose the true relationship between $Y$ and $X$ is given by the solid curve in Figure 1.3, and suppose that we incorrectly propose a straight line, shown as a dashed line, for this relationship. By modeling the relationship with a straight line rather than the appropriate curve, a fixed error, sometimes called the lack of fit error, is the vertical distance between the straight line and the correct curve. For the standard linear regression theory of this chapter, we assume that the lack of fit component to the errors is negligible, although in practice, this assumption requires checking. Often, transformations are used to make lack of fit errors smaller.

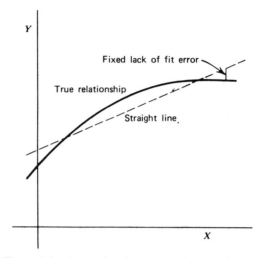

**Figure 1.3**   Approximating a curve by a straight line.

The random component of the errors can arise from several sources. Measurement errors, for now only in $Y$, not $X$, are almost always present, since few quantitative variables can be measured with perfect accuracy. The effects of variables not explicitly included in the model can contribute to the errors. For example, in Forbes' experiments wind speed may have had small effects on the atmospheric pressure, contributing to the variability in the observed values. Also, random errors due to natural variability occur.

Let $e_i$ be the value of the statistical error for the $i$th case, $i = 1, 2, \ldots, n$. Assuming that the fixed component of the errors is negligible, the $e_i$ have zero mean, $E(e_i) = 0$, $i = 1, 2, \ldots, n$. (See Appendix 1A.2 if the symbols $E(\ )$, var$(\ )$, and cov$(\ ,\ )$ are unfamiliar.) An additional convenient assumption is that the errors are mutually uncorrelated (written in terms of the covariance operator, as cov$(e_i, e_j) = 0$, for all $i \neq j$), and have common, though generally unknown, variance var$(e_i) = \sigma^2$, $i = 1, 2, \ldots, n$. Heuristically, uncorrelated means that the value of one of the errors does not depend on or help determine the value of any other error. Little generality is lost if the word *independent* is substituted for *uncorrelated*. For some applications, distributional assumptions concerning the errors are needed also. The usual assumption, and the one that leads most naturally to the use of least squares estimates, is that the errors are normally distributed. In this book, the normality assumption is used primarily to obtain tests and confidence statements. If the errors are thought to follow some different distribution, such as the Poisson or the Gamma, other methods besides least squares may be more appropriate; see Chapter 12 for more discussion.

Making the normality assumption, along with the assumptions concerning means, variances, and covariances just given, we can write

$$e_i \sim \text{NID}(0, \sigma^2) \qquad i = 1, 2, \ldots, n$$

which is read as the $e_i$ are normally and independently distributed with zero mean and common variance $\sigma^2$. In any practical problem, the assumptions made in this section must be examined. This topic is discussed in later chapters.

**The simple regression model.** We have already defined $X$ and $Y$ to be the predictor and the response, respectively, with observed values $(x_i, y_i)$ of $X$ and $Y$ for $i = 1, 2, \ldots, n$. For Forbes' data, the $x_i$'s are given in the second column of Table 1.1 and the $y_i$'s are given in the fifth column. For example, $x_3 = 197.9$ and $y_{12} = 142.44$. We again define $e_i$ to be the statistical error for the $i$th case, $i = 1, 2, \ldots, n$. The simple linear regression model specifies the following:

$$y_i = \beta_0 + \beta_1 x_i + e_i \qquad i = 1, 2, \ldots, n \qquad (1.2)$$

$$\text{with } E(e_i) = 0$$

$$\text{var}(e_i) = \sigma^2$$

$$\text{cov}(e_i, e_j) = 0 \qquad i \neq j$$

In words, the model says that the observed value $y_i$ can be determined from the value of $x_i$ through the specified equation, except that $e_i$, an unknown random quantity, is added on. The three quantities $\beta_0$, $\beta_1$, and $\sigma^2$ are unknown. The $e_i$'s are unobservable quantities introduced into the model to account for the failure of the observed values to fall exactly on a single straight line. Only the $x_i$'s and the $y_i$'s are observed and these data are used to obtain estimates of the unknown parameters, namely, $\beta_0$, $\beta_1$, and $\sigma^2$.

## 1.2   Least squares estimation

Many methods have been suggested for obtaining estimates of parameters in a model. The one discussed here is called _least squares_, in which parameter estimates are chosen to minimize a quantity called the residual sum of squares.

**Notation.** The distinction between parameters and estimates of parameters (statistics) is critical to the use and understanding of statistical models. To keep this distinction clear, parameters are denoted by lowercase Greek letters, usually $\alpha$, $\beta$, $\gamma$, and $\sigma$, and estimates of parameters denoted by putting a "hat" over the corresponding Greek letter; thus, for example, $\hat{\beta}_1$ (read "beta one hat") is the estimator of $\beta_1$. Similarly, $\hat{\sigma}^2$ is the estimator of $\sigma^2$. Although the $e_i$'s are not parameters in the usual sense, we shall use the same hat notation to specify the observed fitting errors or *residuals*: the residual for the $i$th case, denoted by $\hat{e}_i$, is given by the equation

$$\hat{e}_i = y_i - (\hat{\beta}_0 + \hat{\beta}_1 x_i) \qquad i = 1, 2, \ldots, n \tag{1.3}$$

which should be compared to the equation for the statistical errors,

$$e_i = y_i - (\beta_0 + \beta_1 x_i) \qquad i = 1, 2, \ldots, n \tag{1.4}$$

△ The difference between the $e_i$'s and the $\hat{e}_i$'s is important, as the residuals are observable and will be used to check assumptions, while the statistical errors are not observable.

An additional extension of the hat notation is used to identify *fitted values* determined by the estimated regression equation. Thus the $i$th fitted value $\hat{y}_i$ is given by the equation

$$\hat{y}_i = \hat{\beta}_0 + \hat{\beta}_1 x_i \qquad i = 1, 2, \ldots, n \tag{1.5}$$

By comparing (1.5) to (1.3), we see that $\hat{e}_i = y_i - \hat{y}_i$.

All least squares computations for simple regression can be done using only a few summary statistics computed from the data, namely, the sample averages, and corrected sums of squares and corrected cross products. For reference, the definitions of all of these quantities are given in Table 1.2. They have been defined by subtracting the sample average from each of the values before squaring or taking cross products. Appropriate alternative formulas for computing the corrected sums of squares and cross products from uncorrected sums of squares and cross products are also given in the table. Using the uncorrected sums of squares is convenient for computing on a calculator, since many have facilities for accumulating both $\sum x_i$ and $\sum x_i^2$ in a single pass through the data. A formula such as $SXX = \sum x_i^2 - (\sum x_i)^2/n$ can be used to obtain the corrected sum of squares from the uncorrected. However, if computations are to be done on a computer, using the uncorrected sums of squares may lead to severe round-off errors, as illustrated in Appendix 1A.5.

Table 1.2 also lists definitions for the "usual" univariate and bivariate summary statistics, namely, the sample averages ($\bar{x}$, $\bar{y}$), sample variances ($SD_X^2$, $SD_Y^2$), and estimated covariance and correlation ($s_{XY}$, $r_{XY}$). The "hat" rule described earlier would suggest that different symbols should be used for these quantities; for example, $\hat{\rho}_{XY}$ might be more appropriate for the sample correlation if the population correlation is $\rho_{XY}$. However, this inconsistency is deliberate, because in many regression situations these statistics are not estimates of population parameters. For example, in

**Table 1.2  Definitions of symbols**[a]

| Quantity | Definition and Alternative Forms | Description |
|---|---|---|
| $\bar{x}$ | $\sum x_i/n$ | Sample average for the $x_i$'s |
| $\bar{y}$ | $\sum y_i/n$ | Sample average for the $y_i$'s |
| $SXX$ | $\sum (x_i - \bar{x})^2 = \sum x_i^2 - (\sum x_i)^2/n$ $= \sum x_i^2 - n(\bar{x})^2$ | Corrected sum of squares for the $x_i$'s |
| $SD_X^2$ | $SXX/(n-1)$ | Sample variance of the $x_i$'s |
| $SD_X$ | $\sqrt{SXX/(n-1)}$ | Sample standard deviation |
| $SYY$ | $\sum (y_i - \bar{y})^2 = \sum y_i^2 - (\sum y_i)^2/n$ $= \sum y_i^2 - n(\bar{y})^2$ | Corrected sum of squares for the $y_i$'s; also called the total sum of squares |
| $SD_Y^2$ | $SYY/(n-1)$ | Sample variance of the $y_i$'s |
| $SD_Y$ | $\sqrt{SYY/(n-1)}$ | Sample standard deviation |
| $SXY$ | $\sum (x_i - \bar{x})(y_i - \bar{y})$ $= \sum x_i y_i - (\sum x_i)(\sum y_i)/n$ $= \sum x_i y_i - n\bar{x}\bar{y}$ | Corrected sum of cross products |
| $s_{XY}$ | $SXY/(n-1)$ | Sample covariance |
| $r_{XY}$ | $SXY/\sqrt{(SXX)(SYY)} = s_{XY}/s_X s_Y$ | Sample correlation |

[a]The symbol $\sum$ is shorthand for $\sum_{i=1}^{n}$, which means "add for all values of $i$ between 1 and $n$."

Forbes' experiments, data were collected at 17 selected locations. As a result, the sample variance of boiling points, which can be shown to be $SD_X^2 = 33.17$, cannot be expected to be an estimate of any meaningful population variance. Similarly, $r_{XY}$ depends as much on the method of sampling as it does on the population value $\rho_{XY}$, should such a population value make sense.

However, these usual sample statistics are often presented and used in place of the corrected sums of squares and cross products, so alternative formulas are given using both sets of quantities.

**Least squares criterion.**  The criterion function for obtaining estimators is based on the fitting errors or residuals, $\hat{e}_i = y_i - \hat{y}_i$, where $\hat{y}_i = \hat{\beta}_0 + \hat{\beta}_1 x_i$ is the value on the fitted line at $x = x_i$. The residuals give the vertical distances between the fitted line and the actual $y$-values, as illustrated in Figure 1.4. Clearly, one could consider other functions of the data besides vertical errors to derive a criterion for choosing an estimator, but the residuals are a good choice because they reflect the inherent asymmetry in the roles of the response and the predictor in regression problems.

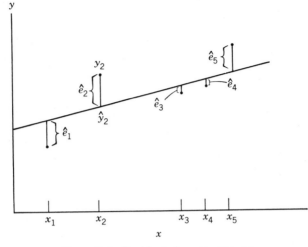

**Figure 1.4** Residuals from the fitted line.

The least squares estimators are those values $\hat{\beta}_0$ of $\beta_0$ and $\hat{\beta}_1$ of $\beta_1$ that minimize the function‡

$$RSS(\beta_0, \beta_1) = \sum [y_i - (\beta_0 + \beta_1 x_i)]^2 \qquad (1.6)$$

When evaluated at $\hat{\beta}_0$, $\hat{\beta}_1$, we call the quantity $RSS(\hat{\beta}_0, \hat{\beta}_1)$ the _residual sum of squares,_ or just _RSS._

Least squares is a purely mathematical formulation that does not depend on any assumptions concerning the $e_i$'s. Least squares estimates can be computed even if the regression model is inappropriate for the data studied.

The least squares estimates can be derived in many ways, one of which is outlined in Appendix 1A.3. They are given by the expressions

$$\hat{\beta}_1 = \frac{SXY}{SXX} = r_{XY} \frac{SD_Y}{SD_X} = r_{XY} \left(\frac{SYY}{SXX}\right)^{1/2}$$

$$\hat{\beta}_0 = \bar{y} - \hat{\beta}_1 \bar{x} \qquad (1.7)$$

The several forms for $\hat{\beta}_1$ are all equivalent.

Sometimes it is convenient to write the simple linear regression model in a different form that is a little easier to manipulate. Taking equation (1.2), and adding $\beta_1 \bar{x} - \beta_1 \bar{x}$, which equals zero, to the right-hand side, and com-

‡Here and elsewhere in this book, we abuse notation by using the symbol for a fixed though unknown quantity like $\beta_1$ as if it were a variable argument. Thus, for example, $RSS(\beta_0, \beta_1)$ is a function of two variables to be evaluated as its arguments $\beta_0$ and $\beta_1$ vary. The same abuse of notation is used in the discussion of confidence intervals.

bining terms, we can write

$$y_i = \beta_0 + \beta_1 \bar{x} + \beta_1 x_i - \beta_1 \bar{x} + e_i$$
$$= (\beta_0 + \beta_1 \bar{x}) + \beta_1 (x_i - \bar{x}) + e_i$$

Define $\alpha = \beta_0 + \beta_1 \bar{x}$ ($\alpha$ does not depend on $i$), and rewrite the last equation in the equivalent form,

$$y_i = \alpha + \beta_1 (x_i - \bar{x}) + e_i \qquad i = 1, 2, \ldots, n \qquad (1.8)$$

This is called the *deviations from the sample average* form for simple regression. The least squares estimates are

$$\hat{\alpha} = \bar{y} \qquad \hat{\beta}_1 \text{ as given by (1.7)} \qquad (1.9)$$

**Forbes' data.**   The four quantities $SXX$, $SXY$, $\bar{x}$, and $\bar{y}$ are needed for computing the least squares estimators. These are

$$\bar{x} = 202.95294118 \qquad SXX = 530.78235294 \qquad SXY = 475.29570589$$
$$\bar{y} = 139.60588235 \qquad SYY = 427.76281177 \qquad (1.10)$$

The quantity $SYY$, although not yet needed, is given for completeness. Also the number of digits given in each of these computations is excessive, since there are at most four significant digits in the original data (the logarithms of the pressures were rounded to the digits shown in Table 1.1 to enable the interested reader to reproduce the preceding calculations). However, since these are intermediate calculations, they should be done as accurately as possible, and rounding should be done only to final results. Using the calculations given, we find

$$\hat{\beta}_1 = \frac{SXY}{SXX} = 0.895$$
$$\hat{\beta}_0 = \bar{y} - \hat{\beta}_1 \bar{x} = -42.131.$$

The reason for multiplying the log(pressure) by 100 may now be evident. If this had not been done, then the estimate of $\beta_1$ would have been 100 times smaller, that is, 0.00895, and such small numbers often lead to mistakes. In the deviations from the average form of the simple regression model, the estimate of the slope is as given earlier, and the estimate of $\alpha$ is $\hat{a} = \bar{y} = 139.606$.

The estimated line, given by either of the equations

$$\hat{y} = -42.131 + 0.895x$$
$$= 139.606 + 0.895(x - 202.953)$$

has been drawn in Figure 1.1. As previously noted, the fit of this line to the data is excellent.

## 1.3    Estimating $\sigma^2$

Ideally, an estimate of $\sigma^2$ that does not depend on the appropriateness of the fitted model is desirable. Generally, such an estimate is obtainable only in data sets with several values of $y$ recorded at each of several values of $x$ or from prior information, as described in Sections 4.2 and 4.3. Lacking these special situations, the usual estimate of $\sigma^2$ is model dependent, as it is a function of the residual sum of squares, $RSS = \sum \hat{e}_i^2$.

Since $\sigma^2$ is essentially the average squared size of the $e_i$'s, we should expect that its estimator, to be called $\hat{\sigma}^2$, is obtained by averaging the $\hat{e}_i^2$. Under the assumption that the $e_i$'s are uncorrelated random variables with zero means and common variance $\sigma^2$, one can show that an unbiased estimate of $\sigma^2$ is obtained by dividing $RSS$ by its degrees of freedom (d.f.), where

residual d.f. = number of cases − number of parameters in model

For simple regression, residual d.f. $= n - 2$, so the estimate of $\sigma^2$ is given by

$$\hat{\sigma}^2 = \frac{RSS}{n-2} \tag{1.11}$$

This quantity is called the residual mean square. In general, any sum of squares divided by its degrees of freedom is called a mean square.

To compute $\hat{\sigma}^2$, it is left to the reader (Exercise 1.6) to show that

$$RSS = SYY - \frac{(SXY)^2}{SXX} = SYY - \hat{\beta}_1^2 SXX \tag{1.12}$$

For Forbes' data,

$$RSS = 427.76281177 - \frac{(475.29570589)^2}{530.78235294}$$

$$= 2.15332 \tag{1.13}$$

(or 2.153 to four digits) and

$$\hat{\sigma}^2 = \frac{2.15332}{17-2} = 0.14355$$

(or 0.144 to three digits—the more accurate figure will be needed later). The square root of this quantity, $\hat{\sigma} = \sqrt{0.144} = 0.379$ is often called the standard error of regression. It is in the same units as is the variable $Y$; for Forbes' data, the units are $100 \times \log(\text{pressure})$.

If, in addition to the assumptions made previously, the $e_i$'s are drawn from

see note

Writing calls $\hat{\sigma}$ standard error of regression

Netor, Wasserman and Kutner call $\hat{\sigma}$ an estimator of the standard deviation of regression (p.47)

$\hat{\sigma}^2$ is mean square error $\sigma$  $RSS/d.f$ error

$\hat{\sigma}$ is called root mean square error  $\sigma$ standard deviation

I think Writing is incorrect!

standard error formula maclay p.22, p 47 reflects particular fitted value a measurement, St. dev. reflects all values a estimator

a normal distribution, then the residual mean square will be distributed as a multiple of a chi-squared random variable with $n-2$ degrees of freedom, or symbolically,

$$(n-2)\frac{\hat{\sigma}^2}{\sigma^2} \sim \chi^2(n-2)$$

This fact, proved in more advanced books on linear models, is used to obtain the distribution of test statistics, and also to make confidence statements concerning $\sigma^2$. In particular, this fact implies that

$$E(\hat{\sigma}^2) = \sigma^2$$

although normality is not required for unbiasedness.

## 1.4   Properties of least squares estimates

The least squares estimates depend on data only through the aggregate statistics given in Table 1.2. This is both an advantage, making computing easy, and a disadvantage, since any two data sets for which these are identical give the same fitted regression, even if a straight line model is appropriate for one but not the other (as in Example 5.1 in Chapter 5). Also, the estimates $\hat{\beta}_0$ and $\hat{\beta}_1$ are linear in $y_1, \ldots, y_n$. From (1.8), the fitted value at $x=\bar{x}$ is $\hat{y} = \bar{y} + \hat{\beta}(\bar{x}-\bar{x}) = \bar{y}$, so the fitted line must pass through the point $(\bar{x}, \bar{y})$, intuitively the center of the data. Finally, it is easy to show that $\sum \hat{e}_i = 0$, so negative and positive residuals balance out. Actually, $\sum \hat{e}_i = 0$ is a consequence of fitting an intercept term $\beta_0$, and models in which no intercept term appears will usually have $\sum \hat{e}_i \neq 0$.

Now, if the $e_i$'s are random variables, then so are the estimates of $\beta_0$ and $\beta_1$, since these depend on the $y_i$'s and hence on the $e_i$'s. If all the $e_i$'s have zero mean, $E(e_i) = 0$, $i = 1, \ldots, n$ and the model is correct, then, as shown in Appendix 1A.4, the least squares estimates are unbiased,

$$E(\hat{\beta}_0) = \beta_0$$
$$E(\hat{\beta}_1) = \beta_1$$

For the variance of the estimators, we now consider only the special case of $\text{var}(e_i) = \sigma^2$, $(i = 1, \ldots, n)$ and $\text{cov}(e_i, e_j) = 0$, $i \neq j$. Then, from Appendix 1A.4,

$$\text{var}(\hat{\beta}_1) = \sigma^2 \frac{1}{SXX}$$

$$\text{var}(\hat{\beta}_0) = \sigma^2 \left( \frac{1}{n} + \frac{\bar{x}^2}{SXX} \right) \qquad (1.14)$$

In the somewhat simpler deviations from the sample average model, the variance of $\hat{\alpha}$ is given by

$$\text{var}(\hat{\alpha}) = \frac{\sigma^2}{n} \qquad (1.15)$$

One important difference between the parameterization of the model (1.2) and that of (1.8) is the fact that the estimates $\hat{\beta}_0$ and $\hat{\beta}_1$ are correlated, as outlined in Appendix 1A.4,

$$\text{cov}(\hat{\beta}_0, \hat{\beta}_1) = -\sigma^2 \frac{\bar{x}}{SXX} \qquad (1.16)$$

but the estimates $\hat{\beta}_1$ and $\hat{\alpha}$ are uncorrelated, a fact that makes other computations, such as that of the variance of a prediction, relatively simple.

With the assumptions that the $e_i$ are uncorrelated random variables with common variance, we can apply the Gauss–Markov theorem: under these conditions, the least squares estimates, which are linear functions of the $y_i$'s, have the smallest possible variance of any linear unbiased estimators. This means that, if one believes the assumptions, and is interested in using linear unbiased estimates, the least squares estimates are the ones to use.

When the errors are normally distributed, the least squares estimates can be justified using a completely different argument, since they are then also *maximum likelihood estimates*. An introduction to maximum likelihood estimation is given by Lindgren (1976).

Normality is also useful in finding the distributions of the estimates $\hat{\beta}_0$ and $\hat{\beta}_1$. Under the assumption that

$$e_i \sim \text{NID}(0, \sigma^2) \qquad i = 1, \ldots, n$$

then $\hat{\beta}_0$ and $\hat{\beta}_1$ are also normally distributed, since they are linear functions of the $y_i$'s and hence of the $e_i$'s, with variances and covariances given by (1.14) and (1.16). These results are used in obtaining confidence intervals.

**Estimated variances.**    Estimates of $\widehat{\text{var}}(\hat{\beta}_0)$ and $\widehat{\text{var}}(\hat{\beta}_1)$ are obtained by substituting $\hat{\sigma}^2$ for $\sigma^2$ in (1.14). We use the symbol $\widehat{\text{var}}(\ )$ to mean estimated variances. Thus

$$\widehat{\text{var}}(\hat{\beta}_0) = \hat{\sigma}^2 \left( \frac{1}{n} + \frac{\bar{x}^2}{SXX} \right)$$

$$\widehat{\text{var}}(\hat{\beta}_1) = \hat{\sigma}^2 \frac{1}{SXX} \qquad (1.17)$$

The square root of an estimated variance is called a *standard error*, for which we use the symbol se( ). The use of this notation is illustrated by

$$se(\hat{\beta}_1) = \sqrt{\widehat{var}(\hat{\beta}_1)}$$

## 1.5 Comparing models: The analysis of variance

The analysis of variance provides a convenient method of comparing the fit of two or more models to the same set of data. The methodology developed here is very useful in multiple regression, although all of the important principles can now be illustrated.

An elementary alternative to the simple regression model suggests fitting the equation

$$y_i = \beta_0 + e_i \qquad i = 1, 2, \ldots, n \tag{1.18}$$

This model asserts that the $y_i$'s depend on a single parameter $\beta_0$ plus random variation, but not on $x_i$. Fitting this model is equivalent to finding the best line parallel to the $x$ axis, as shown in Figure 1.5. The least squares line is $y = \hat{\beta}_0$, where $\hat{\beta}_0$ is chosen to be the value of $\beta_0$ that minimizes $\sum (y_i - \beta_0)^2$. It is easy to show that for this model

$$\hat{\beta}_0 = \bar{y} \tag{1.19}$$

The residual sum of squares is

$$\sum (y_i - \hat{\beta}_0)^2 = \sum (y_i - \bar{y})^2 = SYY \tag{1.20}$$

This residual sum of squares has $n-1$ degrees of freedom ($n$ cases minus 1 parameter in the model).

Next, consider the simple regression model obtained from (1.18) by adding a term that depends on $x_i$,

$$y_i = \beta_0 + \beta_1 x_i + e_i \qquad i = 1, 2, \ldots, n \tag{1.21}$$

Fitting this model is equivalent to finding the best line of arbitrary slope, as shown in Figure 1.5. The least squares estimates for this model are given by (1.7). As an important aside, we see that the estimates of $\beta_0$ under the two models are different, just as the meanings of the parameters in the two models are different. For (1.18), $\hat{\beta}_0$ is the average of the $y$'s, but for (1.21), $\hat{\beta}_0$ is the average when $x_i = 0$.

For (1.21), the residual sum of squares, given in (1.12), is

$$RSS = SYY - \frac{(SXY)^2}{SXX} \tag{1.22}$$

As mentioned earlier, $RSS$ has $n-2$ degrees of freedom.

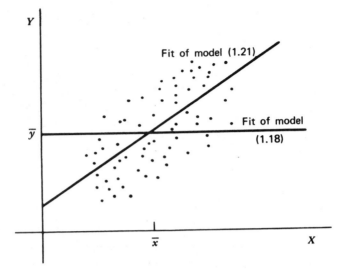

**Figure 1.5**   Two models compared by the analysis of variance.

The difference between the sum of squares at (1.20) and that at (1.22) is the reduction in residual sum of squares due to enlarging the model from that at (1.18) to the simple regression model (1.21). This is the *sum of squares due to regression*, *SSreg*, defined by

$$SSreg = SYY - RSS$$

$$= SYY - \left( SYY - \frac{(SXY)^2}{SXX} \right)$$

$$= \frac{(SXY)^2}{SXX} \tag{1.23}$$

The degrees of freedom associated with *SSreg* is the difference in degrees of freedom for the model (1.18), $n-1$ and the d.f. for model (1.21), $n-2$, so the d.f. for *SSreg* is $(n-1)-(n-2)=1$ for simple regression.

These results are often summarized in an analysis of variance table, abbreviated as ANOVA, given in Table 1.3. In the analysis of variance table the column marked source refers to descriptive labels given to the sums of squares; in more complicated tables, there may be many sources and the labels given are not unique. The degrees of freedom (d.f.) column gives the number of degrees of freedom associated with the named source. The next column gives the associated sum of squares. The mean square column is computed from the sum of squares column by dividing sums of squares by *(handwritten annotations at top: $SYY - RSS$   $SYY - \dfrac{(SXY)^2}{SXX}$)*

**Table 1.3   Analysis of variance**

| Source | Degrees of Freedom (d.f.) | Sum of Squares (SS) | Mean Square (MS) | F |
|---|---|---|---|---|
| Regression on $X$ | 1 | $SSreg$ | $SSreg/1$ | $MSreg/MSE$ |
| Residual for larger model | $n-2$ | $RSS$ | $RSS/(n-2)$ | |
| Total corrected sum of squares | $n-1$ | $SYY$ | | |

*(handwritten: $=\hat{\sigma}^2$ above Mean Square column)*

the corresponding degrees of freedom. The mean square on the residual line is just $\hat{\sigma}^2$, as already discussed.

The analysis of variance for Forbes' data is given in Table 1.4. For this table $SYY$ is given at (1.10), $RSS$ is given at (1.13), and $SSreg$ is computed by subtraction.

The ANOVA is always computed relative to a specific larger model, here given by (1.21), and a smaller model obtained from the full model by setting some parameters to zero. This was done at (1.18), which was obtained from (1.21) by setting $\beta_1 = 0$. The line in the ANOVA table for the total gives the residual sum of squares corresponding to the model with the fewest parameters. In the next chapter, the analysis of variance is applied to a sequence of models, but the reference to a fixed large model remains intact.

In practice, the ANOVA table is computed by finding $SYY$ and $SSreg = SXY^2/SXX$, or by some equivalent formula from Table 1.2. $RSS$ is computed by subtraction.

**The $F$-test for regression.**   If the sum of squares for regression SSreg is large, then the simple regression model $y_i = \beta_0 + \beta_1 x_i + e_i$ should be a significant improvement over the model given by (1.18), $y_i = \beta_0 + e_i$. This is equivalent to saying that the additional parameter in the simple regression

**Table 1.4   Analysis of variance**

| Source | d.f. | SS | MS | F |
|---|---|---|---|---|
| Regression | 1 | 425.610 | 425.610 | 2955 |
| Residual | 15 | 2.153 | 0.144 $=\hat{\sigma}^2$ | |
| Total | 16 | 427.763 | | |

model $\beta_1$ is different from zero, or that $Y$ is indeed related to $X$. To formalize this notion, we need to be able to judge how large is "large." This is done by comparing the regression mean square ($SSreg$ divided by its degree of freedom, which for simple regression is 1) to the residual mean square $\hat{\sigma}^2$. We call this ratio $F$:

$$F = \frac{(SYY - RSS)/1}{\hat{\sigma}^2} = \frac{SSreg/1}{\hat{\sigma}^2} \qquad (1.24)$$

Clearly $F$ is a rescaled version of $SSreg = SYY - RSS$, with larger values of $SSreg$ resulting in larger values of $F$. Formally, we can consider testing the *null hypothesis* (NH) against the *alternative hypothesis* (AH)

$$NH: \quad y_i = \beta_0 + e_i \qquad i = 1, 2, \ldots, n$$
$$AH: \quad y_i = \beta_0 + \beta_1 x_i + e_i \qquad i = 1, 2, \ldots, n \qquad (1.25)$$

If the $e_i$'s are NID(0, $\sigma^2$) then, under NH, (1.24) will follow an $F$ distribution with degrees of freedom associated with the numerator and denominator of (1.24), 1 and $n-2$ for simple regression. This is written $F \sim F(1, n-2)$. Percentage points of the $F$ distribution can be used to assign significance levels, or $p$-values, to the computed $F$-test.

For Forbes' data, we compute

$$F = \frac{425.610}{0.144} = 2955$$

with (1, 15) degrees of freedom. From Table B at the end of the book we see that the 0.01 point of $F(1, 15)$, written as $F(0.01; 1, 15)$, is 8.68, so that the $p$-value for this test is (much) smaller than 0.01, providing very strong evidence against NH and in favor of AH.

**Interpreting $p$-values.** Under the appropriate assumptions, the $p$-value is the conditional probability of observing a value of the computed statistic (here, the value of $F$) as extreme or more extreme (here, as large or larger) than the observed value, given that the NH is true. A small $p$-value provides evidence against the NH. The observed $p$-value will depend on the sample size, the sampling plan, and on how far the correct AH is from the NH. Large (in absolute value) $\beta_1$'s will generally lead to smaller $p$-values than would smaller values of $\beta_1$. Similarly, as sample size increases, $p$-values will generally get smaller since the power of the $F$-test increases with sample size, and the $F$-test will detect less extreme alternative hypotheses. Also, in regression situations, the $p$-value will depend on the sample range for the

$x$'s in the data; if the $x$'s are obtained over a small range, the $p$-value will be larger than if the $x$'s are sampled over a wider range.

There is an important distinction between statistical significance, the observation of a sufficiently small $p$-value, and scientific significance, observing an effect of sufficient magnitude to be meaningful. Judgement of the latter usually will require examination of more than just the $p$-value.

## 1.6   The coefficient of determination, $R^2$

If both sides of (1.23) are divided by $SYY$, we get

$$\frac{SSreg}{SYY} = 1 - \frac{RSS}{SYY} \tag{1.26}$$

The left-hand side of (1.26) is the proportion of variability of $Y$ explained by regression on $X$ or, equivalently, by adding $X$ to the model. The right-hand side consists of 1 minus the remaining unexplained variability. This concept of dividing up the total variability according to whether or not it is explained is of sufficient importance that a special name is given to it. We define $R^2$, the *coefficient of determination*, to be

$$R^2 = \frac{SSreg}{SYY} = 1 - \frac{RSS}{SYY} \tag{1.27}$$

$R^2$ is easily computed from quantities that are available in the analysis of variance table. It is a scale-free one-number summary of the strength of the relationship between the $x_i$ and the $y_i$ in the data. It is a very popular statistic since it generalizes nicely to multiple regression, depends only on the sums of squares, and appears to be easy to interpret. For Forbes' data,

$$R^2 = \frac{SSreg}{SYY} = \frac{425.610}{427.763} = 0.995$$

and thus about 99.5% of the variability in the observed values of $100 \times$ log(pressure) is explained by boiling point.

**Relationship to the correlation coefficient.**   By appealing to (1.27) and to Table 1.2 we can write

$$R^2 = \frac{SSreg}{SYY} = \frac{(SXY)^2}{(SXX)(SYY)} = r_{XY}^2$$

and thus $R^2$ is the same as the square of the sample correlation between $X$ and $Y$.

## 1.7   Confidence intervals and tests

When the errors are NID(0, $\sigma^2$) parameter estimates, fitted values, and predictions will be normally distributed because all of these are linear combinations of the $y_i$'s and hence of the $e_i$'s. Consequently, confidence intervals and tests can be based on the $t$ distribution. Suppose we let $t(\alpha, d)$ be the value that cuts off $\alpha/2 \times 100\%$ in the upper tail of the $t$ distribution with $d$ degrees of freedom. These values are given in Table A at the end of the book.

**The intercept.**   The standard error of the intercept is $\text{se}(\hat{\beta}_0) = \hat{\sigma}(1/n + \bar{x}^2/SXX)^{1/2}$. Hence a $(1-\alpha) \times 100\%$ confidence interval for the intercept is the set of points $\beta_0$ in the interval

$$\hat{\beta}_0 - t(\alpha, n-2)\text{se}(\hat{\beta}_0) \leqslant \beta_0 \leqslant \hat{\beta}_0 + t(\alpha, n-2)\text{se}(\hat{\beta}_0) \tag{1.28}$$

For Forbes' data, $\text{se}(\hat{\beta}_0) = 0.0379(1/17 + (202.95)^2/530.78)^{1/2} = 3.339$. For a 90% confidence interval, $t(0.10, 15) = 1.75$, and the interval is

$$-42.131 - 1.75(3.339) \leqslant \beta_0 \leqslant -42.131 + 1.75(3.339)$$

$$-47.974 \leqslant \beta_0 \leqslant -36.288 \tag{1.29}$$

Ninety percent of such intervals will include the true value.

A hypothesis test of

$$\text{NH:} \quad \beta_0 = \beta_0^*, \quad \beta_1 \text{ arbitrary}$$

$$\text{AH:} \quad \beta_0 \neq \beta_0^*, \quad \beta_1 \text{ arbitrary}$$

is obtained by computing the $t$ statistic

$$t = \frac{\hat{\beta}_0 - \beta_0^*}{\text{se}(\hat{\beta}_0)} \tag{1.30}$$

and referring this ratio to the $t$ distribution with $n-2$ degrees of freedom. For example, in Forbes' data consider testing

$$\text{NH:} \quad \beta_0 = -35, \quad \beta_1 \text{ arbitrary}$$

$$\text{AH:} \quad \beta_0 \neq -35, \quad \beta_1 \text{ arbitrary} \tag{1.31}$$

42.131

The statistic is

$$t = \frac{-42.121 - (-35)}{3.339} = 2.136 \tag{1.32}$$

which has a $p$-value near 0.05, $t(0.05, 15) = 2.13$, providing some evidence against NH. Of course, this hypothesis test for these data is not one that would occur to most investigators and is used only as an illustration.

**Slope.**   The standard error of $\hat{\beta}_1$ is $\text{se}(\hat{\beta}_1) = \hat{\sigma}/\sqrt{SXX} = 0.0164$. A 95% confidence interval for the slope is the set of $\beta_1$ such that

$$0.895 - 2.13(0.0164) < \beta_1 < 0.895 + 2.13(0.0164)$$

$$0.860 < \beta_1 < 0.930 \tag{1.33}$$

As with the intercept, hypothesis tests for the slope are probably not of interest in this example, as there is no obvious value for the parameter for a null hypothesis. The usual test is

$$\text{NH:} \quad \beta_1 = 0, \quad \beta_0 \text{ arbitrary}$$

$$\text{AH:} \quad \beta_1 \neq 0, \quad \beta_0 \text{ arbitrary} \tag{1.34}$$

For the data given, $t = (0.895 - 0)/0.0164 = 54.45$. Although clearly unnecessary here, since $t$ is so large, we could compare this computed value to the $t$-distribution with 15 d.f. The associated $p$-value is very small but this is hardly surprising. If $\beta_1$ were actually near zero, Forbes probably would not ever have done the experiments, or if he had, he would not have published the results.

Compare the hypothesis (1.34) to (1.25). Both appear to be identical. In fact,

$$t^2 = \left(\frac{\hat{\beta}_1}{\text{se}(\hat{\beta}_1)}\right)^2 = \frac{\hat{\beta}_1^2}{\hat{\sigma}^2/SXX} = \frac{\hat{\beta}_1^2 SXX}{\hat{\sigma}^2} = F$$

so the square of a $t$ statistic with $d$ degrees of freedom is equivalent to an $F$ statistic with $(1, d)$ degrees of freedom.

**Prediction.**   The fitted equation is often used to obtain values of the response for given values of the predictor. The two important variants of this problem are prediction and estimation of fitted values. Since prediction is far more important, we discuss it first.

In prediction, we have a new case, not one used to estimate parameters, with observed value of the predictor $x_*$. We would like to know the value $y_*$, the corresponding response, but it has not yet been observed. Since $y_*$ is not available, we will use a model at the observed value of $x_*$ to predict it. We assume that the data used to estimate the fitted line are relevant to the new case, so the fitted model applies to it. In the Forbes example, we could not expect reasonable predictions for water boiled in an airplane or in a submarine, since the experimental conditions are likely to be much different than they were in Forbes' experiments. However, we might expect the model to work for points on Earth at reasonable altitudes. Given this additional assumption, a point prediction of $y_*$, say $\tilde{y}_*$, is just

$$\tilde{y}_* = \hat{\beta}_0 + \hat{\beta}_1 x_* \tag{1.35}$$

Strictly speaking, $\tilde{y}_*$ predicts the expectation of the as yet unobserved $y_*$. As a result, the variability of this predictor will have two sources: the variation in the estimates $\hat{\beta}_0$ and $\hat{\beta}_1$, and the variation due to the fact that $y_*$ will not equal its expectation. One can show, using Appendix 1A.4, that

$$\text{var}(\tilde{y}_*|x_*) = \sigma^2 + \sigma^2 \left[ \frac{1}{n} + \frac{(x_* - \bar{x})^2}{SXX} \right] \tag{1.36}$$

Taking square roots and estimating $\sigma^2$ by $\hat{\sigma}^2$, we get the standard error of prediction (sepred) at $x_*$,

$$\text{sepred}(\tilde{y}_*|x_*) = \hat{\sigma} \left[ 1 + \frac{1}{n} + \frac{(x_* - \bar{x})^2}{SXX} \right]^{1/2} \tag{1.37}$$

A predictive interval for a single prediction is obtained in the usual way, using multipliers from the $t$ distribution. For prediction of $100 \times \log(\text{pressure})$ for a location with $x_* = 200$, the point prediction is $\tilde{y}_* = -42.13 + 0.895(200) = 136.87$, with standard error of prediction $\quad \hat{\beta}_0 + \hat{\beta}_1$

$$\text{sepred}(\tilde{y}_*|x_* = 200) = 0.379 \left[ 1 + \frac{1}{17} + \frac{(200 - 203.0)^2}{530.8} \right]^{1/2} = 0.393$$

Thus a 99% predictive interval is the set of all $y_*$ such that

$$136.87 - 2.95(0.393) \leqslant y_* \leqslant 136.87 + 2.95(0.393)$$

$$135.71 \leqslant y_* \leqslant 138.03$$

Figure 1.6 is a plot of the least squares regression line for Forbes' data along with curves at $\hat{\beta}_0 + \hat{\beta}_1 x_* \pm t(.99; 15)\text{sepred}(\tilde{y}_*|x_*)$, for $x_*$ in the range 180 to 220. The vertical distance between the two curves for any $x_*$ corresponds to a 99% prediction interval for $y_*$ given $x_*$. The interval is wider for $x_*$'s far from $\bar{x}$, as the curves bend outward; this may be hard to see in the figure because the variation about the regression line is so small.

**Fitted values.**   In rare problems one may be interested in obtaining an estimate of the average of $y$ at a given value of $x$. This only makes sense when the model used is known to be correct, except that the values of the parameters are estimated. In Forbes' example this is like asking for the "true" average value of $100 \times \log(\text{pressure})$ corresponding to a fixed value of boiling point. This quantity is estimated by the fitted value $\hat{y} = \hat{\beta}_0 + \hat{\beta}_1 x$, and its standard error is

$$\text{sefit}(\hat{y}|x) = \hat{\sigma} \left[ \frac{1}{n} + \frac{(x - \bar{x})^2}{SXX} \right]^{1/2} \tag{1.38}$$

To obtain confidence intervals, it is more usual to compute a simultaneous

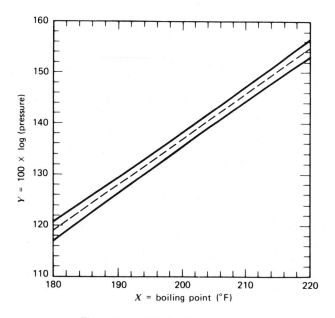

**Figure 1.6** 99% prediction intervals.

interval for all possible values of $x$. This is the same as first computing a joint confidence region for $\beta_0$ and $\beta_1$, and from these, computing the set of all possible regression lines with slope and intercept in the joint confidence set. Although the technique of computing joint confidence sets will be postponed to Section 4.5, the confidence set can be stated simply as the set of all $y$ such that

$$(\hat{\beta}_0 + \hat{\beta}_1 x) - \mathrm{sefit}(\hat{y}|x)[2F(\alpha; 2, n-2)]^{1/2} \leqslant y$$
$$\leqslant (\hat{\beta}_0 + \hat{\beta}_1 x) + \mathrm{sefit}(\hat{y}|x)[2F(\alpha; 2, n-2)]^{1/2}$$

(For multiple regression, replace $2F(\alpha; 2, n-2)$ by $p'F(\alpha; p', n-p')$, where $p'$ is the number of $\beta$'s in the regression equation.) The shape of the simultaneous band for the fitted line is similar to Figure 1.6.

## 1.8 The residuals

The residuals, $\hat{e}_i = y_i - \hat{y}_i$, $i = 1, 2, \ldots, n$, provide information regarding assumptions about error terms and the appropriateness of the model. Any complete data analysis requires examination of the residuals. Here we shall present only the barest outline of analysis of residuals, leaving detailed methods to later chapters.

Plots of residuals versus other quantities are used to find failures of assumptions. The most common plot, especially useful in simple regression, is the plot of $\hat{e}_i$ versus the fitted values $\hat{y}_i$. Systematic features in this plot are of interest. Curvature might indicate that the fitted model is inappropriate and suggest a transformation of the data. Residuals that seem to increase or decrease in average magnitude with $\hat{y}_i$ might indicate nonconstant residual variance. A few relatively large residuals may be indicative of outliers—cases for which the model is somehow inappropriate. On the other hand, if the plot of $\hat{e}_i$ versus $\hat{y}_i$ shows no systematic features, then we would have little reason to suspect that the fitted model was inappropriate for the data.

**Forbes' data (conclusion and summary).**  The fitted values $\hat{y}_i$ and residuals $\hat{e}_i$ for Forbes' data are given in Table 1.5 and are plotted against each other in Figure 1.7. Notice that the residuals are generally small compared to the $\hat{y}_i$'s and that they do not suggest any distinct pattern in Figure 1.7. However, one residual, for case 12, is much larger than the others, as the others are typically less than 0.35 in absolute value, while that for case 12 is about 1.3. This *may* suggest that the assumptions concerning the errors

**Table 1.5   Fitted values and residuals for Forbes' data**

| Case Number | $x_i$ | $y_i$ | $\hat{y}_i$ | $\hat{e}_i$ |
|---|---|---|---|---|
| 1 | 194.50 | 131.79 | 132.04 | −0.25 |
| 2 | 194.30 | 131.79 | 131.86 | −0.07 |
| 3 | 197.90 | 135.02 | 135.08 | −0.06 |
| 4 | 198.40 | 135.55 | 135.53 | 0.02 |
| 5 | 199.40 | 136.46 | 136.42 | 0.04 |
| 6 | 199.90 | 136.83 | 136.87 | −0.04 |
| 7 | 200.90 | 137.82 | 137.77 | 0.05 |
| 8 | 201.10 | 138.00 | 137.95 | 0.05 |
| 9 | 201.40 | 138.06 | 138.22 | −0.16 |
| 10 | 201.30 | 138.05 | 138.13 | −0.08 |
| 11 | 203.60 | 140.04 | 140.19 | −0.15 |
| 12 | 204.60 | 142.44 | 141.08 | 1.36 |
| 13 | 209.50 | 145.47 | 145.47 | 0.00 |
| 14 | 208.60 | 144.34 | 144.66 | −0.32 |
| 15 | 210.70 | 146.30 | 146.54 | −0.24 |
| 16 | 211.90 | 147.54 | 147.62 | −0.08 |
| 17 | 212.20 | 147.80 | 147.89 | −0.09 |

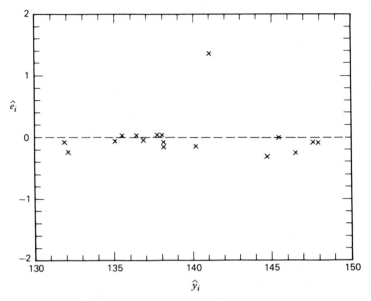

**Figure 1.7** Residual plot for Forbes' data.

are not correct. Either $\sigma^2$ may not be constant, or for case 12 the corresponding error may have a large fixed component. It is possible, for example, that Forbes misread or miscopied the results of his calculations for this case, and the numbers in the data do not correspond to the actual measurements. Forbes noted this possibility himself, by marking this pair of numbers in his paper as being "evidently a mistake" because of the large observed residual.

The problem of what to do with a suspect case must be faced. Lacking a formal procedure for examining such cases, a topic for later chapters, we proceed informally. Since we are concerned with the effects of case 12, we could refit the data, this time without case 12, and then examine the changes that occur in the estimates of parameters, fitted values, residual variance, and so on. This is summarized in Table 1.6, giving for each of the data sets (with 17 and 16 cases, respectively) estimates of parameters, their standard errors, $\hat{\sigma}$, and the coefficient of determination $R^2$. As is clear from the table, for obtaining point estimates of the parameters, case 12 is irrelevant, since the estimates are essentially identical with and without it. In other regression problems, deletion of a single case can change everything. However, the effect of case 12 on standard errors is more marked: if case 12 is deleted, standard errors are decreased by a factor of about 3.1 and variances are decreased by a factor of about $3.1^2 \cong 10$. Inclusion of this case therefore gives the appearance of less reliable results than would be suggested on the

**Table 1.6   Summary for Forbes' data**

| Quantity | Value Using All Data | Value without Case 12 |
|----------|----------------------|-----------------------|
| $\hat{\beta}_0$ | $-42.131$ | $-41.302$ |
| $\hat{\beta}_1$ | 0.895 | 0.891 |
| $se(\hat{\beta}_0)$ | 3.339 | 1.000 |
| $se(\hat{\beta}_1)$ | 0.0164 | 0.00493 |
| $\hat{\sigma}$ | 0.379 | 0.113 |
| $R^2$ | 0.995 | 0.999+ |

basis of the other 16 cases. Figure 1.8, a residual plot obtained when case 12 is deleted before computing, indicates no obvious failures in the model fit to the 16 remaining cases.

Two models have been fit, one using 16 cases and one using 17 cases, which lead to slightly different conclusions, although the results of the two analyses agree more than they disagree. On the basis of the data, there is no real way to choose between the two models, and we have no way of deciding which is the correct least squares analysis of the data. A good approach to this problem is to describe both (or in general all) plausible alternatives.

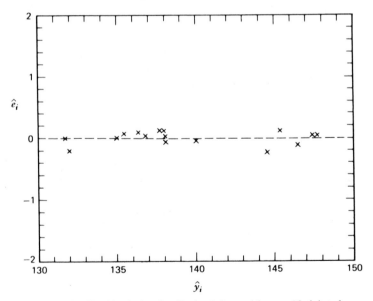

**Figure 1.8**   Residual plot for Forbes' data with case 12 deleted.

In summary, a straight line model of the data collected by Forbes can be used to describe the logarithm of atmospheric pressure as a function of the boiling point of water. For boiling points in the range 180 to 220°F, the data can be described by a straight line, with the equation given by $\hat{y} = -42.131 + 0.895x$, and, at least for any $x$ in this range, this equation will provide a good prediction of $100 \times \log(\text{pressure})$. In the data, one case (number 12) is fit by the line much more poorly than are the other 16 points. Without this case, standard errors of estimates, predictions, and so on are only about one third of what they would be with it.

Even with the excellent fit of the straight line to the data, we cannot assume that the regression of $100 \times \log(\text{pressure})$ on boiling temperature is in fact a straight line; the conclusion should be restricted to the range of values in the observed data. The data provide no information about the usefulness of the model beyond this range.

## Problems

**1.1   Height and weight data.**   These data give $X$=height in centimeters and $Y$=weight in kilograms for a sample of $n=10$ eighteen-year-old girls. The data are taken from a larger study described in problem set 2.1.

| X | Y |
|------|------|
| 169.6 | 71.2 |
| 166.8 | 58.2 |
| 157.1 | 56.0 |
| 181.1 | 64.5 |
| 158.4 | 53.0 |
| 165.6 | 52.4 |
| 166.7 | 56.8 |
| 156.5 | 49.2 |
| 168.1 | 55.6 |
| 165.3 | 77.8 |

The following questions should be answered without the aid of a computer.

**1.1.1.** Draw a scatter plot of $Y$ versus $X$. On the basis of this plot guess plausible values for $\beta_0$, $\beta_1$, and $R^2$ for the regression of $Y$ on $X$ in the simple linear regression model $y_i = \beta_0 + \beta_1 x_i + e_i$, $e_i \sim \text{NID}(0, \sigma^2)$, $i = 1, 2, \ldots, 10$.

## 28  Simple linear regression

**1.1.2.** Show that $\bar{x} = 165.52$, $\bar{y} = 59.47$, $SXX = 472.076$, $SYY = 731.961$, and $SXY = 274.786$. Compute estimates of the slope and the intercept for the regression of $Y$ on $X$. Draw the fitted line on your scatter plot.

**1.1.3.** Estimate $\sigma^2$ and find the estimated standard errors of $\hat{\beta}_0$ and $\hat{\beta}_1$. Also find the estimated covariance between $\hat{\beta}_0$ and $\hat{\beta}_1$. Compute the $t$-tests for the hypotheses that $\beta_0 = 0$ and that $\beta_1 = 0$, and find the appropriate $p$-values for these tests. (Use two-sided tests.)

**1.1.4.** Obtain the analysis of variance table and $F$-test for regression. Show numerically that $F = t^2$, where $t$ was computed in 1.1.3 for testing $\beta_1 = 0$.

**1.1.5.** Obtain the residuals and the fitted values. Show numerically that the sum of the residuals is zero. Draw a graph of the residuals versus the fitted values. Are there any obvious outliers? Are there any clear patterns in the residuals?

**1.1.6.** Interpret the meaning of the parameters $\beta_0$ and $\beta_1$. What units are they measured in? What are the units of $\sigma$?

**1.2  Hooker's data.** In his paper on boiling points and temperatures, Forbes also presented data collected on the same two quantities by Dr. Joseph Hooker. Unlike Forbes, however, Hooker took his measurements in the Himalaya Mountains, generally at higher altitudes. The data below are

| TEMP (°F) | PRES (in. Hg) | TEMP (°F) | PRES (in. Hg) |
|-----------|---------------|-----------|---------------|
| 210.8 | 29.211 | 189.5 | 18.869 |
| 210.2 | 28.559 | 188.8 | 18.356 |
| 208.4 | 27.972 | 188.5 | 18.507 |
| 202.5 | 24.697 | 185.7 | 17.267 |
| 200.6 | 23.726 | 186.0 | 17.221 |
| 200.1 | 23.369 | 185.6 | 17.062 |
| 199.5 | 23.030 | 184.1 | 16.959 |
| 197.0 | 21.892 | 184.6 | 16.881 |
| 196.4 | 21.928 | 184.1 | 16.817 |
| 196.3 | 21.654 | 183.2 | 16.385 |
| 195.6 | 21.605 | 182.4 | 16.235 |
| 193.4 | 20.480 | 181.9 | 16.106 |
| 193.6 | 20.212 | 181.9 | 15.928 |
| 191.4 | 19.758 | 181.0 | 15.919 |
| 191.1 | 19.490 | 180.6 | 15.376 |
| 190.6 | 19.386 | | |

an abstract of Hooker's data giving $n = 31$ pairs of measurements on TEMP = boiling point (degrees Fahrenheit) and PRES = corrected barometric pressure (inches of mercury). Use of packaged computer programs is encouraged, but not necessary, for this problem.

**1.2.1.** Draw the scatter plot of PRES versus TEMP. Would a straight line closely match the data? (A graph of the residuals from the regression PRES = $\beta_0 + \beta_1$ TEMP + $e$ versus fitted values will be useful here.)

**1.2.2.** Draw a scatter plot of $100 \times \log(PRES)$ versus TEMP, and compare to the plot in the last problem. Is this scatter plot more nearly described by a straight line?

**1.2.3.** Fit the simple regression model for $100 \times \log(PRES)$ on TEMP; that is, fit the model $100 \times \log(PRES) = \beta_0 + \beta_1$ TEMP + $e$ and compute the relevant summary statistics (estimates of parameters, tests, analysis of variance table, $R^2$). Draw the fitted line onto the plot in problem 1.2.2. Obtain the residual plot (versus fitted values) and compare to that fit in 1.2.1.

**1.2.4.** Obtain 95% confidence intervals for $\beta_0$ and $\beta_1$.

**1.2.5.** Obtain 90% prediction intervals for $100 \times \log(PRES)$ for predictions at 185 and at 212°F.

**1.2.6.** Qualitatively compare the results of this analysis to the results in the test for Forbes' data. That is, compare the fitted lines, estimates of residual variability, prediction intervals, etc. What do you conclude? In Chapter 7 we will learn tests for comparing regression in different groups.

**1.3  Olympic records.**   The following data give the best foot race running times recorded in the modern Olympic Games up to 1984, taken from the 1985 *World Almanac.*

|  | Men | | Women | |
| --- | --- | --- | --- | --- |
| Distance (m) | Time (sec) | Year | Time (sec) | Year |
| 100 | 9.9 | 1968 | 11.0 | 1984 |
| 200 | 19.8 | 1984 | 21.8 | 1984 |
| 400 | 43.8 | 1968 | 48.8 | 1984 |
| 800 | 103.0 | 1984 | 113.5 | 1980 |
| 1500 | 212.5 | 1984 | 236.6 | 1980 |
| 3000 | | | 516.0 | 1984 |
| 5000 | 785.6 | 1984 | | |
| 10000 | 1658.4 | 1972 | | |
| 42195 | 7761.0 | 1984 | 8692.0 | 1984 |

A computer is not required for this problem.

**1.3.1.** Draw a scatter plot of time versus distance for all the data, using a different symbol for men's times and for women's times. Given such disparate values for both time and distance, you will quickly discover that this display is both hard to draw and to interpret. As an alternative, try plotting transformations of time and distance, perhaps log(time) versus log(distance), or speed = distance/time versus distance or log(distance).

**1.3.2.** Perform relevant calculations that will help to summarize the relationship apparent in these data. You will probably want to do separate analyses for men and for women.

**1.3.3.** Are $F$-tests and $t$-tests relevant in this problem? Why or why not? (*Hint*: What is random in these data?)

**1.4   Regression through the origin.**   Occasionally, a model in which the intercept is known a priori to be zero may be fit. This model is given by

$$y_i = \beta_1 x_i + e_i \qquad i = 1, 2, \ldots, n \qquad (1.39)$$

The residual sum of squares for this model, assuming the $e_i$ are independent with common variance $\sigma^2$, is $RSS = \sum (y_i - \hat{\beta}_1 x_i)^2$.

**1.4.1.** Show that the least sqaures estimate of $\beta_1$ is given by $\hat{\beta}_1 = \sum x_i y_i / \sum x_i^2$. Show that $\hat{\beta}_1$ is unbiased and that $\text{var}(\hat{\beta}_1) = \sigma^2 / \sum x_i^2$. Find an expression for $\hat{\sigma}^2$. How many degrees of freedom does it have?

**1.4.2.** Derive the analysis of variance table with the larger model given by (1.21), but with the smaller model specified in (1.39). Show that the $F$-test derived from this table is numerically equivalent to the square of the $t$-test (1.30) with $\beta_0^* = 0$.

**1.4.3.** The data below give $X$ = water content of snow on April 1 and $Y$ = water yield from April to July (in inches) in the Snake River watershed in Wyoming for $n = 17$ years (1919 to 1935), from Wilm (1950).

| X | Y | X | Y |
|------|------|------|------|
| 23.1 | 10.5 | 37.9 | 22.8 |
| 32.8 | 16.7 | 30.5 | 14.1 |
| 31.8 | 18.2 | 25.1 | 12.9 |
| 32.0 | 17.0 | 12.4 | 8.8 |
| 30.4 | 16.3 | 35.1 | 17.4 |
| 24.0 | 10.5 | 31.5 | 14.9 |
| 39.5 | 23.1 | 21.1 | 10.5 |
| 24.2 | 12.4 | 27.6 | 16.1 |
| 52.5 | 24.9 | | |

Fit a regression through the origin and find $\hat{\beta}_1$ and $\hat{\sigma}^2$. Obtain a 95% confidence interval for $\beta_1$. Test the hypothesis that the intercept is zero.

**1.4.4.** Plot the residuals $(\hat{e}_i = y_i - \hat{\beta}_1 x_i)$ versus the fitted values $(\hat{y}_i = \hat{\beta}_1 x_i)$ and comment on the adequacy of the model. In regression through the origin, $\sum \hat{e}_i \neq 0$.

### 1.5  Scale invariance.

**1.5.1.** In the simple regression model (1.2), suppose each $x_i$ is replaced by $cx_i$, where $c \neq 0$ is a constant. How are $\hat{\beta}_0$, $\hat{\beta}_1$, $\hat{\sigma}^2$, $R^2$, and the $t$-test of NH: $\beta_1 = 0$ affected?

**1.5.2.** Suppose each $y_i$ is replaced by $dy_i$, $d \neq 0$. Repeat 1.5.1.

**1.6**  Using Appendix 1A.3, verify equation (1.12).

**1.7**  Using Appendix 1A.4, verify equation (1.37).

**1.8**  Assuming model (1.2), verify that the sample correlation between the $\hat{e}_i$'s and the $\hat{y}_i$'s is zero. What is the sample correlation between the $\hat{e}_i$'s and the $y_i$'s? In residual plotting, suppose we plotted $\hat{e}_i$ versus $y_i$ rather than $\hat{e}_i$ versus $\hat{y}_i$. Comment on the difference between these plots.

**1.9  Amazon River water levels.**

The Amazon River basin is by far the largest tropical forest on earth, but, like most other natural resources, it has been put under severe pressure because of development. In the 1970s roads were first opened into the upper Amazon area, allowing for rapid population growth and large-scale deforestation. This in turn may cause major climatological and hydrological changes that will reflect on the river as a whole, since both rainfall and runoff are likely to be affected. The data in Table 1.7 for the years 1962 to 1978, give the HIGH and LOW water levels of the Amazon River at Iquitos, Peru (in

**Table 1.7   Amazon River data**

| Year | High (m) | Low (m) | Year | High (m) | Low (m) |
|------|----------|---------|------|----------|---------|
| 1962 | 25.82 | 18.24 | 1971 | 27.36 | 21.91 |
| 1963 | 25.35 | 16.50 | 1972 | 26.65 | 22.51 |
| 1964 | 24.29 | 20.26 | 1973 | 27.13 | 18.81 |
| 1965 | 24.05 | 20.97 | 1974 | 27.49 | 19.42 |
| 1966 | 24.89 | 19.43 | 1975 | 27.08 | 19.10 |
| 1967 | 25.35 | 19.31 | 1976 | 27.51 | 18.80 |
| 1968 | 25.23 | 20.85 | 1977 | 27.54 | 18.80 |
| 1969 | 25.06 | 19.54 | 1978 | 26.21 | 17.57 |
| 1970 | 27.13 | 20.49 | | | |

meters). The data for 1962 to 1969 may be thought of as a control period while 1970 to 1978 represent values obtained after the beginning of development. The exercise is to analyze these data to determine whether deforestation of the upper Amazon has resulted in changes in the water balance of the Amazon basin. Of interest is the study of change in these two characteristics of the river over time. For example, if we fit

$$HIGH = \beta_0 + \beta_1 \times year + e$$

then (1) $\beta_1 = 0$ implies no (linear) change in HIGH water level over time, (2) $\beta_1 > 0$ implies an increase in HIGH water level implying, perhaps, more runoff, and (3) $\beta_1 < 0$ might imply lower runoff over time. See Gentry and Lopez-Parodi (1980) for more discussion.

**1.9.1.** Draw scatter plots of HIGH versus year, LOW versus year, and HIGH versus LOW.

**1.9.2.** Compute the regression of HIGH on year, LOW on year, and HIGH on LOW. Summarize your results for the three regressions and interpret parameters, especially slopes, in terms of the problem.

**1.9.3.** On the basis of these data can we say deforestation is *causing* changes in the water level in the Amazon? What additional information, if available, might be used to infer causality?

**1.10   Computer program.**
Write a computer program that uses the method of updating outlined in Appendix 1A.5 to compute and store the summary statistics consisting of the sample size $n$, the sample averages, and sample corrected sums of squares and cross products. Write the program to allow reading in $n$ cases on *several* variables, perhaps up to 10, computing the summary statistics for each and the corrected cross products for each pair. It is convenient to store the corrected sums of squares and cross products in a two-dimensional array $T(I, J)$, where, for example, $T(J, J)$ is the corrected sum of squares for the $j$th variable, and $T(I, J)$ is the corrected cross product between variable $i$ and variable $j$, so $T(I, J) = T(J, I)$. Given this program, it is an easy matter to compute all the regression statistics, such as estimates and standard errors, outlined in this chapter. Problems are included in other chapters to expand this program.

# 2

# MULTIPLE REGRESSION

In multiple regression, several predictors are used to model a single response variable. For each of the $n$ cases observed, values for the response and for each of the predictors are collected. If the response is called $Y$, and the predictors are called $X_1, X_2, \ldots, X_p$ ($p$ is the number of predictors), then the data will form the $n \times (p+1)$ array:

| Case Number | Values | | | | | |
|---|---|---|---|---|---|---|
| | $Y$ | $X_1$ | $X_2$ | $X_3$ | $\cdots$ | $X_p$ |
| 1 | $y_1$ | $x_{11}$ | $x_{12}$ | $x_{13}$ | | $x_{1p}$ |
| 2 | $y_2$ | $x_{21}$ | $x_{22}$ | $x_{23}$ | | $x_{2p}$ |
| $\vdots$ | $\vdots$ | $\vdots$ | $\vdots$ | $\vdots$ | $\cdots$ | $\vdots$ |
| $n$ | $y_n$ | $x_{n1}$ | $x_{n2}$ | $x_{n3}$ | | $x_{np}$ |

For simple regression, $p = 1$. In this representation of the data, the value $x_{ij}$ refers to the value for the $j$th variable on the $i$th case. The values for one case appear in one row; all the values for a variable appear in one column.

In multiple regression, an equation that expresses the response as a linear function of the $p$ predictors is estimated using the observed data. The model is specified by a linear equation,

$$Y = \beta_0 + \beta_1 X_1 + \beta_2 X_2 + \cdots + \beta_p X_p + e \qquad (2.1)$$

where, as in the previous chapter, the $\beta$'s are unknown parameters, the $e$'s are statistical errors, $Y$ is the response, and $X_1, X_2, \ldots, X_p$ are the predictors.

**33**

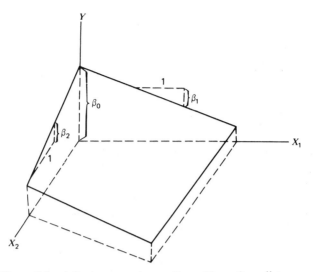

**Figure 2.1**    A linear regression surface with $p = 2$ predictors.

When $p = 2$, (2.1) gives the equation of a two-dimensional plane in the three-dimensional $(X_1, X_2, Y)$ space, as shown in Figure 2.1. Given data, $x_{ij}$ collected on the $X_j$ and $y_i$ collected on $Y$, we rewrite (2.1) as

$$y_i = \beta_0 + \beta_1 x_{i1} + \beta_2 x_{i2} + \cdots + \beta_p x_{ip} + e_i \qquad i = 1, \ldots, n \qquad (2.2)$$

In this chapter, we will be concerned with estimating the $\beta$'s and interpreting these estimates. Most of the results are given in terms of vectors and matrices, for which a brief introduction is given in Appendix 2A.1. With this notation, the results appear simple and elegant. Without it, one can get lost in a sea of subscripts.

### *Example 2.1    Fuel consumption*

Six columns of Table 2.1 list values, for each of the 48 contiguous states, of the following quantities:

POP = 1971 population, in thousands.
TAX = 1972 motor fuel tax rate, in cents per gallon.
NLIC = 1971 thousands of licensed drivers.
INC = 1972 per capita income in thousands of dollars.
ROAD = 1971 thousands of miles of federal-aid primary highways.
FUELC = 1972 fuel consumption, in millions of gallons.

**Table 2.1  Fuel consumption data**

| STATE | POP | $X_1$ TAX | NLIC | $X_3$ INC | $X_4$ ROAD | FUELC | $X_2$ DLIC | $Y$ FUEL |
|---|---|---|---|---|---|---|---|---|
| 1 ME | 1029 | 9.00 | 540 | 3.571 | 1.976 | 557 | 52.5 | 541 |
| 2 NH | 771 | 9.00 | 441 | 4.092 | 1.250 | 404 | 57.2 | 524 |
| 3 VT | 462 | 9.00 | 268 | 3.865 | 1.586 | 259 | 58.0 | 561 |
| 4 MA | 5787 | 7.50 | 3060 | 4.870 | 2.351 | 2396 | 52.9 | 414 |
| 5 RI | 968 | 8.00 | 527 | 4.399 | .431 | 397 | 54.4 | 410 |
| 6 CN | 3082 | 10.00 | 1760 | 5.342 | 1.333 | 1408 | 57.1 | 457 |
| 7 NY | 18366 | 8.00 | 8278 | 5.319 | 11.868 | 6312 | 45.1 | 344 |
| 8 NJ | 7367 | 8.00 | 4074 | 5.126 | 2.138 | 3439 | 55.3 | 467 |
| 9 PA | 11926 | 8.00 | 6312 | 4.447 | 8.577 | 5528 | 52.9 | 464 |
| 10 OH | 10783 | 7.00 | 5948 | 4.512 | 8.507 | 5375 | 55.2 | 498 |
| 11 IN | 5291 | 8.00 | 2804 | 4.391 | 5.939 | 3068 | 53.0 | 580 |
| 12 IL | 11251 | 7.50 | 5903 | 5.126 | 14.186 | 5301 | 52.5 | 471 |
| 13 MI | 9082 | 7.00 | 5213 | 4.817 | 6.930 | 4768 | 57.4 | 525 |
| 14 WI | 4520 | 7.00 | 2465 | 4.207 | 6.580 | 2294 | 54.5 | 508 |
| 15 MN | 3896 | 7.00 | 2368 | 4.332 | 8.159 | 2204 | 60.8 | 566 |
| 16 IA | 2883 | 7.00 | 1689 | 4.318 | 10.340 | 1830 | 58.6 | 635 |
| 17 MO | 4753 | 7.00 | 2719 | 4.206 | 8.508 | 2865 | 57.2 | 603 |
| 18 ND | 632 | 7.00 | 341 | 3.718 | 4.725 | 451 | 54.0 | 714 |
| 19 SD | 579 | 7.00 | 419 | 4.716 | 5.915 | 501 | 72.4 | 865 |
| 20 NE | 1525 | 8.50 | 1033 | 4.341 | 6.010 | 976 | 67.7 | 640 |
| 21 KS | 2258 | 7.00 | 1496 | 4.593 | 7.834 | 1466 | 66.3 | 649 |
| 22 DE | 565 | 8.00 | 340 | 4.983 | .602 | 305 | 60.2 | 540 |
| 23 MD | 4056 | 9.00 | 2073 | 4.897 | 2.449 | 1883 | 51.1 | 464 |
| 24 VA | 4764 | 9.00 | 2463 | 4.258 | 4.686 | 2604 | 51.7 | 547 |
| 25 WV | 1781 | 8.50 | 982 | 4.574 | 2.619 | 819 | 55.1 | 460 |
| 26 NC | 5214 | 9.00 | 2835 | 3.721 | 4.746 | 2953 | 54.4 | 566 |
| 27 SC | 2665 | 8.00 | 1460 | 3.448 | 5.399 | 1537 | 54.8 | 577 |
| 28 GA | 4720 | 7.50 | 2731 | 3.846 | 9.061 | 2979 | 57.9 | 631 |
| 29 FL | 7259 | 8.00 | 4084 | 4.188 | 5.975 | 4169 | 56.3 | 574 |
| 30 KY | 3299 | 9.00 | 1626 | 3.601 | 4.650 | 1761 | 49.3 | 534 |
| 31 TN | 4031 | 7.00 | 2088 | 3.640 | 6.905 | 2301 | 51.8 | 571 |
| 32 AL | 3510 | 7.00 | 1801 | 3.333 | 6.594 | 1946 | 51.3 | 554 |
| 33 MS | 2263 | 8.00 | 1309 | 3.063 | 6.524 | 1306 | 57.8 | 577 |
| 34 AR | 1978 | 7.50 | 1081 | 3.357 | 4.121 | 1242 | 54.7 | 628 |
| 35 LA | 3720 | 8.00 | 1813 | 3.528 | 3.495 | 1812 | 48.7 | 487 |
| 36 OK | 2634 | 6.58 | 1657 | 3.802 | 7.834 | 1695 | 62.9 | 644 |
| 37 TX | 11649 | 5.00 | 6595 | 4.045 | 17.782 | 7451 | 56.6 | 640 |
| 38 MT | 719 | 7.00 | 421 | 3.897 | 6.385 | 506 | 58.6 | 704 |
| 39 ID | 756 | 8.50 | 501 | 3.635 | 3.274 | 490 | 66.3 | 648 |
| 40 WY | 345 | 7.00 | 232 | 4.345 | 3.905 | 334 | 67.2 | 968 |

Table 2.1   (cont.)

| STATE | POP | $X_1$ TAX | NLIC | $X_3$ INC | $X_4$ ROAD | FUELC | $X_2$ DLIC | $Y$ FUEL |
|---|---|---|---|---|---|---|---|---|
| 41 CO | 2357 | 7.00 | 1475 | 4.449 | 4.639 | 1384 | 62.6 | 587 |
| 42 NM | 1065 | 7.00 | 600 | 3.656 | 3.985 | 744 | 56.3 | 699 |
| 43 AZ | 1945 | 7.00 | 1173 | 4.300 | 3.635 | 1230 | 60.3 | 632 |
| 44 UT | 1126 | 7.00 | 572 | 3.745 | 2.611 | 666 | 50.8 | 591 |
| 45 NV | 527 | 6.00 | 354 | 5.215 | 2.302 | 412 | 67.2 | 782 |
| 46 WN | 3443 | 9.00 | 1966 | 4.476 | 3.942 | 1757 | 57.1 | 510 |
| 47 OR | 2182 | 7.00 | 1360 | 4.296 | 4.083 | 1331 | 62.3 | 610 |
| 48 CA | 20468 | 7.00 | 12130 | 5.002 | 9.794 | 10730 | 59.3 | 524 |

These data were collected by Christopher Bingham from the *American Almanac for 1974*, except for fuel consumption, which was given in the 1974 *World Almanac*. We shall use these data to study fuel consumption as a function of the other variables. Of particular interest will be the assessment of the relationship between tax rate and fuel consumption. Although other predictors such as the number of fuel consuming vehicles may be relevant, we limit ourselves to the data in Table 2.1, or transformations of them.

Before beginning the analysis, it is useful to see if we can combine or rescale the variables. The predictors NLIC and FUELC are measured for whole states and will vary with state size, while INC is measured per individual so it is not as sensitive to state size. To make these comparable, we convert both FUELC and NLIC to rates per person by dividing them by POP. The resulting quantities, all given in Table 2.1, are‡

$X_1 = $ TAX (cents per gallon).
$X_2 = $ DLIC $ = 100 \times $ NLIC/POP $ = $ percent of population with driver's licenses.
$X_3 = $ INC, average income (thousands of dollars).
$X_4 = $ ROAD (thousands of miles).
$Y = $ FUEL $ = 1000 \times $ FUELC/POP $ = $ motor fuel consumption (gallons per person).

‡In this book, the names $X_1, X_2, \ldots, X_p$ are generic names of predictors, while $Y$ is the generic name for the response. In a particular problem, however, the variables used may well have other names, such as those used in the example. Most computer programs permit assigning 1 to 10 character names for variables, and that example is followed in this book.

Table 2.2   Basic summary statistics

| Variable | $n$ | Average | Variance | Standard Deviation | Minimum | Maximum |
|----------|-----|---------|----------|--------------------|---------|---------|
| TAX  | 48 | 7.6683 | .90396 | .95077 | 5.0000 | 10.000 |
| DLIC | 48 | 57.033 | 30.770 | 5.5470 | 45.100 | 72.400 |
| INC  | 48 | 4.2418 | .32904 | .57362 | 3.0630 | 5.3420 |
| ROAD | 48 | 5.5654 | 12.191 | 3.4915 | .43100 | 17.782 |
| FUEL | 48 | 576.77 | 12518. | 111.89 | 344.00 | 968.00 |

Matrix of Sample Correlations

| | | | | | |
|------|--------|--------|--------|------|------|
| TAX  | 1.000  |        |        |       |       |
| DLIC | −.2880 | 1.000  |        |       |       |
| INC  | .0127  | .1571  | 1.000  |       |       |
| ROAD | −.5221 | −.0641 | .0502  | 1.000 |       |
| FUEL | −.4513 | .6990  | −.2449 | .0190 | 1.000 |
|      | TAX    | DLIC   | INC    | ROAD  | FUEL  |

The variable POP has been removed from the set of interesting predictors because its effects have been partly modeled by redefining the
other quantities to an individual basis. However, a more thorough
analysis than that to be done here would also include examination
of the additional effects of POP.

The basic summary statistics (sample averages, standard deviations,
and correlations between the five variables of interest) are given in
Table 2.2. For convenience, all the variables have been scaled to be of
roughly the same magnitude. This scaling does not affect the relationships between the measured variables; for example, the correlation
between DLIC and INC, equal to 0.1517, would be the same whether
DLIC is expressed as a fraction or as a percentage. The scaling is useful
because all the estimated coefficients will be of about the same size,
and the use of very small and very large numbers can be avoided.

We will return to this example many times.

## 2.1   Adding a predictor to a simple regression model

Before turning to the general multiple regression model, we study the question
of adding a predictor to a simple regression model. Specifically, in the fuel

consumption data, we will consider adding TAX to a model for FUEL *after* DLIC has already been included. In the end, we will obtain an equation of the form

$$\widehat{\text{FUEL}} = \hat{\beta}_0 + \hat{\beta}_1 \text{TAX} + \hat{\beta}_2 \text{DLIC} \qquad (2.3)$$

where as before $\hat{\beta}_2$, for example, is an estimate of the parameter $\beta_2$. The main idea in adding TAX is to explain the part of FUEL that has not already been explained by DLIC. The fitting of one predictor after another is central to multiple regression and is worth careful study.

Figures 2.2a and 2.2b give scatter plots of FUEL versus DLIC and FUEL versus TAX. These plots display the relationships between FUEL and each of these two predictors without regard to the effects of other predictors. If we fit a simple regression model FUEL on DLIC, we find

$$\widehat{\text{FUEL}} = -227.21 + 14.10 \, \text{DLIC} \qquad (2.4)$$

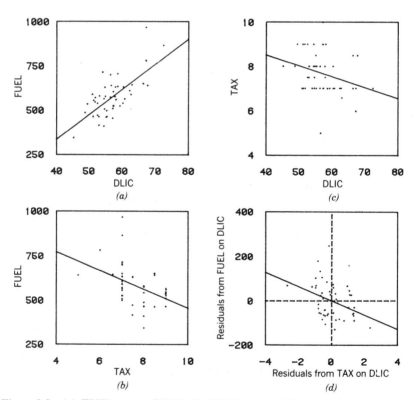

**Figure 2.2**   (*a*) FUEL versus DLIC. (*b*) FUEL versus TAX. (*c*) TAX versus DLIC. (*d*) Added variable plot for TAX after DLIC.

with $R^2 = (0.6990)^2 = 0.4886$, indicating that 48.9% of the variability in FUEL is explained by DLIC, with higher DLIC associated with higher FUEL consumption. Similarly, the regression of FUEL on TAX leads to the fitted model

$$\widehat{\text{FUEL}} = -948.01 - 53.11\,\text{TAX} \tag{2.5}$$

with $R^2 = (-0.4513)^2 = 0.2037$, so TAX explains 20.4% of the variability in FUEL, *ignoring* DLIC, and each penny increase in TAX rate corresponds to an estimated 53.11-gallon decrease per capita in FUEL consumption.

Our goal is to fit a model using both predictors to explain FUEL. What can be said about the proportion of variability in FUEL explained by using DLIC and TAX in a single equation? The answer is, rather little. We can say that the total explained variation must exceed 48.9%, the larger of the two values for variation explained by each variable separately, since knowing both DLIC and TAX must surely be at least as good as knowing just one of them. The total variation will be additive, $48.9\% + 20.4\% = 69.3\%$ *only* if the two variables DLIC and TAX are completely unrelated and measure completely different things. The total can be less than 69.3% if DLIC and TAX are related to each other and are both explaining the same variability. Finally, the total can exceed 69.3% if the two variables interact so that knowing both gives much more information than knowing just one of them. For example, the area of a rectangle may be only poorly determined by either the length or the width alone, but if both length and width are considered in the same model, the area can be determined exactly. It is precisely this inability to predict the relationship between the response and two predictors from the relationships between the response and each predictor separately that makes multiple regression both rich and complicated.

A crucial step in understanding this problem is making a scatter plot of the two predictors, TAX and DLIC. This plot is given in Figure 2.2c, showing that the two predictors are related; states with more drivers per capita tend to have somewhat lower tax rates. The fitted regression equation is

$$\widehat{\text{TAX}} = 10.48 - 0.0494\,\text{DLIC} \tag{2.6}$$

The interesting problem is to find the unique effect of adding TAX to a model that already includes DLIC. We are therefore concerned with modeling that part of FUEL that is not explained by DLIC with the part of TAX that is not explained by DLIC. Graphically, this requires examination of the scatter plot of the residuals from the regression of FUEL on DLIC versus the residuals from the regression of TAX on DLIC or the unexplained part of FUEL versus the unexplained part of TAX. These residuals, obtained from fitting (2.4) and (2.6), are plotted in Figure 2.2d. Figure 2.2b gives the relationship between FUEL and TAX *ignoring* DLIC, while Figure 2.2d *adjusts* for DLIC. If Figure 2.2d shows a stronger relationship than does Figure 2.2b, then TAX and DLIC interact to explain more than 69.3%, while if the relationship is weaker, then the total explained variability is less than 69.3%. The latter seems to be the case here.

If we fit the simple regression line to Figure 2.2d, the fitted line has zero intercept, since the averages of the two plotted variables are zero, and estimated slope $\hat{\beta}_1 = -32.08$. It turns out that this is exactly the estimate of $\beta_1$ in (2.3), the multiple regression model. Figure 2.2d is an example of an *added variable plot*, which will be discussed more generally in Section 2.4.

Thus there are now estimates of $\beta_1$ from two models:

$$\hat{\beta}_1 = -53.11 \quad \textit{ignoring} \text{ DLIC}$$

$$\hat{\beta}_1 = -32.08 \quad \textit{adjusted for} \text{ DLIC} \tag{2.7}$$

While both of these indicate that higher TAX rates are associated with lower per capita FUEL consumption, the latter model, adjusting for DLIC, suggests that the magnitude of this effect is only about 60% as large as one might think if DLIC were ignored. In other regression problems, slope estimates for the same predictor, but in different models, may be even more wildly different, changing signs, magnitude, and significance. This naturally complicates the interpretation of fitted models.

Just as $\hat{\beta}_1$ in (2.3) is an estimate of the effect of TAX adjusted for DLIC, so $\hat{\beta}_2$ is an estimate of the effect of DLIC adjusted for the effect of TAX. Thus $\hat{\beta}_2 = 12.51$ is computed from the regression of the residuals from the regression of FUEL on TAX on the residuals from the regression of DLIC on TAX. In all multiple regression equations, the $\hat{\beta}_j$'s are adjusted for all other predictors in the model.

To complete estimation of (2.3), we can compute $\hat{\beta}_0$ from the formula

$$\hat{\beta}_0 = \bar{y} - \sum_{j=1}^{p} \hat{\beta}_j \bar{x}_j = 108.97 \tag{2.8}$$

where the sample averages are given in Table 2.2, and the $\hat{\beta}_j$'s are given above. The fitted equation for the two predictors is

$$\widehat{\text{FUEL}} = 108.97 - 32.08\,\text{TAX} + 12.51\,\text{DLIC}$$

**Partial correlation.**   In simple regression the relationship between $Y$ and $X_1$ can be measured by the sample correlation $r_{YX_1}$. Similarly, the strength of the relationship between $X_1$ and $Y$ adjusted for $X_2$ can be summarized by the correlation between the residuals for $Y$ on $X_2$ and the residuals for $X_1$ on $X_2$, such as those plotted in Figure 2.2d. This correlation is called a *partial correlation*, symbolically $r_{YX_1 | X_2}$, read as partial correlation between $Y$ and $X_1$ adjusted for $X_2$.

**Orthogonality.** Two variables $X_1$ and $X_2$ are orthogonal if the regression of $Y$ on $X_1$ adjusted for $X_2$ is *identical* to the regression of $Y$ on $X_1$ ignoring $X_2$. This pleasant situation will occur if the sample correlation between $X_1$ and $X_2$, $r_{X_1 X_2}$, is *exactly* zero. When $X_1$ and $X_2$ are orthogonal, then the effect of each is unambiguous, and, consequently, experiments are often designed to have orthogonal variables.

## 2.2 Regression in matrix notation

Matrix notation will simplify most of the results used in multiple regression. In general, a vector or a matrix will be denoted by boldface letters like $\mathbf{X}$, $\mathbf{e}$, and $\boldsymbol{\beta}$. Elements of vectors and matrices are of the form $x_{ij}$, $e_i$, and $\beta_j$.

Let $\mathbf{Y}$ and $\mathbf{e}$ be $n \times 1$ vectors whose elements are given by the $y_i$'s and the $e_i$'s of (2.2), for example,

$$\mathbf{Y} = \begin{bmatrix} y_1 \\ y_2 \\ \vdots \\ y_n \end{bmatrix} \qquad \mathbf{e} = \begin{bmatrix} e_1 \\ e_2 \\ \vdots \\ e_n \end{bmatrix} \tag{2.9}$$

Also, define $\boldsymbol{\beta}$ to be the vector parameter of length $(p+1) \times 1$, including the intercept $\beta_0$,

$$\boldsymbol{\beta} = \begin{bmatrix} \beta_0 \\ \beta_1 \\ \vdots \\ \beta_p \end{bmatrix} \tag{2.10}$$

Next, define $\mathbf{X}$ to be an $n \times (p+1)$ matrix given by

$$\mathbf{X} = \begin{bmatrix} 1 & x_{11} & x_{12} \cdots x_{1p} \\ 1 & x_{21} & x_{22} \cdots x_{2p} \\ \vdots & \vdots & \vdots \\ 1 & x_{n1} & x_{n2} \cdots x_{np} \end{bmatrix} \tag{2.11}$$

As a matter of notational convenience, we will define $p'$ to be the number of columns in $\mathbf{X}$. For problems with an intercept term, $p' = p + 1$, and later we shall see that in problems without an intercept term, $p' = p$.

The matrix $\mathbf{X}$ gives all of the observed values of the predictors, appended to a column of 1's as the leftmost column. The $i$th row of $\mathbf{X}$ (also, the rows of $\mathbf{e}$ and $\mathbf{Y}$) corresponds to values for the $i$th case in the data; the columns of $\mathbf{X}$ correspond to the different predictors. We shall call the leftmost column of $\mathbf{X}$, the column of 1's, the zeroth column, since this will be the column that corresponds to $\beta_0$. The next column, corresponding to the first predictor $X_1$ and parameter $\beta_1$, will be the first column of $\mathbf{X}$, and so on.

Using these quantities, the multiple regression equation (2.2) can be written in matrix terms as

$$\mathbf{Y} = \mathbf{X}\boldsymbol{\beta} + \mathbf{e} \tag{2.12}$$

The reader who is unfamiliar with matrices should use the definitions (2.9) to (2.11) to perform the indicated multiplications and additions in (2.12), and show that the results are, for the $i$th row, exactly the same as equation (2.2).

For the fuel consumption data, the first few and last few rows of the matrices $\mathbf{X}$ and $\mathbf{Y}$ are given by

$$\mathbf{X} = \begin{bmatrix} 1 & 9.0 & 52.5 & 3.571 & 1.976 \\ 1 & 9.0 & 57.2 & 4.092 & 1.250 \\ 1 & 9.0 & 58.0 & 3.865 & 1.586 \\ & & \vdots & & \\ 1 & 9.0 & 57.1 & 4.476 & 3.942 \\ 1 & 7.0 & 62.3 & 4.296 & 4.083 \\ 1 & 7.0 & 59.3 & 5.002 & 9.794 \end{bmatrix} \qquad \mathbf{Y} = \begin{bmatrix} 541 \\ 524 \\ 561 \\ \vdots \\ 510 \\ 610 \\ 524 \end{bmatrix}$$

$\boldsymbol{\beta}$ is a parameter vector (of length $p' = 4 + 1 = 5$), and of course the values in $\boldsymbol{\beta}$ are unknown. The error vector $\mathbf{e}$ is unobserved.

**Variance-covariance matrix of e.**    The error term is a vector of random variables called a random vector (as in Appendix 2A.2). The assumptions concerning the $e_i$'s given in Chapter 1 are summarized in matrix form as $E(\mathbf{e}) = \mathbf{0}$, $\text{var}(\mathbf{e}) = \sigma^2 \mathbf{I}_n$, where $\text{var}(\mathbf{e})$ means the variance-covariance matrix of $\mathbf{e}$, $\mathbf{I}_n$ is the $n \times n$ identity matrix, and $\mathbf{0}$ is an $n \times 1$ vector of zeroes. If we add the assumption that each $e_i$ is normally distributed we will write

$$\mathbf{e} \sim N(\mathbf{0}, \sigma^2 \mathbf{I}_n) \tag{2.13}$$

**Least squares estimators.**    The least squares estimate $\hat{\boldsymbol{\beta}}$ of $\boldsymbol{\beta}$ is chosen to minimize the residual sum of squares. Suppose we let $\mathbf{x}_i^T$ be the $i$th row of $\mathbf{X}$, $i = 1, \ldots, n$; the transpose is required because of the convention that all vectors are column vectors. We choose $\hat{\boldsymbol{\beta}}$ to minimize the function

$$RSS(\boldsymbol{\beta}) = \sum (y_i - \mathbf{x}_i^T \boldsymbol{\beta})^2 = (\mathbf{Y} - \mathbf{X}\boldsymbol{\beta})^T (\mathbf{Y} - \mathbf{X}\boldsymbol{\beta}) \tag{2.14}$$

The least squares estimators can be found from (2.14) by differentiation in a matrix analogue to the development of Appendix 1A.2. They can also be found using an argument that leads to a numerically stable computational method, as in Appendix 2A.3. The basic idea is to change the original problem into an equivalent problem that is easy to solve. The least squares estimate $\hat{\boldsymbol{\beta}}$ of $\boldsymbol{\beta}$ is given by the following formula:

$$\hat{\boldsymbol{\beta}} = (\mathbf{X}^T\mathbf{X})^{-1}\mathbf{X}^T\mathbf{Y} \quad \rightarrow \quad (2A \cdot 13) \tag{2.15}$$

provided that $(\mathbf{X}^T\mathbf{X})^{-1}$ exists. The estimator $\hat{\boldsymbol{\beta}}$ depends only on the sufficient statistics $(\mathbf{X}^T\mathbf{X})$ and $(\mathbf{X}^T\mathbf{Y})$, which are matrices of uncorrected sums of squares and cross products.

As in simple regression, a numerically superior but equivalent formula based on corrected sums of squares and cross products can be derived. Let $\bar{\mathbf{x}} = (\bar{x}_1, \bar{x}_2, \ldots \bar{x}_p)^T$ be the $p \times 1$ vector of sample averages of the $X$'s, and define $\mathscr{X}$ to be the $n \times p$ matrix of the original data with averages subtracted off; the $(i, j)$th element of $\mathscr{X}$ is $x_{ij} - \bar{x}_j$. Similarly, define $\mathscr{Y}$ to be an $n \times 1$ vector with $i$th element $y_i - \bar{y}$. Then the matrices of corrected cross products $\mathscr{X}^T\mathscr{X}$, $\mathscr{Y}^T\mathscr{Y}$ and $\mathscr{X}^T\mathscr{Y}$ are

$$\mathscr{X}^T\mathscr{X} = \begin{bmatrix} \sum (x_{i1} - \bar{x}_1)^2 & \cdots & \sum (x_{i1} - \bar{x}_1)(x_{ip} - \bar{x}_p) \\ \vdots & & \vdots \\ \sum (x_{i1} - \bar{x}_1)(x_{ip} - \bar{x}_p) & \cdots & \sum (x_{ip} - \bar{x}_p)^2 \end{bmatrix} \tag{2.16}$$

$$\mathscr{X}^T\mathscr{Y} = \begin{bmatrix} \sum (x_{i1} - x_1)(y_i - \bar{y}) \\ \vdots \\ \sum (x_{ip} - \bar{x}_p)(y_i - \bar{y}) \end{bmatrix} \qquad \mathscr{Y}^T\mathscr{Y} = \left( \sum (y_i - \bar{y})^2 \right)$$

Generally, these matrices are presented as a single $(p+1) \times (p+1)$ matrix $\mathbf{T}$, where

$$p \text{ columns} \qquad 1 \text{ column}$$

$$\mathbf{T} = \begin{pmatrix} \mathscr{X}^T\mathscr{X} & \mathscr{X}^T\mathscr{Y} \\ \mathscr{Y}^T\mathscr{X} & \mathscr{Y}^T\mathscr{Y} \end{pmatrix} \begin{array}{l} p \text{ rows} \\ 1 \text{ row} \end{array} \tag{2.17}$$

Most computer programs print a matrix $\mathbf{S}$ of sample covariances and variances, obtained from $\mathbf{T}$ by dividing each element by $(n-1)$, $\mathbf{S} = (n-1)^{-1}\mathbf{T}$. The sample correlation matrix given in Table 2.2 is obtained from $\mathbf{S}$ by the rule $r_{ij} = s_{ij}/\sqrt{s_{ii}s_{jj}}$. Since $\mathbf{S}$ is symmetric, only the lower triangular part is usually printed.

Suppose we let

$$\hat{\boldsymbol{\beta}}^* = (\mathscr{X}^T\mathscr{X})^{-1}\mathscr{X}^T\mathscr{Y}$$

Then,

$$\hat{\beta}_0 = \bar{y} - \hat{\beta}^{*T}\bar{x}$$

$$\hat{\beta} = \begin{pmatrix} \hat{\beta}_0 \\ \hat{\beta}* \end{pmatrix} \tag{2.18}$$

In this book, we follow the common practice of presenting formulas in the numerically unstable uncorrected form, because derivations are generally easier to follow. However, the corrected forms can be computed directly, possibly using the method outlined in Appendix 1A.5, and they should be preferred.

**Derived quantities.** Once $\hat{\beta}$ is computed, we can define several related quantities. The vector of *fitted values* $\hat{\mathbf{Y}} = \mathbf{X}\hat{\beta}$ has $i$th element equal to $\hat{y}_i = \mathbf{x}_i^T\hat{\beta}$. The vector of residuals is $\hat{\mathbf{e}} = \mathbf{Y} - \hat{\mathbf{Y}}$, with $i$th element $\hat{e}_i = y_i - \hat{y}_i = y_i - \mathbf{x}_i^T\hat{\beta}$. The function (2.14) when evaluated at $\hat{\beta}$ is called the _residual sum of squares_, abbreviated RSS, and has the form

$$RSS = \hat{\mathbf{e}}^T\hat{\mathbf{e}} = (\mathbf{Y} - \mathbf{X}\hat{\beta})^T(\mathbf{Y} - \mathbf{X}\hat{\beta}) \tag{2.19}$$

**Properties of the estimates.** Additional properties of the least squares estimates are derived in Appendix 2A.3 and are only summarized here. Assuming that $E(\mathbf{e}) = \mathbf{0}$ and $\text{var}(\mathbf{e}) = \sigma^2\mathbf{I}_n$, then $\hat{\beta}$ is unbiased, $E(\hat{\beta}) = \beta$, and

$$\text{var}(\hat{\beta}) = \sigma^2(\mathbf{X}^T\mathbf{X})^{-1} \tag{2.20}$$

An estimate of $\sigma^2$ is obtained, according to the rule in Section 1.3, by

$$\hat{\sigma}^2 = \frac{RSS}{n - p'} \tag{2.21}$$

Several formulas for RSS can be computed by substituting the value for $\hat{\beta}$ into (2.19) and simplifying.

$$\begin{aligned} RSS &= \mathbf{Y}^T\mathbf{Y} - \hat{\beta}^T(\mathbf{X}^T\mathbf{X})\hat{\beta} \\ &= \mathbf{Y}^T\mathbf{Y} - \hat{\beta}^T\mathbf{X}^T\mathbf{Y} \\ &= \mathscr{Y}^T\mathscr{Y} - \hat{\beta}^{*T}(\mathscr{X}^T\mathscr{X})\hat{\beta}* \\ &= \mathscr{Y}^T\mathscr{Y} - \hat{\beta}^T(\mathbf{X}^T\mathbf{X})\hat{\beta} + n\bar{y}^2 \end{aligned} \tag{2.22}$$

As in Section 1.3, if $\mathbf{e}$ is normally distributed, then $(n - p')\hat{\sigma}^2/\sigma^2$ has a $\chi^2(n - p')$ distribution.

By substituting $\hat{\sigma}^2$ for $\sigma^2$ in (2.20), we find the estimated variance of $\hat{\beta}$, $\widehat{\text{var}}(\hat{\beta})$, to be

$$\widehat{\text{var}}(\hat{\beta}) = \hat{\sigma}^2(\mathbf{X}^T\mathbf{X})^{-1} \tag{2.23}$$

**Simple regression in matrix terms.**   For simple regression, $\mathbf{X}$ and $\mathbf{Y}$ are given by

$$\mathbf{X} = \begin{bmatrix} 1 & x_1 \\ 1 & x_2 \\ \vdots & \\ 1 & x_n \end{bmatrix} \qquad \mathbf{Y} = \begin{bmatrix} y_1 \\ y_2 \\ \vdots \\ y_n \end{bmatrix}$$

and thus

$$(\mathbf{X}^T\mathbf{X}) = \begin{bmatrix} n & \sum x_i \\ \sum x_i & \sum x_i^2 \end{bmatrix} \qquad \mathbf{X}'\mathbf{Y} = \begin{bmatrix} \sum y_i \\ \sum x_i y_i \end{bmatrix} \tag{2.24}$$

$(\mathbf{X}^T\mathbf{X})^{-1}$ can be shown to be

$$(\mathbf{X}^T\mathbf{X})^{-1} = \frac{1}{SXX}\begin{pmatrix} \sum x_i^2/n & -\bar{x} \\ -\bar{x} & 1 \end{pmatrix} \tag{2.25}$$

so that

$$\hat{\boldsymbol{\beta}} = \begin{pmatrix} \hat{\beta}_0 \\ \hat{\beta}_1 \end{pmatrix} = (\mathbf{X}^T\mathbf{X})^{-1}\mathbf{X}^T\mathbf{Y} = \frac{1}{SXX}\begin{pmatrix} \sum x_i^2/n & -\bar{x} \\ -\bar{x} & 1 \end{pmatrix}\begin{bmatrix} \sum y_i \\ \sum x_i y_i \end{bmatrix}$$

$$= \begin{bmatrix} \bar{y} - \hat{\beta}_1\bar{x} \\ \dfrac{SXY}{SXX} \end{bmatrix}$$

as found previously. Also, since $\sum x_i^2/(nSXX) = 1/n + \bar{x}^2/SXX$, the variances and covariances for $\hat{\beta}_0$ and $\hat{\beta}_1$ found in the previous chapter are identical to those given by $\sigma^2(\mathbf{X}^T\mathbf{X})^{-1}$.

In the deviations from the sample average form, the results are simpler, since

$$\boldsymbol{\mathscr{X}}^T\boldsymbol{\mathscr{X}} = SXX \qquad \boldsymbol{\mathscr{X}}^Y\boldsymbol{\mathscr{y}} = SXY$$

and

$$\hat{\beta}_1 = (\boldsymbol{\mathscr{X}}^T\boldsymbol{\mathscr{X}})^{-1}\boldsymbol{\mathscr{X}}^T\boldsymbol{\mathscr{y}} = \frac{SXY}{SXX}$$

$$\hat{\beta}_0 = \bar{y} - \hat{\beta}_1\bar{x}$$

**Fuel consumption data (continued).**   We shall now fit the model with all $p = 4$ predictors to the fuel consumption data. We will write FUEL on TAX DLIC INC ROAD to mean "fit the model FUEL $= \hat{\beta}_0 + \hat{\beta}_1$TAX $+ \hat{\beta}_2$DLIC $+ \hat{\beta}_3$INC $+ \hat{\beta}_4$ROAD." For these data and this model, $(\mathbf{X}^T\mathbf{X})^{-1}$

**Table 2.3   $(\mathbf{X}^T\mathbf{X})^{-1}$ for the fuel consumption data. $(\mathscr{X}^T\mathscr{X})^{-1}$ is given by the 4 × 4 lower right submatrix.**

| Intercept | 7.8301941 | −.4265133 | −.0611076 | −.1495090 | −.0753492 |
|---|---|---|---|---|---|
| TAX | −.4265133 | .0382636 | .0022158 | −.0059137 | .0057148 |
| DLIC | −.0611076 | .0022158 | .0008411 | −.0014500 | .0004127 |
| INC | −.1495090 | −.0059137 | −.0014500 | .0674600 | −.0015445 |
| ROAD | −.0753492 | .0057148 | .0004127 | −.0015445 | .0026126 |
| | Intercept | TAX | DLIC | INC | ROAD |

is given in Table 2.3, and the matrix **T** [equation (2.17)] is given by

$$\mathbf{T} = \begin{pmatrix} 42.48627 & -71.39733 & 0.32465 & -81.46385 & -2256.28833 \\ -71.39733 & 1446.16667 & 23.48977 & -58.37527 & 20388.66667 \\ 0.32465 & 23.48977 & 15.46508 & 4.72193 & -738.62083 \\ -81.46385 & -58.37527 & 4.72193 & 572.95925 & 349.62058 \\ -2256.28833 & 20388.66667 & -738.62083 & 349.62058 & 588366.47917 \end{pmatrix}$$

Using the formulas based on corrected sums of squares in this chapter, the estimate $\hat{\boldsymbol{\beta}}^*$ is computed to be

$$\hat{\boldsymbol{\beta}}^* = (\mathscr{X}^T\mathscr{X})^{-1}\mathscr{X}^T\mathscr{Y} = \begin{pmatrix} \hat{\beta}_1 \\ \hat{\beta}_2 \\ \hat{\beta}_3 \\ \hat{\beta}_4 \end{pmatrix} = \begin{pmatrix} -34.790149 \\ 13.364494 \\ -66.588752 \\ -2.425889 \end{pmatrix}$$

The estimated intercept is

$$\hat{\beta}_0 = \bar{y} - \hat{\boldsymbol{\beta}}^{*T}\bar{\mathbf{x}} = 377.2911$$

and the residual sum of squares is

$$RSS = \mathscr{Y}^T\mathscr{Y} - \hat{\boldsymbol{\beta}}^{*T}(\mathscr{X}^T\mathscr{X})\hat{\boldsymbol{\beta}}^*$$
$$= 588{,}366.48 - 399{,}316.51 = 189{,}049.97$$

The residual mean square is

$$\hat{\sigma}^2 = \frac{RSS}{n-p'} = \frac{189{,}049.97}{48-(4+1)} = 4396.5 \text{ (43 d.f.)}$$

The standard errors and estimated covariances of the $\hat{\beta}_j$'s are found from $\hat{\sigma}^2$ and $(\mathbf{X}^T\mathbf{X})^{-1}$. For example,

$$\text{se}(\hat{\beta}_0) = \hat{\sigma}\sqrt{7.83019} = 185.54$$
$$\text{se}(\hat{\beta}_4) = \hat{\sigma}\sqrt{0.0026126} = 3.3892$$
$$\text{cov}(\hat{\beta}_1, \hat{\beta}_2) = \hat{\sigma}^2(0.0022158) = 0.1469$$

**Table 2.4   Computer program regression summary** *change regs order*
**for the regression of FUEL on TAX DLIC INC** *won't Δ βⱼ se or t-value*
**ROAD**

| Variable | Estimate | Standard error | $t$-value |
|---|---|---|---|
| Intercept | 377.2911 | 185.5412 | 2.03 |
| TAX | −34.79015 | 12.97020 | −2.68 |
| DLIC | 13.36449 | 1.922981 | 6.95 |
| INC | −66.58875 | 17.22175 | −3.87 |
| ROAD | −2.425889 | 3.389174 | −.72 → take out first |

$\hat{\sigma}^2 = 4396.511$, d.f. $= 43$, $R^2 = 0.6787$

*BS*

In most computer programs, the usual output obtained is somewhat less than that given here. The results in Table 2.4 are more typical of what might be expected. The first column gives the labels of the predictors. The second column gives the corresponding $\hat{\beta}_j$'s. The third column gives $\hat{\sigma}$ times the square root of the appropriate diagonal entry of $(\mathbf{X}^T\mathbf{X})^{-1}$. The last column ($t$-value) is the ratio $\hat{\beta}_j/\mathrm{se}(\hat{\beta}_j)$, to be discussed shortly. Various other summary statistics, such as the number of d.f. for error, $\hat{\sigma}^2$ and/or $\hat{\sigma}$, and $R^2$ are generally also reported. Sometimes an excessive number of digits is printed, as is the case here. In these data, rounding results to at most four significant digits is appropriate.

**Predictions and fitted values.**   Consider first the question of prediction. We have observed a new $p' \times 1$ vector of predictors, $\mathbf{x}_*$, for which the response $y_*$ is as yet unobserved. The problem is to use the data to predict $y_*$. In exactly the same way as was done in simple regression, the point prediction is just $\tilde{y}_* = \mathbf{x}_*^T\hat{\boldsymbol{\beta}}$. The standard error of prediction, $\mathrm{sepred}(\tilde{y}_*|\mathbf{x}_*)$, using Appendix 2A.2, is

$$\mathrm{sepred}(\tilde{y}_*|\mathbf{x}_*) = \hat{\sigma}\sqrt{(1 + \mathbf{x}_*^T(\mathbf{X}^T\mathbf{X})^{-1}\mathbf{x}_*)} \tag{2.26}$$

Similarly, the estimated average of all possible units with value $\mathbf{x}$ for the predictors is given by $\hat{y} = \mathbf{x}^T\hat{\boldsymbol{\beta}}$, with standard error $\mathrm{sefit}(\hat{y}|\mathbf{x})$ given by

$$\mathrm{sefit}(\hat{y}|\mathbf{x}) = \hat{\sigma}\sqrt{(\mathbf{x}^T(\mathbf{X}^T\mathbf{X})^{-1}\mathbf{x})} \tag{2.27}$$

**Computational considerations.**   Least squares estimators could be computed directly using formulas (2.15) or (2.18), and related statistics could be computed as outlined in this section. However, the use of this approach will often lead to numerical problems. The main computational problems come in actually forming the matrix $\mathbf{X}^T\mathbf{X}$ and finding its inverse. This is generally handled in one of three ways. The most common method uses a technique called Gaussian elimination and is implemented in the Sweep algorithm outlined in Problem 2.7. This method assumes that the matrix $\mathbf{T}$, equation (2.17), has been accurately computed. The Sweep algorithm does, however, compute the inverse of $\mathbf{X}^T\mathbf{X}$. The second method begins with the matrix $\mathbf{T}$, but uses a matrix decomposition to avoid calculation of $(\mathbf{X}^T\mathbf{X})^{-1}$. This method is called the *Cholesky decomposition* of $\mathbf{T}$. The third method avoids computing $\mathbf{T}$, and performs all calculations directly on the original data. These last two methods are very closely related, and some aspects of both are outlined in Problem 2.4. The *Linpack Users Guide* (Dongarra et al., 1979) provides a good source of computational algorithms for least squares problems.

## 2.3   The analysis of variance

For multiple regression, the analysis of variance is a very rich technique that is used to divide variability and to compare models that include different sets of variables. In the overall analysis of variance, the full model

$$\mathbf{Y} = \mathbf{X}\boldsymbol{\beta} + \mathbf{e} \qquad (2.28)$$

is compared to the model with no $X$ variables,

$$\mathbf{Y} = \beta_0 \mathbf{1} + \mathbf{e} \qquad (2.29)$$

where $\mathbf{1}$ is an $n \times 1$ vector of ones. These correspond to (1.21) and (1.18), respectively. Thus, for model (2.29), $\hat{\beta}_0 = \bar{y}$ and the residual sum of squares is $SYY$. For model (2.28), on the other hand, the estimate of $\boldsymbol{\beta}$ is given in (2.15) and $RSS$ is given in (2.19). Clearly, we must have $RSS < SYY$, and the difference between these two,

$$SSreg = SYY - RSS \qquad (2.30)$$

corresponds to the sum of squares of $Y$ explained by the larger model that is not explained by the smaller model. The number of degrees of freedom associated with $SSreg$ is equal to the number of d.f. in $SYY$ minus the number of d.f. in $RSS$, which equals $p$.

These results are summarized in the analysis of variance table:

| Source | d.f. | SS | MS |
|---|---|---|---|
| | | Analysis of Variance (Overall) | |
| Regression on $X_1, \ldots, X_p$ | $p$ | $SSreg$ | $SSreg/p$ |
| Residual | $n - p'$ | $RSS$ | $RSS/(n - p') = \hat{\sigma}^2$ |
| Total | $n - 1$ | $SYY$ | |

We can judge the importance of the regression on the $X$'s by determining if $SSreg$ is sufficiently large by comparing the ratio of the mean square for regression to $\hat{\sigma}^2$ to the $F(p, n-p')$ distribution. If the computed $F$ exceeds a convenient critical value, then we would judge that knowledge of the $X$'s provides a significantly better model than does no knowledge of them. The ratio computed will have an exact $F$ distribution if the errors are NID(0, $\sigma^2$) and the NH is true. The hypothesis tested by this $F$-test is

NH:   model (2.29) applies,   $\boldsymbol{\beta}^* = 0$
AH:   model (2.28) applies,   $\boldsymbol{\beta}^* \neq 0$

**The coefficient of determination.**   As with simple regression the ratio

$$R^2 = \frac{SYY - RSS}{SYY} = \frac{SSreg}{SYY} \tag{2.31}$$

gives the proportion of variability in $Y$ explained by regression on the $X$'s. In addition, one can show that the value $R^2$ is the square of the *multiple correlation coefficient* between $Y$ and the $X$'s: it is the square of the maximum correlation between $Y$ and any linear function of the $X$'s.

**Fuel consumption data.**   The overall analysis of variance table is given by

| Source | d.f. | SS | MS | F |
|--------|------|------|------|------|
| Regression | 4 | 399,316 | 99,829 | 22.70 |
| Residual | 43 | 189,050 | $4397 = \hat{\sigma}^2$ | |
| Total | 47 | 588,366 | | |

Since $F = 22.7$ exceeds $F(0.01; 4, 43) = 3.79$ by a large margin, one would be led to suspect that at least some of the $X$'s are in fact related to fuel consumption. The value of $R^2 = 399316/588366 = 0.68$, indicating that about 68% of the observed variability in the response is modeled by the $X$'s. Without experience in problems like this one, it is not easy to decide if 68% is a lot or a little.

Incidentally, this example points out that the computation of the overall $F$ statistic is not always interesting. Often it is known a priori that the variables are related so very large values of the test statistic are expected. Of more interest is examination of other hypotheses concerning some of the variables.

**Hypotheses concerning one of the predictors.**   In many problems, obtaining information on the usefulness of one of the predictors is of interest.

Can we do as well modeling fuel consumption from, for example, just DLIC, ROAD, and INC as we do from all four variables? This question can be rephrased in a more suggestive manner: if DLIC, ROAD, and INC are known, will knowledge of TAX represent a significant improvement? The following procedure can be used: fit the model that excludes TAX, and obtain the residual sum of squares for that model. Then, fit a second model including TAX, and get the residual sum of squares for this model. Subtracting the residual sum of squares for the larger model from the residual sum of squares for the smaller model will give the sum of squares for regression on TAX after adjusting for the variables that are already included in the model (DLIC, ROAD, and INC). This computation can be done exactly as outlined. Begin by performing the regression of FUEL on ROAD, DLIC, and INC. The residual sum of squares for this model is 220,682 (and $\hat{\sigma}^2 = 5015$). The residual sum of squares for the full model has already been given as 189,050 (and $\hat{\sigma}^2 = 4397$). The sum of squares for regression on TAX after the others is $220,682 - 189,050 = 31,632$ (and the estimated $\hat{\sigma}^2$ is reduced by about 7%). This can be summarized in the following analysis of variance table:

| Source | d.f. | SS | MS | F |
|---|---|---|---|---|
| Regression on ROAD, INC, DLIC | 3 | 367,684 | 122,561 | |
| TAX after others | 1 | 31,632 | 31,632 | 7.19 |
| Residual | 43 | 189,050 | 4,397 | |

The *SS* for regression on ROAD, INC, DLIC is found by subtracting the $SSreg(\text{ROAD, INC, DLIC, TAX}) - SSreg(\text{TAX after others}) = 399,316 - 31,632 = 367,684$. The ratio of mean squares $F = 31,632/4,397 = 7.19$ is the statistic used to test the usefulness of TAX after the other variables are already included in the model; it says nothing about the usefulness of the other variables. It is compared to the *F* distribution with 1 and $n - p' = 43$ degrees of freedom. In the example, TAX appears to be a significant predictor after adjusting for the others, since $F(0.01; 1, 43) = 7.26$, giving a *p*-value near 0.01. We call this a *partial F*-test. Specifically, this *F* tests the hypothesis

$$\text{NH:} \quad \beta_1 = 0; \quad \beta_0, \beta_2, \beta_3, \beta_4 \text{ arbitrary}$$
$$\text{AH:} \quad \beta_1 \neq 0; \quad \beta_0, \beta_2, \beta_3, \beta_4 \text{ arbitrary}$$
(2.32)

**Relationship to the *t* statistic.** Another reasonable procedure for testing the importance of TAX is simply to compare the estimate of the coefficient divided by its standard error to the *t* distribution with 43 degrees of freedom. It can be shown that the square of the *t* ratio is the same number as the *F* ratio just computed, so these two procedures are identical. Therefore, the $t$ statistic tests hypotheses concerning the importance of variables adjusted for all the other variables in the model, *not* ignoring them.

For example, the $t$ statistic for TAX is, from Table 2.4,

$$t = \frac{-34.79015}{12.97020} = -2.68$$

which would be compared to $t(43)$ to find critical values. The hypothesis tested by this statistic is given in (2.32). Also, one finds

$$t^2 = (-2.68)^2 = 7.18$$

numerically identical within rounding error to the value obtained for the $F$-test for this hypothesis. A $t$-test that any of the $\beta_j$'s has a specific value (given that all other $\beta$'s are arbitrary) can be carried out as in Section 1.7.

**Other tests of hypotheses.**   We have obtained a test of a hypothesis concerning the effects of TAX adjusted for all the other variables in the problem. Equally well, we could obtain tests for the effect of TAX adjusting for *some* of the other variables, or for none of the other variables. In general, these tests will not be equivalent: a variable can be judged to be a useful predictor ignoring other variables, but judged as useless when adjusted for them. Furthermore, a predictor that is useless by itself may become important when considered in concert with the other variables. The outcome of these tests depends on the relationship between the $X$'s as reflected in $\mathbf{X}^T\mathbf{X}$, or usually more clearly in the sample correlations. Therefore, a problem of order of fitting the various $X$'s is apparent in multiple regression.

**Sequential analysis of variance tables.**   By separating TAX from the other three predictors, $SSreg$ is divided into two pieces, one for fitting the first three variables and one for fitting TAX after the other three. This subdivision can be continued by dividing the sum of squares for regression into pieces for each variable. Unless all the predictors are orthogonal, this breakdown is not unique. For example, we could first fit DLIC, then TAX adjusted for DLIC, then INC adjusted for DLIC and TAX, and finally ROAD adjusted for the other three. The resulting table is given in Table 2.5(a). Alternatively, we could fit in the order ROAD, INC, DLIC, and then TAX as in Table 2.5(b). As can be seen, the resulting associated sums of squares are quite different: the sum of squares for ROAD when fit first is 213, but adjusted for the other three it has a sum of squares of 2252, larger by a factor of 10, but still not very large compared to $\hat{\sigma}^2$.

**Table 2.5   Two analysis of variance tables
with different orders of fitting**

|  | Source | d.f. | SS | MS |
|---|---|---|---|---|
| | *(a) First analysis* | | | |
| First | DLIC | 1 | 287,448 | 287,448 |
| Then | TAX | 1 | 40,084 | 40,084 |
| Then | INC | 1 | 69,532 | 69,532 |
| Then | ROAD | 1 | 2,252 | 2,252 |
| | Residual | 43 | 189,050 | 4,397 |
| | *(b) Second analysis* | | | |
| First | ROAD | 1 | 213 | 213 |
| Then | INC | 1 | 35,642 | 35,642 |
| Then | DLIC | 1 | 331,828 | 331,829 |
| Then | TAX | 1 | 31,632 | 31,632 |
| | Residual | 43 | 189,050 | 4,397 |

## 2.4   Added variable plots

In simple regression, the relationship between the response $Y$ and the predictor $X$ is displayed by a scatter plot. In multiple regression, the situation is complicated by the relationship between the several predictors, so a scatter plot between $Y$ and any one of the $X$'s need not reflect the relationship when adjusted for the other $X$'s. The *added variable plot* is a graphical device that allows the display of just this relationship.

We have already seen an example of an added variable plot in Figure 2.2*d*, which shows the relationship between FUEL and TAX adjusted for DLIC. The general procedure is as follows:

**1.** Consider the model

$$Y = \beta_0 + \beta_1 X_1 + \cdots + \beta_p X_p + e \qquad (2.33)$$

We are interested in a graphical summary of the relationship between $X_k$ and $Y$, adjusted for the other $X$'s $(1 \leqslant k \leqslant p)$.

**2.** Fit the regression of $Y$ on all the $X$'s *except* $X_k$, and save the residuals from this regression. Call the residuals $\hat{e}_Y(X_k)$. The notation is complicated, unfortunately, but we need to keep reference both to the response variable $Y$ and to the predictor variable $X_k$ not used in computing the residuals. The $\hat{e}_Y(X_k)$ are the part of $Y$ not explained by all of the $X$'s except $X_k$.

**3.** Fit the regression of $X_k$ on the other $X$'s. Save the resulting residuals and call them $\hat{e}_k$. This is the part of $X_k$ not explained by the other $X$'s. Computing and saving residuals is easy in many computer programs. The relationship of interest, between $Y$ and $X_k$ adjusted for the other $X$'s is exactly the relationship between the two sets of residuals.

**4.** Plot $\hat{e}_Y(X_k)$ versus $\hat{e}_k$. This is the added variable plot. A strong linear relationship between the plotted quantities corresponds to a strong adjusted relationship between $Y$ and $X_k$. If the plot does not exhibit a strong trend, then the adjusted relationship is weak. Overall, this plot may be interpreted similarly to a scatter plot in simple regression.

**Properties of added variable plots.**  If the $\hat{e}_Y(X_k)$ are regressed on $\hat{e}_k$ via ordinary least squares, one will find an intercept of exactly zero as long as the intercept was included as one of the variables in the original model (2.33). The slope will be the same as the coefficient for $X_k$ in (2.33). The residuals from this regression are the same as those for (2.33). The standard error of $\hat{\beta}_k$ and the residual mean square $\hat{\sigma}^2$ would agree with that obtained in fitting (2.33) if the degrees of freedom are corrected to $n-p'$ in place of $n-2$. In a real sense, then, the added variable plot does summarize the relationship between $Y$ and $X_k$ adjusted for the other $X$'s. More details are given by Cook and Weisberg (1982, Sec. 2.3.2).

Figure 2.3 gives added variable plots for each of the four predictors in the fuel data. In each plot, the regression line for one set of residuals on the other is shown. Several facts are apparent from these plots. The plot for DLIC shows the strongest trend, since the data matches the fitted line most closely. The weakest adjusted predictor is ROAD, since almost no trend is apparent in Figure 2.3d. Added variable plots allow the analyst to look for non-linearity and any points that either seem to be determining the fitted line or else are isolated away from the line. Although there are a few straggling points, no nonlinearity or important isolated points are apparent.

**Partial residual plots.**  As an alternative to the added variable plots, Ezekiel (1924), Larsen and McCleary (1975), and Wood (1975) have advocated the use of *partial residual plots*, also called *residual plus-component plots*, in which one plots $\hat{e}_i + \hat{\beta}_k x_{ik}$ versus $x_{ik}$, where $\hat{e}_i = y_i - \mathbf{x}_i^T \hat{\boldsymbol{\beta}}$ is a residual from fitting model (2.33). This plot also has slope equal to $\hat{\beta}_k$, but the scatter of points away from the fitted line is not the same as in the added variable plot. The advantage of the added variable plot is that it is directly interpretable as displaying the relationship between $Y$ and $X_k$ adjusted for the other $X$'s. On the other hand, the need to transform $X_k$ to another scale seems to be better reflected in the partial residual plot. Each seems to have a role in regression modeling.

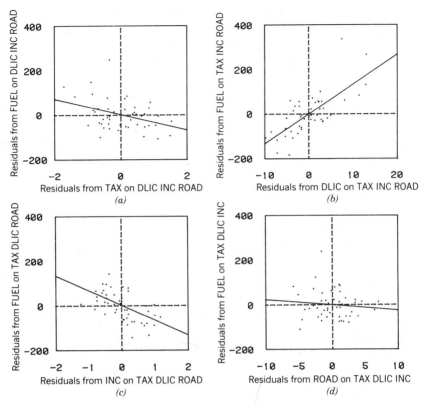

**Figure 2.3**   Added variable plots for (*a*) TAX, (*b*) DLIC, (*c*) INC, and (*d*) ROAD.

## 2.5   Regression through the origin

Models with intercept $\beta_0 = 0$ can arise in several ways. The most obvious circumstance is if $\mathbf{x} = \mathbf{0}$ must imply that $y = 0$. Also, if an experiment is carried out with the sum of a subset of the predictors always equal to a constant, then fitting a nonzero intercept will lead to a rank deficient model. For example, *mixture experiments* often consist of choosing proportions $x_1, \ldots, x_p$ of several chemicals, such that $\sum x_p = 1$ for all cases. For such a model, fitting an intercept is possible only if one of the $x_j$'s is deleted from the model. Interpretation of results is usually simplified if the intercept rather than one of the other predictors is removed.

Regression through the origin can be handled in the same framework that we have discussed so far by defining $\mathbf{X}$ to be a $n \times p$ matrix *without* the column of ones, and $\boldsymbol{\beta}$ to be $p \times 1$ without $\beta_0$. With these modifications, model (2.12) is valid, and all the matrix results following (2.12) apply if $p' = p$ rather than $p + 1$.

**Notational convention.**   Since the same matrix results apply to regression through the origin, let $p' = p + 1$ if the intercept is in the model and $p' = p$ if regression is through the origin, and view $\mathbf{X}$ as $n \times p'$ and $\boldsymbol{\beta}$ as $p' \times 1$.

## Problems

**2.1   Berkeley Guidance Study.**   The data for this example are excerpted from the Berkeley Guidance Study, a longitudinal monitoring of boys and girls born in Berkeley, California, between January 1928 and June 1929. The variables included in the data are:

WT2 = weight at age 2 (kg).
HT2 = height at age 2 (cm).
WT9 = weight at age 9.
HT9 = height at age 9.
LG9 = leg circumference at age 9 (cm).
ST9 = a composite measure of strength at age 9 (high values = stronger).
WT18 = weight at age 18.
HT18 = height at age 18.
LG18 = leg circumference at age 18.
ST18 = strength at age 18.
SOMA = somatotype, a seven-point scale, as a measure of fatness (1 = slender, 7 = fat), determined using a photograph taken at age 18.

Data for 26 boys and for 32 girls are given in Tables 2.6 and 2.7, respectively (the complete study consisted of larger sample sizes and of more variables; see Tuddenham and Snyder, 1954, for details).

**2.1.1.**   For the girls, obtain the matrix of sample correlations, the sample averages, and the sample standard deviations.

**2.1.2.**   Fit the model

$$\text{SOMA} = \beta_0 + \beta_1 \, \text{HT2} + \beta_2 \, \text{WT2} + \beta_3 \, \text{HT9} + \beta_4 \, \text{WT9} + \beta_5 \, \text{ST9} + e$$

Find $\hat{\sigma}$, $R^2$, the overall analysis of variance and $F$-test, and state the conclusion for the $F$-test. Compute $t$-statistics to be used to test each of the $\beta_j$ in the model equal to zero. Explicitly state the hypothesis tested by each and the conclusion.

**Table 2.6   Berkeley Guidance Study: Boys**

| Identi-fication Number | WT2 | HT2 | WT9 | HT9 | LG9 | ST9 | WT18 | HT18 | LG18 | ST18 | SOMA |
|---|---|---|---|---|---|---|---|---|---|---|---|
| 201 | 13.6 | 90.2 | 41.5 | 139.4 | 31.6 | 74.0 | 110.2 | 179.0 | 44.1 | 226.0 | 7.0 |
| 202 | 12.7 | 91.4 | 31.0 | 144.3 | 26.0 | 73.0 | 79.4 | 195.1 | 36.1 | 252.0 | 4.0 |
| 203 | 12.6 | 86.4 | 30.1 | 136.5 | 26.6 | 64.0 | 76.3 | 183.7 | 36.9 | 216.0 | 6.0 |
| 204 | 14.8 | 87.6 | 34.1 | 135.4 | 28.2 | 75.0 | 74.5 | 178.7 | 37.3 | 220.0 | 3.0 |
| 205 | 12.7 | 86.7 | 24.5 | 128.9 | 24.2 | 63.0 | 55.7 | 171.5 | 31.0 | 200.0 | 1.5 |
| 206 | 11.9 | 88.1 | 29.8 | 136.0 | 26.7 | 77.0 | 68.2 | 181.8 | 37.0 | 215.0 | 3.0 |
| 207 | 11.5 | 82.2 | 26.0 | 128.5 | 26.5 | 45.0 | 78.2 | 172.5 | 39.1 | 152.0 | 6.0 |
| 209 | 13.2 | 83.8 | 30.1 | 133.2 | 27.6 | 70.0 | 66.5 | 174.6 | 37.3 | 189.0 | 4.0 |
| 210 | 16.9 | 91.0 | 37.9 | 145.6 | 29.0 | 61.0 | 70.5 | 190.4 | 33.9 | 183.0 | 3.0 |
| 211 | 12.7 | 87.4 | 27.0 | 132.4 | 26.0 | 74.0 | 57.3 | 173.8 | 33.3 | 193.0 | 3.0 |
| 212 | 11.4 | 84.2 | 25.9 | 133.7 | 25.8 | 68.0 | 50.3 | 172.6 | 31.6 | 202.0 | 3.0 |
| 213 | 14.2 | 88.4 | 31.1 | 138.3 | 27.3 | 59.0 | 70.8 | 185.2 | 36.6 | 208.0 | 4.0 |
| 214 | 17.2 | 87.7 | 34.6 | 134.6 | 30.6 | 87.0 | 73.7 | 178.4 | 39.2 | 227.0 | 3.0 |
| 215 | 13.7 | 89.6 | 34.6 | 139.0 | 28.9 | 71.0 | 75.2 | 177.6 | 36.8 | 204.0 | 2.5 |
| 216 | 14.2 | 91.4 | 43.1 | 146.0 | 32.4 | 98.0 | 83.1 | 183.5 | 38.0 | 226.0 | 4.0 |
| 217 | 15.9 | 90.0 | 33.2 | 133.2 | 28.5 | 82.0 | 74.3 | 178.1 | 37.8 | 233.0 | 2.5 |
| 218 | 14.3 | 86.4 | 30.7 | 133.3 | 27.3 | 73.0 | 72.2 | 177.0 | 36.5 | 237.0 | 2.0 |
| 219 | 13.3 | 90.0 | 31.6 | 130.3 | 27.5 | 68.0 | 88.6 | 172.9 | 40.4 | 230.0 | 7.0 |
| 221 | 13.8 | 91.4 | 33.4 | 144.5 | 27.0 | 92.0 | 75.9 | 188.4 | 36.5 | 250.0 | 1.0 |
| 222 | 11.3 | 81.3 | 29.4 | 125.4 | 27.7 | 70.0 | 64.9 | 169.4 | 35.7 | 236.0 | 3.0 |
| 223 | 14.3 | 90.6 | 30.2 | 135.8 | 26.7 | 70.0 | 65.6 | 180.2 | 35.4 | 177.0 | 4.0 |
| 224 | 13.4 | 92.2 | 31.1 | 139.9 | 27.2 | 63.0 | 66.4 | 189.0 | 35.3 | 186.0 | 4.0 |
| 225 | 12.2 | 87.1 | 27.6 | 136.8 | 25.8 | 73.0 | 59.0 | 182.4 | 33.5 | 199.0 | 3.0 |
| 226 | 15.9 | 91.4 | 32.3 | 140.6 | 27.9 | 69.0 | 68.1 | 185.8 | 34.2 | 227.0 | 1.0 |
| 227 | 11.5 | 89.7 | 29.0 | 138.6 | 24.6 | 61.0 | 67.7 | 180.7 | 34.3 | 164.0 | 4.0 |
| 228 | 14.2 | 92.2 | 31.4 | 140.0 | 28.2 | 74.0 | 68.5 | 178.7 | 37.0 | 219.0 | 2.0 |

**2.1.3.** Now fit the model

$$SOMA = \beta_0 + \beta_3 HT9 + \beta_4 WT9 + \beta_5 ST9 + e$$

and compare this model to that fit in 2.1.2 (i.e., compute an $F$-test).

**2.1.4.** Repeat 2.1.1 to 2.1.3, except for the boys. Qualitatively describe the difference between the fitted models for boys and for girls (formal procedures will be studied in Chapter 7).

**2.1.5.** For the boys, repeat the derivation in Section 2.1 when adding the variable WT9 to the model $HT18 = \beta_0 + \beta_2 WT2$.

**2.2   (Matrix manipulation).** Define the following matrices:

$$A = \begin{bmatrix} 1 & 1 \\ -1 & 0 \end{bmatrix} \quad B = \begin{bmatrix} 3 & 1 \\ 2 & 1 \end{bmatrix} \quad I = \begin{bmatrix} 1 & 0 \\ 0 & 1 \end{bmatrix} \quad D = \begin{bmatrix} -2 \\ 3 \end{bmatrix}$$

Table 2.7 Berkeley Guidance Study: Girls

| Identi-fication Number | WT2 | HT2 | WT9 | HT9 | LG9 | ST9 | WT18 | HT18 | LG18 | ST18 | SOMA |
|---|---|---|---|---|---|---|---|---|---|---|---|
| 331 | 12.6 | 83.8 | 33.0 | 136.5 | 29.0 | 57.0 | 71.2 | 169.6 | 38.8 | 107.0 | 6.0 |
| 334 | 12.0 | 86.2 | 34.2 | 137.0 | 27.3 | 44.0 | 58.2 | 166.8 | 34.3 | 130.0 | 5.0 |
| 335 | 10.9 | 85.1 | 28.1 | 129.0 | 27.4 | 48.0 | 56.0 | 157.1 | 37.8 | 101.0 | 5.0 |
| 351 | 12.7 | 88.6 | 27.5 | 139.4 | 25.7 | 68.0 | 64.5 | 181.1 | 34.2 | 149.0 | 4.0 |
| 352 | 11.3 | 83.0 | 23.9 | 125.6 | 24.5 | 22.0 | 53.0 | 158.4 | 32.4 | 112.0 | 5.0 |
| 353 | 11.8 | 88.9 | 32.2 | 137.1 | 28.2 | 59.0 | 52.4 | 165.6 | 33.8 | 136.0 | 4.0 |
| 354 | 15.4 | 89.7 | 29.4 | 133.6 | 26.6 | 58.0 | 56.8 | 166.7 | 32.7 | 118.0 | 4.5 |
| 355 | 10.9 | 81.3 | 22.0 | 121.4 | 24.4 | 44.0 | 49.2 | 156.5 | 33.5 | 110.0 | 4.0 |
| 356 | 13.2 | 88.7 | 28.8 | 133.6 | 26.5 | 58.0 | 55.6 | 168.1 | 34.1 | 104.0 | 4.5 |
| 357 | 14.3 | 88.4 | 38.8 | 134.1 | 31.1 | 57.0 | 77.8 | 165.3 | 39.8 | 138.0 | 6.5 |
| 358 | 11.1 | 85.1 | 36.0 | 139.4 | 28.2 | 64.0 | 69.6 | 163.7 | 38.6 | 108.0 | 5.5 |
| 359 | 13.6 | 91.4 | 31.3 | 138.1 | 27.6 | 64.0 | 56.2 | 173.7 | 34.2 | 134.0 | 3.5 |
| 361 | 13.5 | 86.1 | 33.3 | 138.4 | 29.4 | 73.0 | 64.9 | 169.2 | 36.7 | 141.0 | 5.0 |
| 362 | 16.3 | 94.0 | 36.2 | 139.5 | 28.0 | 52.0 | 59.3 | 170.1 | 32.8 | 122.0 | 4.5 |
| 364 | 10.2 | 82.2 | 23.4 | 129.8 | 22.6 | 60.0 | 49.8 | 164.2 | 30.0 | 128.0 | 4.0 |
| 365 | 12.6 | 88.2 | 33.8 | 144.8 | 28.3 | 107.0 | 62.6 | 176.0 | 35.8 | 168.0 | 5.0 |
| 366 | 12.9 | 87.5 | 34.5 | 138.9 | 30.5 | 62.0 | 66.6 | 170.9 | 38.8 | 126.0 | 5.0 |
| 367 | 13.3 | 88.6 | 34.4 | 140.3 | 31.2 | 88.0 | 65.3 | 169.2 | 39.0 | 142.0 | 5.0 |
| 368 | 13.4 | 86.9 | 38.2 | 143.8 | 29.8 | 78.0 | 65.9 | 172.0 | 35.7 | 132.0 | 5.5 |
| 369 | 12.7 | 86.4 | 31.7 | 133.6 | 27.5 | 52.0 | 59.0 | 163.0 | 32.7 | 116.0 | 5.5 |
| 370 | 12.2 | 80.9 | 26.6 | 123.5 | 27.2 | 40.0 | 47.4 | 154.5 | 32.2 | 112.0 | 4.0 |
| 371 | 15.4 | 90.0 | 34.2 | 139.9 | 29.1 | 71.0 | 60.4 | 172.5 | 35.7 | 137.0 | 4.0 |
| 372 | 12.7 | 94.0 | 27.7 | 136.1 | 26.7 | 30.0 | 56.3 | 175.6 | 34.0 | 114.0 | 3.0 |
| 373 | 13.2 | 89.7 | 28.5 | 135.8 | 25.5 | 76.0 | 61.7 | 167.2 | 35.5 | 122.0 | 4.5 |
| 374 | 12.4 | 86.4 | 30.5 | 131.9 | 28.6 | 59.0 | 52.4 | 164.0 | 34.8 | 121.0 | 5.0 |
| 376 | 13.4 | 86.4 | 39.0 | 130.9 | 29.3 | 38.0 | 58.4 | 161.6 | 33.0 | 107.0 | 6.5 |
| 377 | 10.6 | 81.8 | 25.0 | 126.3 | 25.0 | 50.0 | 52.8 | 153.6 | 33.4 | 140.0 | 5.0 |
| 380 | 12.7 | 91.4 | 29.8 | 135.5 | 27.0 | 57.0 | 67.4 | 173.5 | 34.5 | 123.0 | 5.0 |
| 382 | 11.8 | 88.6 | 27.0 | 134.0 | 26.5 | 54.0 | 56.3 | 166.2 | 36.2 | 135.0 | 4.5 |
| 383 | 13.3 | 86.4 | 41.4 | 138.2 | 32.5 | 44.0 | 82.8 | 162.8 | 42.5 | 125.0 | 7.0 |
| 384 | 13.2 | 94.0 | 41.6 | 142.0 | 31.0 | 56.0 | 68.1 | 168.6 | 38.4 | 142.0 | 5.5 |
| 385 | 15.9 | 89.2 | 42.4 | 140.8 | 32.6 | 74.0 | 63.1 | 169.2 | 37.9 | 142.0 | 5.5 |

$$E = \begin{bmatrix} 2 \\ 1 \end{bmatrix} \qquad H = \begin{bmatrix} \dfrac{1}{\sqrt{2}} & \dfrac{-1}{\sqrt{2}} \\ \dfrac{1}{\sqrt{2}} & \dfrac{1}{\sqrt{2}} \end{bmatrix} \qquad C = \begin{bmatrix} 1 & 2 \\ 3 & 4 \\ 5 & 6 \end{bmatrix}$$

**2.2.1.** Find $A^T$, $B^T$, $C^T$, $D^T$, $E^T$.
**2.2.2.** Find $A + B$.
**2.2.3.** Find $AB$ and $BA$. Does $AB = BA$?
**2.2.4.** Show that $(AB)^T = B^T A^T$.

**2.2.5.** Compute $C^T C$ and $C C^T$. Are they equal?

**2.2.6.** Find $D E^T$, $D^T E$.

**2.2.7.** Show that $H$ is orthogonal (that is, $H H^T = H^T H = I$).

**2.3  Partitioned matrices.**    An $n \times p$ matrix $C$ may be partitioned by columns into $C = (C_1 \ C_2)$, where $C_1$ is an $n \times p_1$ matrix and $C_2$ is an $n \times (p - p_1)$ matrix, and $C_1$ is the first $p_1$ columns of $C$, and $C_2$ is the last $p - p_1$ columns of $C$. With this definition

$$C^T C = (C_1 \ C_2)^T (C_1 \ C_2)$$

$$= \binom{C_1^T}{C_2^T}(C_1 \ C_2)$$

$$= \begin{pmatrix} C_1^T C_1 & C_1^T C_2 \\ C_2^T C_1 & C_2^T C_2 \end{pmatrix}$$

and

$$C C^T = (C_1 \ C_2)(C_1 \ C_2)^T = C_1 C_1^T + C_2 C_2^T$$

**2.3.1.** Show, by direct multiplication, that if $(C^T C)$ is of full rank, we can write

$$(C^T C)^{-1} = \begin{pmatrix} (C_1^T C_1)^{-1} + F E^{-1} F^T & -F E^{-1} \\ -E^{-1} F^T & E^{-1} \end{pmatrix}$$

where

$$E = C_2^T C_2 - C_2^T C_1 (C_1^T C_1)^{-1} C_1^T C_2,$$

and

$$F = (C_1^T C_1)^{-1} C_1^T C_2$$

**2.3.2.** Suppose that the correct linear model is $Y = X\beta + e$, with $X$ an $n \times p'$ matrix, but we fit the model $Y = X\beta + Z\gamma + e$, so that, unknown to us, $\gamma = 0$. Use the result of 2.3.1 to show that the estimate of $\beta$ in the latter model is unbiased, and find its variance. Compare the variance of $\hat{\beta}$ from the correct linear model to the variance that would be obtained if the larger model were fit, and comment. Find conditions on $X$ and $Z$ such that the estimator for $\beta$ is numerically identical for either model.

**2.4  QR factorization.**    Suppose we have an $n \times p'$ matrix $Q$ and a $p' \times p'$ upper triangular matrix $R$ such that $Q^T Q = I$ and $Q R = X$ as in Appendix 2A.3.

**2.4.1.** Show that $R^T R = X^T X$.

**2.4.2.** Show that if $(X^T X)^{-1}$ exists then $(X^T X)^{-1} X^T Y = R^{-1} Q^T Y$. Thus, to compute $\hat{\beta}$, first compute $z = Q^T Y$, then solve the set of linear equations $R\hat{\beta} = z$ by back substitution; see Appendix 2A.3.

**2.4.3.** Show that the vector of fitted values $\hat{\mathbf{Y}} = \mathbf{QQ}^T\mathbf{Y}$. This, incidentally, will show that $\mathbf{QQ}^T = \mathbf{X}(\mathbf{X}^T\mathbf{X})^{-1}\mathbf{X}^T$, which will be an important matrix in Chapters 5 and 6. Also, find $\hat{\mathbf{e}}$ in terms of $\mathbf{Y}$ and $\mathbf{Q}$.

**2.4.4.** The variance of a fitted value $\mathbf{c}^T\hat{\boldsymbol{\beta}}$ at any vector $\mathbf{c}$ is given by $\sigma^2\mathbf{c}^T(\mathbf{X}^T\mathbf{X})^{-1}\mathbf{c}$. If $\mathbf{c}$ is a vector with $j$th element equal to one and the other elements all zero, then $\sigma^2\mathbf{c}(\mathbf{X}^T\mathbf{X})^{-1}\mathbf{c} = \text{var}(\hat{\beta}_j)$. Thus if we can find $\mathbf{c}^T(\mathbf{X}^T\mathbf{X})^{-1}\mathbf{c}$ for any $p' \times 1$ vector $\mathbf{c}$, we can find the estimated variance of any fitted value, prediction, linear combination of coefficients, or any individual coefficient. From 2.4.1, since $\mathbf{X}^T\mathbf{X} = \mathbf{R}^T\mathbf{R}$, if the inverses exist, then $(\mathbf{X}^T\mathbf{X})^{-1} = (\mathbf{R}^T\mathbf{R})^{-1} = \mathbf{R}^{-1}\mathbf{R}^{-T}$, where $-T$ means the inverse of the transpose. Show that

$$\mathbf{c}^T(\mathbf{X}^T\mathbf{X})^{-1}\mathbf{c} = (\mathbf{R}^{-T}\mathbf{c})^T(\mathbf{R}^{-T}\mathbf{c})$$

Suppose we let $\mathbf{d} = \mathbf{R}^{-T}\mathbf{c}$. Then $\mathbf{c}^T(\mathbf{X}^T\mathbf{X})^{-1}\mathbf{c} = \mathbf{d}^T\mathbf{d}$. To find $\mathbf{d}$, write

$$\mathbf{d} = \mathbf{R}^{-T}\mathbf{c}$$

or

$$\mathbf{R}^T\mathbf{d} = \mathbf{c}$$

and use back substitution, as in Appendix 2A.3, to solve for $\mathbf{d}$. This avoids explicit inversion of $(\mathbf{X}^T\mathbf{X})^{-1}$ or of $\mathbf{R}$.

**2.4.5.** Suppose $p' = 3$ and

$$\mathbf{R} = \begin{bmatrix} 2 & 4 & 3 \\ 0 & 1 & 5 \\ 0 & 0 & 8 \end{bmatrix}$$

Assuming $\hat{\sigma}^2 = 1$, find $\text{var}(\hat{\beta}_0)$ (set $\mathbf{c} = (1, 0, 0)^T$), $\text{var}(\hat{\beta}_1)$ (set $\mathbf{c} = (0, 1, 0)^T$) and the covariance between $\hat{\beta}_0$ and $\hat{\beta}_1$. The latter will require a slight extension of the above results.

## 2.5 Computational example.
Consider the linear model $\mathbf{Y} = \mathbf{X}\boldsymbol{\beta} + \mathbf{e}$, where

$$\mathbf{X} = \begin{bmatrix} 1 & 1 \\ 1 & 3 \\ 1 & 5 \\ 1 & 7 \end{bmatrix} \qquad \mathbf{Y} = \begin{bmatrix} 34 \\ 47 \\ 55 \\ 64 \end{bmatrix}$$

**2.5.1.** Compute $\mathbf{X}^T\mathbf{X}$, $\mathbf{X}^T\mathbf{Y}$, $\mathbf{Y}^T\mathbf{Y}$. Using the rule that if $\mathbf{A}$ is a $2 \times 2$ symmetric matrix, then, if $ac \neq b^2$,

$$\mathbf{A}^{-1} = \begin{pmatrix} a & b \\ b & c \end{pmatrix}^{-1} = \frac{1}{ac - b^2}\begin{pmatrix} c & -b \\ -b & a \end{pmatrix}$$

find $(\mathbf{X}^T\mathbf{X})^{-1}$, $\hat{\boldsymbol{\beta}}$, and $\text{var}(\hat{\boldsymbol{\beta}})$. Find $\hat{\mathbf{Y}}$ and $\hat{\mathbf{e}}$.

**2.5.2.** Define

$$Q = \begin{bmatrix} -\dfrac{1}{2} & \dfrac{3\sqrt{5}}{10} \\[2ex] -\dfrac{1}{2} & \dfrac{\sqrt{5}}{10} \\[2ex] -\dfrac{1}{2} & -\dfrac{\sqrt{5}}{10} \\[2ex] -\dfrac{1}{2} & -\dfrac{3\sqrt{5}}{10} \end{bmatrix} \qquad R = \begin{pmatrix} -2 & -8 \\ 0 & -2\sqrt{5} \end{pmatrix}$$

Show that $Q^T Q = I$ and $QR = X$. Show that $R^T R = X^T X$. Find $z = Q^T Y$ and $\hat{Y} = Qz$. Compute $\hat{e}$. Find $\hat{\beta}$ via back substitution.

**2.6**  Use appropriate methods to study the addition of POP to the fuel consumption model, and summarize the results.

**2.7  Sweep algorithm.**  Define $A$ to be a $(p'+1) \times (p'+1)$ matrix

$$A = \begin{pmatrix} X^T X & X^T Y \\ Y^T X & Y^T Y \end{pmatrix}$$

**Definition.**  We will say that the matrix $A$ with elements $a_{ij}$ has been *swept on pivot* k when it has been transformed into a matrix $B$ with elements $b_{ij}$ according to the following rules:

$$b_{kk} = \frac{1}{a_{kk}}$$

$$b_{ik} = -\frac{a_{ik}}{a_{kk}} \qquad i \neq k$$

$$b_{kj} = \frac{a_{kj}}{a_{kk}} \qquad j \neq k$$

$$b_{ij} = a_{ij} - \frac{a_{ik}a_{kj}}{a_{kk}} \qquad i \neq k, j \neq k$$

The sweep algorithm is easily programmed to overwrite the matrix $A$ by the matrix $B$. First, replace $a_{kk}$ by $b_{kk} = 1/a_{kk}$. Then the remaining elements in column $k$ are replaced by $b_{ik} = -a_{ik}b_{kk}$. The elements in neither row $k$ nor column $k$ are replaced by $b_{ij} = a_{ij} + a_{kj}b_{ik}$. Finally, the $k$th row is replaced by $b_{kj} = a_{kj}b_{kk}$. Sweep is described in a slightly different notation by Goodnight (1979). The name was apparently coined by A. Beaton (1964), but the algorithm was used earlier (Ralston, 1960).

**2.7.1.** Show that if the first column of **X** is a column of 1's, the first row of **A** is given by the vector

$$(n, n\bar{x}_1, \ldots, n\bar{x}_p, n\bar{y})^T = (n, n\bar{\mathbf{x}}^T, n\bar{y})$$

**2.7.2.** Let the notation **B** = Sweep **A**[$i, j, \ldots, m$] mean "**B** is the result of sweeping the matrix **A** on pivot $i$, then sweeping the result on pivot $j, \ldots,$ then sweeping the result on pivot $m$." Show that

$$\text{Sweep } \mathbf{A}[1] = \begin{pmatrix} \dfrac{1}{n} & \bar{\mathbf{x}}^T & \bar{y} \\ -\bar{\mathbf{x}} & \mathscr{X}^T\mathscr{X} & \mathscr{X}^T\mathscr{Y} \\ -\bar{y} & \mathscr{Y}^T\mathscr{X} & \mathscr{Y}^T\mathscr{Y} \end{pmatrix} = \begin{pmatrix} \dfrac{1}{n} & \bar{\mathbf{x}}^T & \bar{y} \\ -\bar{\mathbf{x}} & & \\ -\bar{y} & & \mathbf{T} \end{pmatrix}$$

Thus, except for sign changes, sweeping on the first pivot turns **A** into the matrix **T** of corrected cross products, augmented by a new first row and column of the averages of the $X$'s and of $Y$. A computer program for regression calculations using Sweep would usually start by computing Sweep **A**[1] directly, possibly using the algorithm described in Appendix 1A.5, as this will lead to increased numerical precision in most problems.

**2.7.3.** Write a computer program to implement the Sweep algorithm for regression calculations. (You should read the rest of this problem before writing your program.) The program should begin by taking as input the matrix Sweep **A**[1], as defined in problem 2.7.2. This matrix can be computed using the program written for problem 1.10. Suppose we call this matrix **C**. Two points concerning the program should be made.

a. Include storage for two copies of the matrix **C**, perhaps calling them **C** and **CC**. It is desirable for numerical reasons to be able to recover the original matrix. This can be done by sweeping only the matrix **C**, leaving **CC** untouched, and copying **CC** to **C** as needed.

b. One of the operations in the algorithm is to replace a diagonal element $b_{kk}$ by its inverse $1/b_{kk}$, which can cause numerical problems if $b_{kk}$ is nearly zero. Aside from rounding error, if $b_{kk}$ is exactly zero, then the matrix **B** is not invertible, and unique least squares estimates do not exist. Because of rounding errors intrinsic to digital computers, a tolerance check is required: declare $b_{kk}$ to be zero if $b_{kk} < tol/SX_kX_k$ where $tol$ is a predetermined value that depends on the word length of the computer and $SX_kX_k$ is the corrected sum of squares of the values in the $k$th column of **X**. $tol = .001$ is a common choice. If the tolerance check fails, variable $X_k$ is essentially a linear combination of variables with pivots already swept, and it must be deleted from the model.

Berk (1977) has reported that this tolerance check is not sufficient to guarantee numerically accurate computations, as the addition of the $k$th pivot may modify one of the previously entered pivots to make it too large and hence numerically unstable. Suppose the $j$th pivot has already been performed, and we now consider the $k$th pivot. Berk suggests the following procedure. If $b_{kk}$ passes the tolerance test, sweep on pivot $k$. Next, check the pivots corresponding to each of the previous sweeps. If the $j$th of these pivots is too large, exceeding $SX_jX_j/tol$, then reject the pivot of column $k$, and declare $X_k$ to be a linear combination of columns already entered into the equation. This will probably require recomputing the swept matrix by copying **CC** to **C**, and repeating the sweeps up to but not including the $k$th.

**2.7.4.** Either prove mathematically or demonstrate using your program that sweep is reversible: if **B** = Sweep **A**$[i]$ then **A** = Sweep **B**$[i]$, except for rounding error.

**2.7.5.** Either prove mathematically or demonstrate using your program that Sweep is commutative: Sweep **A**$[i, j]$ = Sweep **A**$[j, i]$, except for rounding error.

**2.7.6.** Either prove mathematically or demonstrate using your program that if **A** is invertible, then **A**$^{-1}$ can be computed by sweeping **A** once on each pivot, except for rounding error.

**2.7.7.** Label the pivots of **A** by $0, 1, \ldots, p, p+1$. The zeroth pivot corresponds to the constant column of ones while the $(p+1)$st pivot corresponds to the response $Y$. Then show that

$$\text{Sweep } \mathbf{A}[0, 1, 2, \ldots, p] = \begin{pmatrix} (\mathbf{X}^T\mathbf{X})^{-1} & \hat{\boldsymbol{\beta}} \\ -\hat{\boldsymbol{\beta}}^T & RSS \end{pmatrix}$$

where, as usual, $\hat{\boldsymbol{\beta}} = (\mathbf{X}^T\mathbf{X})^{-1}\mathbf{X}^T\mathbf{Y}$, and $RSS$ is the residual sum of squares for the regression of $Y$ on the $X$'s, including the intercept. (If regression is through the origin, do not sweep on pivot 0.)

**2.7.8.** Describe how Sweep could be used to estimate parameters for the regression of $Y$ on any subset of the $X$'s.

**2.7.9.** Describe an algorithm using Sweep for getting a sequential analysis of variance table like Table 2.5.

**2.7.10.** Using the results of this problem, extend your Sweep program to compute and print all summary statistics for multiple regression described in this chapter. In particular, the program should compute $\hat{\boldsymbol{\beta}}$, $\hat{\sigma}^2$, $R^2$, and standard errors of the components of $\hat{\boldsymbol{\beta}}$.

**2.8**   For the period 1965 to 1977, the number of days on which ozone levels exceeded federal standards for 1 hour or more in the San Francisco Bay area showed a decline of about 5% per year, but with large unexplained fluctuations. One possible cause of the fluctuations might be the weather from the previous year or two years. For example, winter precipitation might influence summer ozone levels.

The following data are from Sandberg, Basso, and Okin (1978):

YEAR = Year of ozone measurement.

RAIN = Average winter precipitation in centimeters in the San Francisco Bay area for the preceding two winters.

SF = Summer quarter maximum hourly average ozone reading in parts per million at San Francisco.

SJ = Summer quarter maximum hourly average ozone reading in parts per million at San Jose, at the southern end of the Bay.

**2.8.1.** Use SF as the response variable. Fit the regression of SF on YEAR. Then obtain the added variable plot for RAIN after YEAR. Is RAIN a useful predictor?

**2.8.2.** Repeat the last problem, but fit RAIN first, and comment on fitting YEAR second.

**2.8.3.** Compute two new variables, SUM = SF + SJ and DIFF = SF − SJ. Study the regression models of SUM on YEAR and RAIN and of DIFF on YEAR and RAIN. Interpret the results.

| YEAR | RAIN | SF | SJ | YEAR | RAIN | SF | SJ |
|------|------|-----|-----|------|------|-----|-----|
| 1965 | 18.9 | 4.3 | 4.2 | 1972 | 19.0 | 3.1 | 4.6 |
| 1966 | 23.7 | 4.2 | 4.8 | 1973 | 30.6 | 3.4 | 5.1 |
| 1967 | 26.2 | 4.6 | 5.3 | 1974 | 34.1 | 3.4 | 3.7 |
| 1968 | 26.6 | 4.7 | 4.8 | 1975 | 23.7 | 2.1 | 2.7 |
| 1969 | 39.6 | 4.1 | 5.5 | 1976 | 14.6 | 2.2 | 2.1 |
| 1970 | 45.5 | 4.6 | 5.6 | 1977 | 7.6 | 2.0 | 2.5 |
| 1971 | 26.7 | 3.7 | 5.4 |      |      |     |     |

# 3

# DRAWING CONCLUSIONS

After calculations are completed, the results must be interpreted. Although the outline of the analysis may be similar in many problems, the conclusions that may be drawn are varied. Often a fitted model is an approximation, either because important variables are not included or are incorrectly measured, or because the functional form used is not exactly correct.

## 3.1   Interpreting parameter estimates

**Do parameters exist?**   The model $Y = X\beta + e$ is often a useful fiction. The model is chosen more for its ease of fitting than from any underlying theory that connects the response to the predictors. As a result, $\hat{\beta}$ may not estimate a real quantity. If data were collected on the same variables but over different ranges or from a different sampling scheme, the computed $\hat{\beta}$ could "estimate" something else. Estimation of parameters, unknown constants that characterize a process, makes sense only when the presumed functional form of the model is nearly exact.

*Example 3.1   Estimating the area of a rectangle*

Suppose we have a sample of $n$ rectangles from which we want to model ln(area) as a function of ln(length), perhaps through the simple regres-

64

sion equation

$$\ln(\text{area}) = \beta_0 + \beta_1 \ln(\text{length}) + e$$

Ln(length) alone will not be a good predictor of ln(area) for arbitrary rectangles since both length and width are needed to determine area. For a specific population of rectangles, however, the model with only ln(length) as a predictor may work very well. If the rectangles are the tops of tables, each with width of about .5 × length, then $\ln(\text{area}) \cong \ln(\text{length}) + \ln(.5 \times \text{length}) = \ln(.5) + 2\ln(\text{length})$. The linear model is likely to fit reasonably well with intercept near $\ln(.5) \cong -0.07$ and slope near 2. The values of the slope and intercept have little to do with the properties of rectangles, but rather they are determined by the sampled population. If a different population of rectangles were sampled, different parameters might result.

---

**Interpreting estimates.**    In the fuel consumption data of Example 2.1, the fitted model was

$$\widehat{\text{FUEL}} = 377.29 - 34.79 \text{ TAX} + 13.36 \text{ DLIC} - 66.59 \text{ INC} - 2.43 \text{ ROAD}$$

The usual interpretation of an estimated coefficient is as a rate of change: increasing TAX rate by 1 cent should decrease consumption, all other factors being held constant, by 34.79 gallons per person. This assumes that a predictor can in fact be changed by one unit without affecting the other predictors and that the model fit to the available data will apply when the predictor is so changed. In the example, the data are observational since the assignment of values for the predictors was not under the control of the analyst, so whether fuel consumption would be decreased by increasing taxes cannot be directly assessed. Rather, a more conservative interpretation is in order: states with higher tax rates are observed to have lower fuel consumption. To draw conclusions concerning the effects of changing tax rates, the rates must in fact be changed, and the results observed.

**Signs of estimates.**    The sign of a parameter estimate indicates the direction of the relationship between the predictor and the response. In multiple regression, if the predictors are correlated, the sign of a coefficient may change depending on the other predictors in the model. While this is mathematically possible and, occasionally, scientifically reasonable, it certainly makes interpretation more difficult. Sometimes this problem can be removed by redefining the predictors into new linear combinations that are easier to interpret.

*Example 3.2   Berkeley Guidance Study*

---

Data from the Berkeley Guidance Study on the growth of boys and girls are given in problem 2.1. In studying these data, suppose we wish to model somatotype (SOMA) by weights at ages 2, 9, and 18 (WT2, WT9, WT18) for $n = 32$ girls. The correlation matrix for these four variables is given in Table 3.1. All the variables are positively correlated, as one might expect. Yet, the regression of SOMA on WT2, WT9, and WT18, Table 3.2, leads to the unexpected conclusion that heavier girls at age 2 tend to be thinner (have lower somatotype) at age 18. This result may be due to the correlations between the predictors. In place of the preceding variables, consider the following:

WT2 = Weight at age 2.

DW9 = WT9 − WT2 = weight gain from age 2 to 9.

DW18 = WT18 − WT9 = weight gain from age 9 to 18.

Since all three variables measure weights, combining them in this way is reasonable. If the variables measured different quantities, then combining them could lead to conclusions that are even less useful

**Table 3.1   Correlation matrix for weight variables for Berkeley Guidance Study girls**

| | | | | |
|---|---|---|---|---|
| WT2 | 1.0000 | | | |
| WT9 | .5969 | 1.0000 | | |
| WT18 | .3508 | .7108 | 1.0000 | |
| SOMA | .1234 | .6508 | .6865 | 1.0000 |
| | WT2 | WT9 | WT18 | SOMA |

**Table 3.2   Regression of SOMA on WT2 WT9 and WT18**

| Variable | Coefficient | Standard Error | $t$-Value |
|---|---|---|---|
| Intercept | 2.03686 | 1.08218 | 1.88 |
| WT2 | − .217635 | .088180 | − 2.47 |
| WT9 | .094583 | .031871 | 2.97 |
| WT18 | .043269 | .018394 | 2.35 |

$\hat{\sigma}^2 = 0.332625$, d.f. $= 28$, $R^2 = 0.610$.

Table 3.3   Regression of SOMA on WT2 DW9 and DW18

| Variable | Coefficient | Standard Error | $t$-Value |
|----------|-------------|----------------|-----------|
| Intercept | 2.03686 | 1.08218 | 1.88 |
| WT2 | $-.079782$ | .076185 | $-1.05$ |
| DW9 | .137853 | .023908 | 5.77 |
| DW18 | .043269 | .018394 | 2.35 |

$\hat{\sigma}^2 = 0.332625$, d.f. $= 28$, $R^2 = 0.610$.

than those originally obtained. The fitted regression for SOMA on WT2, DW9, and DW18 is given in Table 3.3.

Compare this regression to that fit to the weights themselves. The estimates $\hat{\beta}_0$, se$(\hat{\beta}_0)$, $\hat{\sigma}^2$, and $R^2$ are identical in each. In fact, since the three variables WT2, DW9, and DW18 can be obtained from WT2, WT9, and WT18 via a linear transformation, the two sets of variables carry exactly the same information concerning SOMA. However, the estimated coefficient for WT2 depends on which set of variables is used. Table 3.2 shows that $\hat{\beta}_{WT2} = -0.22$, with $t = -2.47$, while in Table 3.3, $\hat{\beta}_{WT2} = -0.08$, with $t = -1.05$. In the former case, the effect of WT2 appears substantial, while in the latter it does not. Although $\hat{\beta}_{WT2}$ is negative in each, we would be led in the latter case to conclude that the effect of WT2 is negligible. Thus interpretation of the effect △ of a variable depends not only on the other variables in a model, but also upon which linear transformation of those variables is used.

The linear transformation used above is not unique, and, depending on the context, many others might be preferred. For example, another set might be

AVE $=$ (WT2 $+$ WT9 $+$ WT18)/3.
LIN $=$ WT18 $-$ WT2.
QUAD $=$ WT2 $-$ 2WT9 $+$ WT18.

This transformation focuses on the fact that WT2, WT9, and WT18 are ordered in time and are more or less equally spaced. Pretending that the weight measurements are equally spaced, AVE, LIN, and QUAD are, respectively, the average, linear, and quadratic time trends in weight gain.

**Rank deficient and overparameterized models.**    In the last example, several combinations of the basic predictors WT2, WT9, and WT18 were studied. One might naturally ask what would happen if more than three combinations of these predictors were used in the same regression model. As long as we use *linear* combinations of the predictors, we cannot use more than three, the number of measured, linearly independent quantities.

To see why this is true, consider adding QUAD to the model SOMA on WT2 DW9 DW18. As in Chapter 2, we can examine this by study of the residuals from the regression of SOMA on WT2 DW9 DW18 versus the residuals from the regression of QUAD on WT2 DW9 DW18. But since QUAD can be written as an exact linear combination of the other predictors, QUAD = DW18 − DW9, the residuals from this second regression are all exactly zero. A slope coefficient for QUAD is thus not defined after adjusting for the other three predictors. We would say that the four predictors WT2, DW9, DW18, and QUAD are linearly dependent, since one can be determined exactly from the others. The maximum number of predictors that could be included in a model is called the *rank* of the model or of the data matrix **X**. Most of the models in this book are of *full rank*, with all predictors linearly independent, but singular models, or models with linearly related predictors, are frequently used. The simplest example is the one-way design. Suppose that a unit is assigned to one of three treatment groups, and let $X_1 = 1$ if the unit is in group 1 and zero otherwise, $X_2 = 1$ if the unit is in group 2 and zero otherwise, and $X_3 = 1$ if the unit is in group 3 and zero otherwise. Thus, for each unit, we must have $X_1 + X_2 + X_3 = 1$ since each unit is in only one of the three groups. We therefore cannot fit the model

$$Y = \beta_0 + \beta_1 X_1 + \beta_2 X_2 + \beta_3 X_3 + e$$

because the sum of the $X$'s is equal to the column of ones, and the model is singular. To fit a model, we must do something else. The options are (1) Place a constraint like $\beta_1 + \beta_2 + \beta_3 = 0$ on the parameters; (2) exclude one of the $X$'s from the model; or (3) leave out the intercept and force the regression through the origin. All of these options will in some sense be equivalent, since the same $R^2$, $\hat{\sigma}^2$ and overall $F$-test and predictions will result. Of course, some care must be taken in using parameter estimates, since these will surely depend on the parameterization used to get a full rank model. For further reading on matrices and models of less than full rank, see Searle (1971, 1982).

**How good is best? (Ehrenberg, 1982)**    The least squares estimates give the best fitted line in the sense that the residual sum of squares function is minimized. However, how much better than competitors is this best line? As we shall see, for many estimators the value of the residual sum of squares function is very close to that given by the least squares estimates.

Consider only simple regression, with least squares estimates $\hat{\beta}_0$ and $\hat{\beta}_1$, and with residual sum of squares given by $RSS$. As alternative estimates, consider $\tilde{\beta}_0$ and $\tilde{\beta}_1$ defined for each $k$ by

$$\tilde{\beta}_1 = k\hat{\beta}_1 \quad \text{and} \quad \tilde{\beta}_0 = \bar{y} - \tilde{\beta}_1 \bar{x}$$

As $k$ varies, these estimates give a fitted line through the point $(\bar{x}, \bar{y})$, but of arbitrary slope. The residual sum of squares function for these estimates, $RSS(\tilde{\beta}_0, \tilde{\beta}_1)$, is

$$RSS(\tilde{\beta}_0, \tilde{\beta}_1) = \sum (y_i - \tilde{\beta}_0 - \tilde{\beta}_1 x_i)^2$$

$$= \sum [(y_i - \bar{y}) - \hat{\beta}_1(x_i - \bar{x}) - (\tilde{\beta}_1 - \hat{\beta}_1)(x_i - \bar{x})]^2$$

$$= RSS + (\tilde{\beta}_1 - \hat{\beta}_1)^2 SXX$$

Since $RSS = SYY(1 - r^2)$, where $r$ is the sample correlation between $x$ and $y$, and $\hat{\beta}_1 = SXY/SXX$,

$$RSS(\tilde{\beta}_0, \tilde{\beta}_1) = SYY[(1 - r^2) + r^2(1 - k)^2] \tag{3.1}$$

If we divide both sides of (3.1) by $RSS$ and take square roots, we will have the ratio of the standard error of regression using the alternative estimates to that using the least squares estimates,

$$\left[\frac{RSS(\tilde{\beta}_0, \tilde{\beta}_1)}{RSS}\right]^{1/2} = \left[1 + \frac{r^2}{1 - r^2}(1 - k)^2\right]^{1/2}$$

This ratio must always be at least 1 since the least squares estimates minimize the residual sum of squares function. But how big can $k$ be and still have the standard error of regression within 5% of its minimum value? The answer is obtained by setting the right-hand side of the last expression equal to 1.05 and solving for $k$. We find

$$k \leqslant 1 \pm 0.32 \left[\frac{1 - r^2}{r^2}\right]^{1/2}$$

Any value of $k$ in the shaded area in Figure 3.1 will satisfy the bound. For example, if $r^2 = 0.49$, any value of $\tilde{\beta}_1$ in the range $0.67\hat{\beta}_1$ to $1.33\hat{\beta}_1$ gives an estimated standard error of regression within 5% of that given by least squares. If $r^2$ is not large, the least squares estimates will not be much better than a whole range of other possible estimates.

**Tests.**   Even if the fitted model were correct and errors were normally distributed, tests and confidence statements for parameters are difficult to interpret because nonorthogonality in the data leads to a multiplicity of possible tests. Sometimes, tests of effects adjusted for other variables are

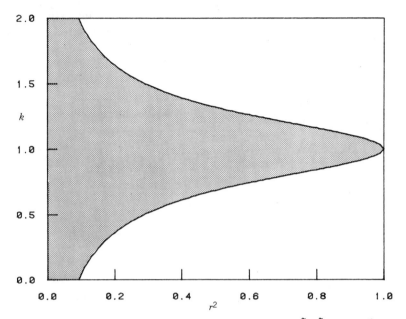

**Figure 3.1**    Shaded region gives values of $k$ for which $[RSS(\tilde{\beta}_0, \tilde{\beta}_1)/RSS]^{1/2} \leqslant 1.05$.

clearly desirable, such as in assessing a treatment effect after adjusting for other variables to reduce variability. At other times, the order of fitting is not clear, and the analyst must expect ambiguous results.

If the model that is fit depends on the data, the situation is further complicated, since testing a fictional parameter equal to zero is not necessarily equivalent to testing the significance of a variable in a regression. However, the usual test statistics do provide a useful guide, even if strict probability interpretations are not reliable. For example, the $t$-value for WT2 in Table 3.3 indicates that WT2 has little effect on the regression, and essentially the same results would be obtained without that variable in the model.

All that has been said so far should serve to point out the futility of using accept/reject rules to determine significance of predictors. In most situations the only true test of significance is repeated experimentation.

## 3.2    Sampling models

**Experimentation versus observation.**    There are fundamentally two types of predictors in a regression analysis: experimental and observational. For the former type, the values of the predictors are under the control of the experimenter, while for the latter type, the values of the predictors are observed, rather than set. Consider, for example, a hypothetical study of factors determining the yield of a certain crop. Experimental variables might include the amount and type of fertilizers used, the spacing of plants, and the amount of irrigation, since each of these can be assigned by the investigator to the units, which are plots of land. Observational predictors might include characteristics of the plots in the study, such as drainage, exposure, soil fertility, and weather variables. All of these are beyond the control of the experimenter, yet may have important effects on the observed yields.

Some experimental designs, including those that use randomization, are constructed so that the effects of observational factors can be ignored or used in analysis of covariance (see, e.g., Cox, 1958). For data from designed experiments, fitted models may lead to useful results concerning the effects of experimental factors, and these models can be used in predicting future values of the response.

At the other extreme, purely observational studies not under the control of the analyst can only be used to predict or model the events that were observed in the data, as in the fuel consumption example. To apply observational results to predict future values, additional assumptions about the behavior of future values compared to the behavior of the existing data must be made.

**Sampling from a normal population.**   Much of the intuition for the use of least squares estimation is based on the assumption that the observed data are a sample from a multivariate normal population. While the assumption of multivariate normality is almost never tenable in practical regression problems, it is worthwhile to explore the relevant results for normal data, first assuming random sampling and then removing that assumption.

*Example 3.3*   *Multivariate normality*

Suppose that all of the observed variables are normal random variables, and the observations on each case are independent of the observations on each other case. In a two-variable problem, for the $i$th case observe $(x_i, y_i)$, and suppose that

$$\begin{pmatrix} x_i \\ y_i \end{pmatrix} \sim N\left( \begin{pmatrix} \mu_X \\ \mu_Y \end{pmatrix}, \begin{pmatrix} \sigma_X^2 & \rho_{XY}\sigma_X\sigma_Y \\ \rho_{XY}\sigma_X\sigma_Y & \sigma_Y^2 \end{pmatrix} \right) \qquad i = 1, 2, \ldots, n \quad (3.2)$$

Equation (3.2) says that $x_i$ and $y_i$ are each realizations of normal random variables with means $\mu_X$ and $\mu_Y$, variances $\sigma_X^2$ and $\sigma_Y^2$, and correlation coefficient $\rho_{XY}$. Now, suppose we consider the *conditional distribution* of $y_i$ given that we have already observed the value of $x_i$. It can be shown (see, e.g., Lindgren, 1976) that the conditional distribution of $y_i$ given $x_i$, written as $y_i|x_i$, is normal, and

$$y_i|x_i \sim N\left(\mu_y + \rho_{XY}\frac{\sigma_Y}{\sigma_X}(x_i - \mu_X),\ \sigma_Y^2(1 - \rho_{XY}^2)\right) \qquad i = 1, 2, \ldots, n \qquad (3.3)$$

If we define

$$\beta_0 = \mu_Y - \beta_1\mu_X; \qquad \beta_1 = \rho_{XY}\frac{\sigma_Y}{\sigma_X}; \qquad \sigma^2 = \sigma_Y^2(1 - \rho_{XY}^2) \qquad (3.4)$$

then the conditional distribution of $y_i$ given $x_i$ is simply

$$y_i|x_i \sim N(\beta_0 + \beta_1 x_i, \sigma^2) \qquad i = 1, 2, \ldots, n \qquad (3.5)$$

which is essentially the same as the simple regression model.

Given random sampling, the five parameters in (3.2) are estimated, using the notation of Table 1.2, by

$$\hat{\mu}_X = \bar{x} \qquad \hat{\sigma}_X^2 = SD_X^2 \qquad \hat{\rho}_{XY} = r_{XY}$$

$$\hat{\mu}_Y = \bar{y} \qquad \hat{\sigma}_Y^2 = SD_Y^2 \qquad (3.6)$$

Estimates of $\beta_0$ and $\beta_1$ are obtained by substituting estimates from (3.6) for parameters in (3.4), so that $\hat{\beta}_1 = r_{XY}SD_Y/SD_X$, and so on, as derived in Chapter 1. (However, $\hat{\sigma}^2 = [(n-1)/(n-2)]SD_Y^2(1 - r_{XY}^2)$ to correct for degrees of freedom.)

If the observations on the $i$th case are $y_i$ and a $p \times 1$ vector $\mathbf{x}_i$ not including a constant, *multivariate* normality is shown symbolically by

$$\begin{pmatrix} y_i \\ \mathbf{x}_i \end{pmatrix} \sim N\left(\begin{pmatrix} \mu_y \\ \boldsymbol{\mu}_x \end{pmatrix},\ \begin{pmatrix} \sigma_Y^2 & \boldsymbol{\Sigma}_{XY}^T \\ \boldsymbol{\Sigma}_{XY} & \boldsymbol{\Sigma}_{XX} \end{pmatrix}\right)$$

where $\boldsymbol{\Sigma}_{XX}$ is a $p \times p$ matrix of variances and covariances between the $X$'s, and $\boldsymbol{\Sigma}_{XY}$ is a $p \times 1$ vector of covariances between the $X$'s and $Y$. The conditional distribution of $y_i$ given $\mathbf{x}_i$ is then

$$y_i|\mathbf{x}_i \sim N((\mu_Y - \boldsymbol{\beta}^{*T}\boldsymbol{\mu}_X) + \boldsymbol{\beta}^{*T}\mathbf{x}_i, \sigma^2) \qquad (3.7)$$

and, if $\mathcal{R}^2$ is the population multiple correlation,

$$\boldsymbol{\beta}^* = \boldsymbol{\Sigma}_{XX}^{-1}\boldsymbol{\Sigma}_{XY}; \qquad \sigma^2 = \sigma_Y^2 - \boldsymbol{\Sigma}_{XY}^T\boldsymbol{\Sigma}_{XX}^{-1}\boldsymbol{\Sigma}_{XY} = \sigma^2(1 - \mathcal{R}^2)$$

The formulas for $\boldsymbol{\beta}^*$ and $\sigma^2$ and the formulas for their least squares estimators differ only by the substitution of estimates for parameters, with $n^{-1}(\mathscr{X}^T\mathscr{X})$ estimating $\Sigma_{XX}$, and $n^{-1}(\mathscr{X}^T\mathscr{Y})$ estimating $\Sigma_{XY}$.

---

***Example 3.4   Nonrandom sampling of a population (or how to get bigger $R^2$)***

---

The reader may have noticed that the conditional distribution in (3.3) or (3.7) does not depend on random sampling, but only on normal distributions. Thus whenever multivariate normality seems to be a reasonable model for the variables, a linear regression model is suggested for the conditional distribution of one variable given the others. However, if random sampling is not used, some of the usual summary statistics, including $R^2$, lose meaning. This is illustrated with artificial data.

Figure 3.2 gives a bivariate pseudorandom sample of $n = 250$ pairs $(x_i, y_i)$ drawn on a computer as if each $(x_i, y_i)$ was independently drawn

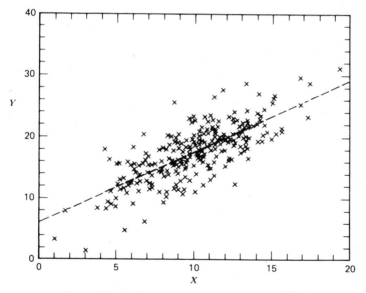

**Figure 3.2**   A bivariate normal sample, $n = 250$.

from

$$\binom{x_i}{y_i} \sim N\left(\binom{10.0}{17.5}, \begin{pmatrix} 9.0000 & 11.2510 \\ 11.2510 & 23.0625 \end{pmatrix}\right) \qquad i=1, 2, \ldots, 250 \qquad (3.8)$$

From (3.3), the conditional distribution of $y_i|x_i$ is

$$y_i|x_i \sim N(5+1.25x_i, 9) \qquad i=1, 2, \ldots, 250$$

In the figure, the cloud of points is generally elliptical, as is characteristic of normal data. The fitted regression line, given in Table 3.4, and drawn in Figure 3.2, is not far from the actual line $y=5+1.25x$. The computed $R^2=0.584$ is close to the true $\rho_{XY}^2=0.610$. All of the usual summaries and tests will be useful because random sampling was used.

Now, consider the same set of points, but select $n$ cases according to their value $x_i$. Let $S_X^2=\sum(x_i-\bar{x})^2/n$ be the variance of the values of $X$ that are actually in the sample; $S_X^2$ is not an estimate of $\sigma_X^2$ because the values of $X$ were selected, not sampled. Often, the experimenter can choose units to make $S_X^2$ take on any value; usually, in designed experiments $S_X^2$ is as large as possible. Two selections are shown graphically in Figures 3.3 and 3.4. These two figures were obtained from Figure 3.2. In Figure 3.3, only cases with $x_i \leqslant 7$ or $x_i \geqslant 13$ were chosen, while in Figure 3.4, the range for $X$ was restricted to $7 < x_i < 13$. Thus in Figure 3.3, $S_X^2$ is large, while in Figure 3.4, $S_X^2$ is small. For all three data sets (Figures 3.2 to 3.4) the fitted equations are nearly identical, and $\hat{\sigma}^2$ is fairly constant, as given in Table 3.4. However, note the very large change in $R^2$. In Figure 3.3, where $S_X^2 > \sigma_X^2$, $R^2=0.762$ is much too big, while in Figure 3.4, where $S_X^2 < \sigma_X^2$, $R^2=0.279$ is much too small. While all three graphs lead to nearly the same fitted equation, the usual summary of the fit—$R^2$—leads to very different conclusions.

**Table 3.4   Regression summaries**

|   |            |             | Estimates from |           |
|---|------------|-------------|-----------------------|-----------|
|   | True Values | Full Sample | $X \leqslant 7$ or $X \geqslant 13$ | $7 < X < 13$ |
| $n$ | $\infty$ | 250 | 90 | 160 |
| $\beta_0$ | 5.000 | 5.816 | 5.829 | 5.376 |
| $\beta_1$ | 1.250 | 1.167 | 1.157 | 1.216 |
| $\sigma^2$ | 9.000 | 9.010 | 9.166 | 9.024 |
| $R^2$ | 0.610 | 0.584 | 0.762 | 0.279 |

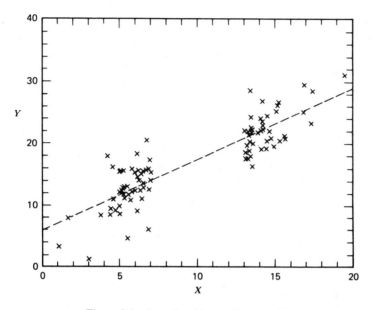

**Figure 3.3** Sample with $x_i \leqslant 7$ or $x_i \leqslant 13$.

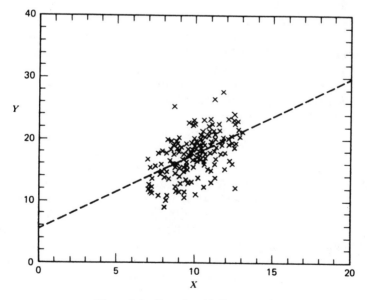

**Figure 3.4** Sample with $7 \leqslant x_i \leqslant 13$.

The reason for this is apparent from the three graphs. In Figure 3.4, it appears that, because the range of $X$ is so narrow, the variation in $Y$ ignoring $X$ is not much greater than the variation of $Y$ given $X$. Thus the regression appears to explain little—and $R^2$ is small. A similar argument applied to Figure 3.3 suggests large $R^2$.

This example points out that even in the unusual event of analyzing data drawn from a multivariate normal population, if sampling of the population is not random, the interpretation of summary statistics such as $R^2$ may be completely misleading, as this statistic will be strongly influenced by the method of sampling. In particular, a few cases with unusual values for the predictors can largely determine the observed value of this statistic.

---

## 3.3  Predictors measured with error

The desirable properties of least squares estimators depend on the assumptions. A major topic in this book is the study of problems when assumptions fail. One assumption is that the predictors are fixed values measured without error. When errors do occur in the predictors as well as in the response, the usual least squares criterion may not make much sense, since only errors in the response are used to determine the estimator. Alternative assumptions that allow for errors in the predictors are much more complicated, and there is no agreed correct way to proceed.

Suppose that $X$ is the observed matrix of predictors, with elements possibly measured with error. The "true values" that should have been observed are given by $\tilde{X}$, where

$$X = \tilde{X} + D$$

and $D$ is an $n \times p'$ matrix of errors; usually the first column of $D$ is all zeros, since the column of ones in $X$ is known without error, and any other column of $X$ that is known exactly will have corresponding column in $D$ equal to all zeroes. The other elements of $D$ may represent rounding error, always present when using computing machinery, or measurement errors. We assume here that any row of $D$, say $d_i^T$, representing the errors for one case, is independent of the errors for any other case. We assume $E(d_i) = 0$, and var$(d_i)$ is a $p' \times p'$ diagonal matrix S,

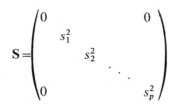

The zero in the upper left corner corresponds to an assumption of no error in the indicator variable for the intercept.

Suppose first that the problem of interest is the linear model

$$Y = \tilde{X}\beta + e \tag{3.9}$$

but estimation of $\beta$ must be based on the erroneous model

$$Y = X\beta + e = (\tilde{X} + D)\beta + e \tag{3.10}$$

The estimator $\hat{\beta} = (X^T X)^{-1} X^T Y$ from fitting (3.10) need not be a reasonable estimate of $\beta$ for the model (3.9). Hodges and Moore (1972) show that

$$E(\hat{\beta}) - \beta \approx -(n - p')(X^T X)^{-1} S\beta \tag{3.11}$$

and thus fitting (3.10) gives biased estimates of $\beta$ in (3.9). The biases may be large, either positive or negative, and they do not disappear as sample size increases. For simple regression, the expectation of the slope coefficient is

$$E(\hat{\beta}_1 \,|\, \text{simple regression}) \approx \beta_1 \left[ 1 - \frac{s_1^2}{SXX/(n-2)} \right]$$

and the estimated slope will, on the average, have magnitude that is too small, depending on the ratio of measurement error $s_1^2$ to $SXX/(n-2)$. In the presence of measurement error, if (3.9) is of interest, least squares regression can be very misleading.

Fortunately, (3.9) is *not* always of interest. Berkson (1950) has considered the very important problem of prediction. If predictors are measured with error, both now and in the future, fitting of (3.10), not (3.9), would be of interest, since the future predictors are not the true values but the imperfectly measured values. Thus measurement errors may not be important in problems if the relationship of interest is based on the observed values, not on the unobserved true values.

In parameter estimation problems where the relationship between the response and the unobserved true values is of interest, we are in the situation described earlier of fitting (3.10), but wanting to fit (3.9). Reasonable approaches to this more difficult problem require further assumptions and information. A lucid treatment of the two-variable problem is given by

Madansky (1959). A more recent account and bibliography is given by Anderson (1976).

Another approach to the problem of errors in variables is to find diagnostic statistics to decide if fitting (3.10) is very different from fitting (3.9). These diagnostics are generally derived using the methods of numerical analysis; for example, one could expand $\hat{\beta}$ in a Taylor series about $\mathbf{D} = \mathbf{0}$ and use the size of higher-order terms to judge the effects of measurement errors; see, for example, Hodges and Moore (1972); Davies and Hutton (1975); and Beaton, Rubin, and Barone (1976). These diagnostics are intimately related to the question of collinearity between the predictors discussed in Chapter 8; no method yet proposed seems to be completely adequate for deciding if measurement errors can be neglected.

## Problems

**3.1** Fit the regression of SOMA on AVE LIN and QUAD as defined in Section 3.1 for the girls in the Berkeley Guidance Study data, and compare to the results in Section 3.1.

**3.2** For a bivariate normal, find the conditional distribution of $x_i|y_i$. From this you will find that the regression line for the regression of $y$ on $x$ is *different* from the regression line for the regression of $x$ on $y$. Under what conditions is it the same? For the bivariate normal given at (3.8) find the conditional distribution of $x_i|y_i$ and draw the two regression lines ($x$ on $y$ and $y$ on $x$) on the same graph.

**3.3** The data given in Table 3.5 were first given by Longley (1967) to demonstrate inadequacies of regression computer programs then available. The seven variables are

$X_1 = $ GNP price deflator, in percent.

$X_2 = $ GNP, in millions of dollars.

$X_3 = $ Unemployment, in thousands of persons.

$X_4 = $ Size of armed forces, in thousands.

$X_5 = $ Noninstitutional population 14 years of age and over, in thousands.

$X_6 = $ Year.

$Y = $ Total derived employment in thousands.

**3.3.1.** Fit the regression of $Y$ on all the $X$'s.

**3.3.2.** The predictors in these data are clearly measured with error. Obtain estimates of the measurement errors, $s_1^2, \ldots, s_6^2$ for the predictors. (*Hint*: The data themselves contain little information concerning the measurement errors. One way to proceed that will give a *lower bound* on the

Table 3.5 Longley's data

| $X_1$ | $X_2$ | $X_3$ | $X_4$ | $X_5$ | $X_6$ | $Y$ |
|---|---|---|---|---|---|---|
| 83.0 | 234289. | 2356. | 1590. | 107608. | 1947. | 60323. |
| 88.5 | 259426. | 2325. | 1456. | 108632. | 1948. | 61122. |
| 88.2 | 258054. | 3682. | 1616. | 109773. | 1949. | 60171. |
| 89.5 | 284599. | 3351. | 1650. | 110929. | 1950. | 61187. |
| 96.2 | 328975. | 2099. | 3099. | 112075. | 1951. | 63221. |
| 98.1 | 346999. | 1932. | 3594. | 113270. | 1952. | 63639. |
| 99.0 | 365385. | 1870. | 3547. | 115094. | 1953. | 64989. |
| 100.0 | 363112. | 3578. | 3350. | 116219 | 1954. | 63761. |
| 101.2 | 397469. | 2904. | 3048. | 117388. | 1955. | 66019. |
| 104.6 | 419180. | 2822. | 2857. | 118734. | 1956. | 67857. |
| 108.4 | 442769. | 2936. | 2798. | 120445. | 1957. | 68169. |
| 110.8 | 444546. | 4681. | 2637. | 121950. | 1958. | 66513. |
| 112.6 | 482704. | 3813. | 2552. | 123366. | 1959. | 68655. |
| 114.2 | 502601. | 3931. | 2514. | 125368. | 1960. | 69564. |
| 115.7 | 518173. | 4806. | 2572. | 127852. | 1961. | 69331. |
| 116.9 | 554894. | 4007. | 2827. | 130081. | 1962. | 70551. |

$s_j^2$ is to assume that the last digit of each number is subject to round-off error. This would suggest that the "true value" of $X_2$ for 1947 is any number between 234,288.5 and 234,289.5. If we assume that all the possible outcomes are equally likely to be the true one, then rounding error is uniformly distributed on the interval $(-.5, +.5)$. Using the fact that a random variable that is uniform on the interval $(a, b)$ has variance $(b-a)^2/12$, we can estimate $s_2^2 = [.5 - (-.5)]^2/12 = \frac{1}{12}$. The same method can be applied to the other predictors, except possibly year. These estimated $s_j^2$ are probably much too small, but they do provide a starting point. Rounding error for year could correspond to the variation in the length of a year, leap years are longer, or else to the fact that the other quantities are not measured exactly a year apart.)

3.3.3. To examine the effects of errors of the magnitudes estimated in the last section of this problem, perform a simulation experiment as follows: (1) Begin with the Longley data. (2) On a computer, generate random numbers that are uniform on the interval you think reasonable for rounding errors. For $X_2$, for example, you may generate random variables on $(-.5, .5)$. (3) Add the random numbers to the predictors; do not change $Y$. (4) Compute the regression of $Y$ on the modified $X$'s, and record the coefficient estimates. (5) Repeat steps 2 to 4 several times, perhaps 100, and summarize results. This simulation will give a good idea of the effects of rounding error on the estimates of coefficients in this problem.

# 4

# WEIGHTED LEAST SQUARES, TESTING FOR LACK OF FIT, GENERAL *F*-TESTS, AND CONFIDENCE ELLIPSOIDS

This chapter begins with a discussion of the use of additional information about error variances and covariances. This information will sometimes be used to obtain generalized least squares estimators rather than the ordinary least squares estimators previously discussed. Alternatively, this information may be used to test for lack of fit of a model applied to observed data. Next, we turn to a heuristic discussion of the general methodology for obtaining test statistics that will follow $F$ distributions when errors are normally distributed. These tests are used in many situations in regression, including those outlined in Chapters 1 and 2, as well as others to be described in later chapters. Finally, ellipsoids used as simultaneous confidence regions for more than one parameter are discussed. These regions, which depend on the $F$ distribution, will be of use in assessing the influence of cases on regression estimates, a topic in Chapter 5.

## 4.1  Generalized and weighted least squares

The assumptions of earlier chapters that error variances are unknown and equal and that the errors are independent are usually made out of necessity since specific knowledge of these variances is exceptional. In a few problems, however, additional information concerning error variances will be available, with variances either known or known up to some multiplicative constant. The methodology for incorporating this information into the analysis is not difficult and is presented in this section.

**Generalized least squares.** Suppose that we know the value of a symmetric positive definite matrix $\Sigma$, such that the covariance matrix for the error vector $\mathbf{e}$ is given by $\mathrm{var}(\mathbf{e}) = \sigma^2\Sigma$, with $\sigma^2 > 0$, but not necessarily known. We might reasonably expect that in these circumstances the ordinary least squares estimator of $\boldsymbol{\beta}$, although still unbiased, will no longer be the minimum variance estimator, since it ignores obviously useful information. Formally, consider the model

$$\mathbf{Y} = \mathbf{X}\boldsymbol{\beta} + \mathbf{e} \qquad \mathbf{X}: n \times p' \text{ rank } p' \qquad (4.1a)$$

$$\mathrm{var}(\mathbf{e}) = \sigma^2\Sigma \qquad \Sigma \text{ known, } \sigma^2 > 0 \text{ not necessarily known} \qquad (4.1b)$$

We will continue to use the symbol $\hat{\boldsymbol{\beta}}$ for the estimator of $\boldsymbol{\beta}$, even though the estimate will be obtained via generalized, not ordinary, least squares. Once $\hat{\boldsymbol{\beta}}$ is determined, the residuals $\hat{\mathbf{e}}$ are given by $\hat{\mathbf{e}} = \mathbf{Y} - \hat{\mathbf{Y}} = \mathbf{Y} - \mathbf{X}\hat{\boldsymbol{\beta}}$. The estimator $\hat{\boldsymbol{\beta}}$ is chosen to minimize the generalized residual sum of squares function,

$$RSS(\boldsymbol{\beta}) = (\mathbf{Y} - \mathbf{X}\boldsymbol{\beta})^T\Sigma^{-1}(\mathbf{Y} - \mathbf{X}\boldsymbol{\beta}) \qquad (4.2)$$

Roughly speaking, the use of the generalized residual sum of squares recognizes that some of the residuals, or fitting errors, are more important than others. In particular, residuals corresponding to errors with a larger variance will be less important in the computation of the generalized residual sum of squares. The generalized least squares estimator is given by

$$\hat{\boldsymbol{\beta}} = (\mathbf{X}^T\Sigma^{-1}\mathbf{X})^{-1}\mathbf{X}^T\Sigma^{-1}\mathbf{Y} \qquad (4.3)$$

While this last equation can be found directly, it is convenient to transform the problem specified by model (4.1) to one that can be solved by ordinary least squares. Then, all of the results for ordinary least squares can be applied to generalized least squares problems.

Model (4.1) differs from the ordinary least squares model only in that $\mathrm{var}(\mathbf{e}) = \sigma^2\Sigma$. Now, suppose that we could find an $n \times n$ matrix $\mathbf{C}$, such that $\mathbf{C}$ is symmetric and $\mathbf{C}^T\mathbf{C} = \mathbf{C}\mathbf{C}^T = \Sigma^{-1}$ (and $\mathbf{C}^{-1}\mathbf{C}^{-T} = \Sigma$). Such a matrix $\mathbf{C}$ will be called a *square root* of $\Sigma^{-1}$. Then, using Appendix 2A.2 on random vectors, the variance-covariance matrix of $\mathbf{Ce}$ is given by

$$\mathrm{var}(\mathbf{Ce}) = \mathbf{C}(\sigma^2\Sigma)\mathbf{C}^T$$

$$= \sigma^2\mathbf{C}(\mathbf{C}^{-1}\mathbf{C}^{-T})\mathbf{C}^T$$

$$= \sigma^2\mathbf{C}\mathbf{C}^{-1}\mathbf{C}^{-T}\mathbf{C}^T$$

$$= \sigma^2\mathbf{I}_n \qquad (4.4)$$

Multiplying both sides of equation (4.1a) by $C$ gives

$$CY = CX\beta + Ce \tag{4.5}$$

Now, define $Z = CY$, $M = CX$, and $d = Ce$. Then equation (4.5) becomes

$$Z = M\beta + d \tag{4.6}$$

where, from (4.4), $var(d) = \sigma^2 I_n$, and in (4.6) $\beta$ is exactly the same as $\beta$ in (4.1). Model (4.6) can be solved using ordinary least squares. For example, the estimator $\hat{\beta}$ in terms of $Z$ and $M$ is from (2.15),

$$\hat{\beta} = (M^T M)^{-1} M^T Z$$

Substituting $X$'s, $Y$'s and $C$'s for $M$ and $Z$, this last equation becomes

$$
\begin{aligned}
\hat{\beta} &= [(CX)^T(CX)]^{-1}(CX)^T(CY) \\
&= (X^T C^T C X)^{-1}(X^T C^T C Y) \\
&= (X^T \Sigma^{-1} X)^{-1}(X^T \Sigma^{-1} Y)
\end{aligned}
$$

which is the estimator given at (4.3).

A <u>practical procedure</u> for <u>generalized least squares</u> is to obtain $C$, a square root of $\Sigma^{-1}$, multiply the observed data vector $Y$ and matrix $X$ on the left by $C$, and solve the resulting regression problem using ordinary least squares. A numerical difficulty arises in actually finding $C$. However, in the special case of weighted least squares, computing $C$ from $\Sigma$ is simple.

**Weighted least squares.**   When the errors may have different variances, but are all uncorrelated, $\Sigma$ will be a diagonal matrix and we have a *weighted least squares* problem. The variances are usually expressed using weights, $w_i > 0$, and $var(e_i) = \sigma^2 / w_i$. Cases with large weight have small variance and are more important in the regression problem. The matrix $\Sigma$ has the form $\Sigma = W^{-1}$, where

$$
W^{-1} = \begin{pmatrix} \dfrac{1}{w_1} & & & 0 \\ & \dfrac{1}{w_2} & & \\ & & \ddots & \\ 0 & & & \dfrac{1}{w_n} \end{pmatrix} \tag{4.7}
$$

For this choice of $\Sigma$ ($= W^{-1}$), the matrix $C$ is easily found:

$$
C = \begin{pmatrix} \sqrt{w_1} & & & 0 \\ & \sqrt{w_2} & & \\ & & \ddots & \\ 0 & & & \sqrt{w_n} \end{pmatrix} \tag{4.8}
$$

and

$$
\mathbf{M} = \begin{pmatrix} \sqrt{w_1} & \sqrt{w_1}x_{11} & \cdots & \sqrt{w_1}x_{1p} \\ \sqrt{w_2} & \sqrt{w_2}x_{21} & \cdots & \sqrt{w_2}x_{2p} \\ & & \vdots & \\ \sqrt{w_n} & \sqrt{w_n}x_{n1} & \cdots & \sqrt{w_n}x_{np} \end{pmatrix} \qquad \mathbf{Z} = \begin{pmatrix} \sqrt{w_1}y_1 \\ \sqrt{w_2}y_2 \\ \vdots \\ \sqrt{w_n}y_n \end{pmatrix} \tag{4.9}
$$

Even the column of ones gets multiplied by the $\sqrt{w_i}$. The regression problem is then solved using $\mathbf{M}$ and $\mathbf{Z}$ in place of $\mathbf{X}$ and $\mathbf{Y}$. (*Exception*: the residuals and fitted values are still computed from $\hat{\mathbf{e}} = \mathbf{Y} - \hat{\mathbf{Y}}$, and $\hat{\mathbf{Y}} = \mathbf{X}\hat{\boldsymbol{\beta}}$, not from the equivalent formulas using $\mathbf{M}$ and $\mathbf{Z}$.)

Most computer programs are designed to allow use of case weights in regression. The $w_i$ usually appear as a column of data, and the user simply specifies that column as weights. The program computes $\mathbf{M}^T\mathbf{M}$ and $\mathbf{M}^T\mathbf{Z}$ or their deviation from the average equivalents and proceeds as if unweighted least squares were used (see problem 4.8).

**Applications of weighted least squares.** If the $i$th response is an average of $n_i$ equally variable observations, then $\mathrm{var}(y_i) = \sigma^2/n_i$, and $w_i = n_i$. If $y_i$ is a total of $n_i$ observations, $\mathrm{var}(y_i) = n_i\sigma^2$, and $w_i = 1/n_i$. If variance is proportional to some predictor $x_i$, $\mathrm{var}(y_i) = x_i\sigma^2$, then $w_i = 1/x_i$.

*Example 4.1  Strong interaction*

---

The purpose of the experiment described here is to study the interactions of certain kinds of elementary particles in collision with proton targets (Weisberg et al., 1978). The particles studied include the $\pi^-$ meson and its antiparticle $\pi^+$. These are unstable and can be produced by high-energy accelerators. They are in a group of particles called hadrons, and have a positive or negative electrical charge equal in magnitude to the charge of the electron. Hadrons interact via the electromagnetic force that holds atoms together. In addition, and unlike the electron, they also interact via the so-called strong interaction force that holds nuclei together. Although the electromagnetic force is well understood, the strong interaction is still somewhat mysterious to physicists, and this experiment was designed to test certain theories of the nature of the strong interaction.

The scattering processes studied are denoted

$$
a + p \rightarrow c + x \tag{4.10}
$$

and may be represented by a diagram:

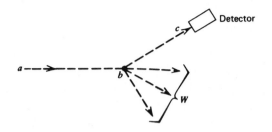

This diagram shows a beam of particles of type $a$ aimed at a target $b$ containing protons $p$. A detector then detects outgoing particles of type $c$. The experiment is carried out for $a$ and $c$ equal to various particles; only $\pi^-$ and $\pi^+$ are considered here. For a typical collision at high energy, both $a$ and $p$ will break up and, by the process of the conversion of energy into matter, fragment into many hadrons. The experiment measures the rate of production of a particular particle type $c$, no matter what other particles (labeled $W$ in the diagram) are produced.

The quantity measured is the scattering cross section $y$ (usually denoted by $\Delta\sigma$), given by the equation

$$y = \frac{N_c}{N_a \rho l} \tag{4.11}$$

where $N_a$ is the number of incoming beam particles per second, $\rho$ is the target density in particles per unit volume, $l$ is the target length, and $N_c$ is the number of particles $c$ detected per second. The cross section $y$ has the dimensions of area and is conventionally measured in millibarns (mb), where 1 mb $= 1 \times 10^{-27}$ cm$^2$.

The experiment is carried out with beam $a$ having various values of incident momentum $p_a^{\text{lab}}$, measured in the laboratory frame of reference. A quantity of more basic theoretical significance than $p_a^{\text{lab}}$ is $s$, the square of the total energy in the center-of-mass frame of reference system. For the high momenta used in this experiment, particle $a$ travels nearly at the speed of light, and the relationship

$$s = 2m_p p_a^{\text{lab}} \tag{4.12}$$

holds to a good approximation. The units of $s$ are (GeV)$^2$, where 1 GeV $= 1 \times 10^9$ electron volts is the energy that an elementary particle reaches on being accelerated by an electric potential of one billion volts. The momentum $p_a^{lab}$ and the mass $m_p$ are measured in GeV, and $m_p = 0.938$ GeV for a proton.

Theoretical physicists have constructed various models of the strong interaction force. Certain models predict that, in the high-energy limit $s \to \infty$, the cross section $y$ should approach a constant limit. In addition, it is predicted that the functional form of the approach to this limit should be

$$y = \beta_0 + \beta_1 s^{-1/2} + \text{relatively small terms} \qquad (4.13)$$

The theory makes quantitative predictions about $\beta_0$ and $\beta_1$ and their dependence on particle types $a$ and $c$. Of interest, therefore, are: (1) estimation of $\beta_0$ and $\beta_1$, given (4.13) as a model, for each choice of $a$ and $c$; (2) assessment of whether the model (4.13) provides an accurate description of the observed data; and (3) comparison of $\beta_0$ and $\beta_1$ to theoretical predictions if (4.13) is in fact appropriate.

The data given in Table 4.1 summarize the results of experiments with $a = c = \pi^-$. At each $p_a^{lab}$, a very large number of particles $N_a$ was used so that the variance of the observed $y$ values could be accurately obtained from theoretical considerations. The square roots of these variances (i.e., the $\sigma/\sqrt{w_i}$) are given in column 4 of Table 4.1.

For continuity of notation, set $x = s^{-1/2}$, and $e = $ smaller terms, so

Table 4.1    Data for the physics example

| $p_a^{lab}$ GeV/c | $s^{-1/2}$ GeV/c$^{-1}$ | $y$ ($\mu$b) | $\sigma/\sqrt{w_i} =$ Estimated Standard Deviation |
|---|---|---|---|
| 4 | 0.345 | 367 | 17 |
| 6 | 0.287 | 311 | 9 |
| 8 | 0.251 | 295 | 9 |
| 10 | 0.225 | 268 | 7 |
| 12 | 0.207 | 253 | 7 |
| 15 | 0.186 | 239 | 6 |
| 20 | 0.161 | 220 | 6 |
| 30 | 0.132 | 213 | 6 |
| 75 | 0.084 | 193 | 5 |
| 150 | 0.060 | 192 | 5 |

(4.13) can be written as

$$y_i = \beta_0 + \beta_1 x_i + e_i \qquad i = 1, 2, \ldots, n \qquad (4.14)$$

with the $e_i$ independent and $\text{var}(e_i) = \sigma^2/w_i$ as given in Table 4.1. Without any loss of generality, we can set $\sigma^2 = 1$, so the numbers in column 4 of Table 4.1 correspond to $1/\sqrt{w_i}$, $i = 1, 2, \ldots, n$.

The estimated values of $\beta_0$ and $\beta_1$ must be obtained by weighted least squares. This can be done either by directly minimizing the generalized residual sum of squares (4.2), which in scalar form is given by

$$RSS(\boldsymbol{\beta}) = \sum w_i (y_i - \mathbf{x}_i^T \boldsymbol{\beta})^2 \qquad (4.15)$$

or by rescaling the data, and then applying ordinary least squares. Table 4.2 lists the results of the weighted least squares calculations, and the fitted line is plotted in Figure 4.1. The usual summaries such as $R^2$ and $t$-tests indicate that the fitted model matches the observed data reasonably well, and the parameter estimates themselves are well determined. The next question is whether (4.13) does in fact fit the data. This question of fit or lack of fit of a model is the subject of the next section.

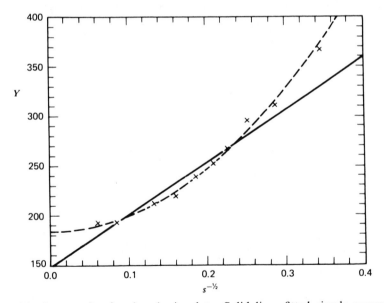

**Figure 4.1**   Scatter plot for the physics data. Solid line: fitted simple regression; dashed line: fitted quadratic regression.

Table 4.2  Weighted least squares estimates for
the physics data

| Variable | Estimate | Standard Error | $t$-value |
|---|---|---|---|
| Intercept | 148.473 | 8.079 | 18.38 |
| Slope | 530.835 | 47.550 | 11.16 |

$\hat{\sigma}^2 = 2.744$    d.f. $= 8$    $R^2 = 0.940$

*Analysis of variance*

| Source | d.f. | SS | MS | $F$ |
|---|---|---|---|---|
| Regression | 1 | 341.991 | 341.991 | 124.63 |
| Residual | 8 | 21.953 | 2.744 | |

**Additional comments.** Many statistical models, including variance components, time series, and some econometric models, will specify that $\Sigma$ depends on a small number of parameters. This may lead to use of *iteratively reweighted least squares*, defined as follows:

1. Set $\tilde{\Sigma} = I$, or some other convenient starting value.
2. Estimate coefficients as if $\Sigma = \tilde{\Sigma}$.
3. Use the residuals from step 2 to estimate $\Sigma$ by a new $\tilde{\Sigma}$.
4. Repeat steps 2 and 3 until $\tilde{\Sigma}$ does not change much.

This method can be justified by appeal to asymptotic theory, which guarantees that in large samples, as long as the method of estimation in step 3 has good properties, estimates have the same properties they would if $\Sigma$ were known (see Carroll, 1982). In small samples, estimates of parameters are likely to be unbiased if the errors have a symmetric distribution, but estimated standard errors of coefficients may be much too small; Freedman and Peters (1984) report underestimates by a factor of 3 in one problem. Using estimated weights in iteratively reweighted least squares may mislead the investigator concerning precision of estimates.

Weights may sometimes be estimated if each value of $x$ in the data is replicated many times. The sample variance of the responses for the fixed $x$ may then provide an estimate of var$(y|x)$, and the inverses of these sample variances may be used as weights. This method was used to get weights in the physics data, where the number of cases per value of $x$ was extremely large. Problem 4.7 provides another example using estimated weights as if they were true weights. The usefulness of this method depends on having a large sample size at each value of **x**.

## 4.2    Testing for lack of fit, variance known

A model specifies a shape for the relationship between the predictors and the response. When the hypothesized shape is correct, then the residual mean square from the fitted model, $\hat{\sigma}^2$, will provide an unbiased estimate of the residual variance. If the hypothesized shape is incorrect, then $\hat{\sigma}^2$ will estimate a quantity larger than $\sigma^2$, since its size will depend both on the errors and on systematic biases due to fitting the wrong shape. If $\sigma^2$ is known, or if a model-free estimate of it is available, a test for lack of fit of the model can be obtained by comparing $\hat{\sigma}^2$ to the model-free value. If $\hat{\sigma}^2$ is too large, we may have evidence that the fitted model is not adequate.

In the physics data of Example 4.1, we want to know if the straight-line model (4.13) or (4.14) provides an adequate description. As outlined in Section 4.1, the inverses of the squares of the values in column 4 of Table 4.1 are used as weights, with $\sigma^2 = 1$, a known value. From Table 4.2, $\hat{\sigma}^2 = 2.744$. Evidence against the simple regression model will be obtained if we judge $\hat{\sigma}^2 = 2.744$ large when compared to the known value of $\sigma^2 = 1$. To assign a $p$-value to this comparison, we use the following result:

If the $e_i \sim \text{NID}(0, \sigma^2/w_i)$, $i = 1, 2, \ldots, n$, with the $w_i$ and $\sigma^2$ known, and the linear model has parameters estimated using weighted least squares with the $w_i$'s as weights, and the model is correct, then

$$\chi^2 = \frac{RSS}{\sigma^2} = \frac{(n-p')\hat{\sigma}^2}{\sigma^2} \qquad (4.16)$$

is distributed as a chi-squared random variable with $n-p'$ degrees of freedom. As usual, $RSS$ is the residual sum of squares.

For the example, from Table 4.2,

$$\chi^2 = \frac{21.953}{1} = 21.953$$

From Table C at the back of the book, $\chi^2(0.01, 8) = 20.09$, so the $p$-value associated with the test is less than 0.01, which suggests that the model may not be adequate.

When this test indicates the lack of adequate fit, it is usual to fit alternative models, either by transforming some of the predictors or the response or by adding polynomial terms in the predictors. The available physical theory suggests this latter approach, and the model

$$y = \beta_0 + \beta_1 s^{-1/2} + \beta_2 (s^{-1/2})^2 + \text{smaller terms}$$

Table 4.3   Model (4.17) for the physics data

| Variable | Estimate | Standard Error | t-value |
|----------|----------|----------------|---------|
| Intercept | 183.831 | 6.459 | 28.46 |
| X | .971 | 85.369 | .01 |
| $X^2$ | 1597.505 | 250.586 | 6.38 |

| $\hat{\sigma}^2 = 0.461$ | d.f. $= 7$ | $R^2 = 0.991$ |
|---|---|---|

*Analysis of variance*

| Source | d.f. | SS | MS | F |
|--------|------|-----|-----|---|
| Regression | 2 | 360.719 | 180.359 | 391.41 |
| Residual | 7 | 3.226 | .461 | |

or

$$y_i = \beta_0 + \beta_1 x_i + \beta_2 x_i^2 + e_i \tag{4.17}$$

should be fit to the data. This model requires that the relationship between $y$ and $x$ be a quadratic curve rather than a straight line.

Model (4.17) is a multiple regression model with two predictors, $x$ and $x^2$. Fitting must again use weighted least squares with the same weights as before. The fitted equation and the analysis of variance are given in Table 4.3, and the fitted curve is graphed in Figure 4.1. The curve matches the data very closely. We can test for lack of fit of this model by computing

$$\chi^2 = \frac{RSS}{\sigma^2} = \frac{3.226}{1} = 3.226$$

Comparing this value to the percentage points of $\chi^2(7)$ indicates no evidence of lack of fit for model (4.17).

Although (4.13) does not describe the data, (4.17) does result in an adequate fit. Judgment of the success or failure of the model for the strong interaction force requires analysis of data for other choices of incidence and product particles, as well as the data analyzed here. Based on this further analysis, Weisberg et al. (1978) concluded that the theoretical model for the strong interaction force is consistent with the observed data.

## 4.3   Testing for lack of fit, variance unknown

When $\sigma^2$ is unknown, a test for lack of fit requires a model-free estimate of the residual variance. The most common model-free estimate makes use of the variation between cases with the same values on all of the predictors. For example, consider the artificial data with $n=10$ given in Table 4.4. The data were generated by first choosing the values of $x_i$ and then computing $y_i = 2.0 + 0.5x_i + e_i$, $i=1, 2, \ldots, 10$, where the $e_i$ were taken from a table of standard normal random deviates. If we consider only the values of $y_i$ corresponding to $x=1$, we can compute the average response $\bar{y}$ and the standard deviation SD with $3-1=2$ degrees of freedom, as shown in the table. If we assume that the true residual variance is the same for all values of $x$, a pooled estimate of the common variance is obtained by pooling the SD into a single estimate. If $n$ is the number of cases at a value of $x$, then the *sum of sauares for pure error*, symbolically SS(pe), is given by

$$SS(pe) = \sum (n-1)SD^2 = \sum \sum (y_i - \bar{y})^2 \tag{4.18}$$

where the sum is over all groups of cases. For the example, SS(pe) is simply the sum of the numbers in the fourth column of Table 4.4,

$$SS(pe) = 0.0243 + 0.0000 + 0.1301 + 2.2041 = 2.3585$$

Associated with SS(pe) is its degrees of freedom, d.f.(pe) $= \sum (n-1) = 2+0 +1+3=6$. The pooled, or pure error estimate of variance is $\hat{\sigma}^2_{pooled} = SS(pe)/$ d.f.(pe) $= 0.3931$. This is the same estimate that would be obtained for the residual variance if the data were analyzed using a one-way analysis of variance, grouping according to the value of $x$.

The pure error estimate of variance makes no reference to the linear regression model. It uses the assumptions that the residual variance for the example is the same for each $x$ and that all observations are independent.

**Table 4.4   A hypothetical example**

| X | Y | $\bar{y}$ | $\sum(y_i - \bar{y})^2$ | SD | d.f. |
|---|---|---|---|---|---|
| 1 | 2.55 | | | | |
| 1 | 2.75 | 2.6233 | 0.0243 | 0.1102 | 2 |
| 1 | 2.57 | | | | |
| 2 | 2.40 | 2.4000 | 0 | 0 | 0 |
| 3 | 4.19 | 4.4450 | 0.1301 | 0.3606 | 1 |
| 3 | 4.70 | | | | |
| 4 | 3.81 | | | | |
| 4 | 4.87 | | | | |
| 4 | 2.93 | 4.0325 | 2.2041 | 0.8571 | 3 |
| 4 | 4.52 | | | | |

SS(pe) $= 2.3585$,    d.f.(pe) $= 6$

d.f. $(pf) = M - p'$   $4 - 2 = 2$

Table 4.5   Analysis of variance

| Source | d.f. | SS | MS | F |
|---|---|---|---|---|
| Regression | 1 | 4.5693 | 4.5693 | |
| Residual | 8 | 4.2166 | 0.5271 | |
| { Lack of fit | { 2 | { 1.8581 | { 0.9291 | 2.36 |
| { Pure error | { 6 | { 2.3585 | { 0.3931 | |

Now suppose we fit a linear regression model to the data. The analysis of variance is given in Table 4.5. The residual mean square in Table 4.5 provides an estimate of $\sigma^2$, but this estimate does depend on the model. Thus we have two estimates of $\sigma^2$, and if the latter is much larger than the former, the model is inadequate.

We can obtain an $F$-test if the residual sum of squares in Table 4.5 is divided into two parts, the sum of squares for pure error, as given in Table 4.4, and the remainder, called the sum of squares for lack of fit, or $SS(lof) = RSS - SS(pe) = 4.2166 - 2.3585 = 1.8581$ with degrees of freedom $n - p' - df(pe)$. The $F$-test is the ratio of the mean square for lack of fit to the mean square for pure error. The observed $F = 2.36$ is considerably smaller than $F(0.05; 2, 6) = 5.14$, suggesting no lack of fit of the model to these data.

Although all the examples in this section have a single predictor, the ideas used to get a model-free estimate of $\sigma^2$ are perfectly general. The pure error estimate of variance is based on the sum of squares between the values of the response for all cases with the same values on all of the predictors.

### Example 4.2   Apple shoots

Many types of trees produce two types of morphologically different shoots. Some branches remain vegetative year after year and contribute considerably to the size of the tree. Called long shoots or leaders, they may grow as much as 15 or 20 cm over a single growing season.

On the other hand, other shoots will seldom exceed 1 cm in total length. Called short, dwarf, or spur shoots, these usually produce flowers from which fruit may arise. To complicate the issue further, long shoots occasionally change to short in a new growing season and vice versa. The mechanism that the tree uses to control the long and short shoots is not well understood.

Bland (1978) has done a descriptive study of the difference between long and short shoots of McIntosh apple trees. Using healthy trees of clonal stock planted in 1933 and 1934, he took samples of long and short shoots from the trees every few days throughout the 1971 growing season (about 106 days). The shoots sampled are presumed to be a sample of available shoots at the sampling dates. The sampled shoots were removed from the tree, marked, and taken to the laboratory for analysis.

Among the many measurements taken, Bland counted the number of stem units in each shoot. The long and the short shoots could differ because of the number of stem units, the average size of stem units, or both. An abstract of Bland's data is given in Table 4.6, for both long and short shoots. For now we will consider only the long shoots, leaving the short shoots to the Problems section.

Our goal is to find an equation that can adequately describe the relationship between DAY = days from dormancy and $Y$ = number of stem units. Lacking a theoretical form for this equation, we first examine Figure 4.2, a scatter plot of average number of stem units versus DAY. The apparent linearity of this plot should encourage us to fit a straight

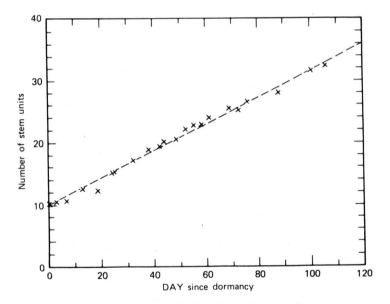

**Figure 4.2**  Scatter plot for the apple shoot data.

**Table 4.6   Bland's data for long and short apple shoots**

| | Long Shoots | | | | Short Shoots | | |
|---|---|---|---|---|---|---|---|
| DAY | $n$ | $\bar{y}$ | SD | DAY | $n$ | $\bar{y}$ | SD |
| 0 | 5 | 10.20 | 0.83 | 0 | 5 | 10.00 | 0.00 |
| 3 | 5 | 10.40 | 0.54 | 6 | 5 | 11.00 | 0.72 |
| 7 | 5 | 10.60 | 0.54 | 9 | 5 | 10.00 | 0.72 |
| 13 | 6 | 12.50 | 0.83 | 19 | 11 | 13.36 | 1.03 |
| 18 | 5 | 12.00 | 1.41 | 27 | 7 | 14.29 | 0.95 |
| 24 | 4 | 15.00 | 0.82 | 30 | 8 | 14.50 | 1.19 |
| 25 | 6 | 15.17 | 0.76 | 32 | 8 | 15.38 | 0.51 |
| 32 | 5 | 17.00 | 0.72 | 34 | 5 | 16.60 | 0.89 |
| 38 | 7 | 18.71 | 0.74 | 36 | 6 | 15.50 | 0.54 |
| 42 | 9 | 19.22 | 0.84 | 38 | 7 | 16.86 | 1.35 |
| 44 | 10 | 20.00 | 1.26 | 40 | 4 | 17.50 | 0.58 |
| 49 | 19 | 20.32 | 1.00 | 42 | 3 | 17.33 | 1.52 |
| 52 | 14 | 22.07 | 1.20 | 44 | 8 | 18.00 | 0.76 |
| 55 | 11 | 22.64 | 1.76 | 48 | 22 | 18.46 | 0.75 |
| 58 | 9 | 22.78 | 0.84 | 50 | 7 | 17.71 | 0.95 |
| 61 | 14 | 23.93 | 1.16 | 55 | 24 | 19.42 | 0.78 |
| 69 | 10 | 25.50 | 0.98 | 58 | 15 | 20.60 | 0.62 |
| 73 | 12 | 25.08 | 1.94 | 61 | 12 | 21.00 | 0.73 |
| 76 | 9 | 26.67 | 1.23 | 64 | 15 | 22.33 | 0.89 |
| 88 | 7 | 28.00 | 1.01 | 67 | 10 | 22.20 | 0.79 |
| 100 | 10 | 31.67 | 1.42 | 75 | 14 | 23.86 | 1.09 |
| 106 | 7 | 32.14 | 2.28 | 79 | 12 | 24.42 | 1.00 |
| | | | | 82 | 19 | 24.79 | 0.52 |
| | | | | 85 | 5 | 25.00 | 1.01 |
| | | | | 88 | 27 | 26.04 | 0.99 |
| | | | | 91 | 5 | 26.60 | 0.54 |
| | | | | 94 | 16 | 27.12 | 1.16 |
| | | | | 97 | 12 | 26.83 | 0.59 |
| | | | | 100 | 10 | 28.70 | 0.47 |
| | | | | 106 | 15 | 29.33 | 1.74 |

line,

$$Y = \beta_0 + \beta_1 \mathrm{DAY} + \text{error} \qquad (4.19)$$

If this model were adequate, we would have the interesting result that the observed rate of production of stem units per day is constant over the growing season.

For each sampled day, Table 4.6 reports $n$ = number of shoots sampled,

$\bar{y}$ = average number of stem units on that day, and $SD$ = within-day standard deviation. Assuming that residual variance is constant from day to day, we can do the regression in two ways. First, since var($\bar{y}$) = $\sigma^2/n$, we can compute the weighted regression of $\bar{y}_i$ on DAY with weights equal to $n$. This is summarized in Table 4.7. Alternatively, if the original 189 data points were available, we could compute the unweighted regression of the original data on DAY. This is summarized in Table 4.8. Both methods give identical intercept, slope, and regression sum of squares. They differ on any calculation that uses the residual sum of squares, because in Table 4.8 the residual sum of squares is the sum of SS(pe) and SS(lof). For example, the standard errors of the

Table 4.7    **Weighted regression of $\bar{y}$ on DAY**

| Variable | Estimate | Standard Error | $t$-value |
|---|---|---|---|
| Intercept | 9.9738 | .31427 | 31.74 |
| DAY | 0.2173 | .00534 | 40.71 |

$\hat{\sigma}^2 = 3.7196,$     d.f. = 20,     $R^2 = 0.988$

*Analysis of variance*

| Source | d.f. | SS | MS | F |
|---|---|---|---|---|
| Regression | 1 | 6164.28 | 6164.28 | 1657.2 |
| Residual | 20 | 74.39 | 3.72 | |

Table 4.8    **Unweighted regression of $y_i$ on DAY**

| Variable | Estimate | Standard Error | $t$-value |
|---|---|---|---|
| Intercept | 9.9738 | .21630 | 46.11 |
| DAY | 0.2173 | .00367 | 59.12 |

$\hat{\sigma}^2 = 1.7621,$     d.f. = 187,     $R^2 = 0.949$

*Analysis of variance*

| Source | d.f. | SS | MS | F |
|---|---|---|---|---|
| Regression | 1 | 6164.28 | 6164.28 | 3498.35 |
| Residual | 187 | 329.50 | 1.76 | |
| Lack of fit | 20 | 74.39 | 3.72 | 2.43 |
| Pure error | 167 | 255.12 | 1.53 | |

coefficients in the two tables differ because in Table 4.7 the apparent estimate of variance is 3.7196 with 20 d.f., while in Table 4.8 it is 1.7621 with 187 d.f. Often it will be more appropriate to use pure error alone to estimate $\sigma^2$, especially if the model is doubtful; this would lead to a third set of standard errors.

The SS(pe) can be computed directly from Table 4.6 using (4.18), SS(pe) $= \sum (n-1)SD^2 = 255.11$ with $\sum (n-1) = 167$ d.f. The *F*-test for lack of fit is $F = 2.43$. Since $F(0.01; 20, 167) = 1.99$, the *p*-value for this test is less than 0.01, indicating that the straight-line model (4.19) does not appear to be adequate. However, an *F*-test with this many degrees of freedom is very powerful and will detect very small deviations from the null hypothesis. Thus, while the result here is statistically significant, it may not be scientifically important, and for purposes of describing the growth of apple shoots, the model (4.19) may be adequate.

---

**Additional comments.** The pure error test requires replicated observations to obtain an estimate of variance. Without replication or additional information, there is no test for lack of fit. Daniel and Wood (1981) have suggested an *approximate* lack of fit test. Their idea is to use a clustering algorithm to find cases that are *almost* replicates, and use the variation in the responses for the near replicates to compute a lack of fit test. The usefulness of this test will depend on the success of the clustering algorithm. Different clusters will give different tests. At the moment, both experience and supporting theory for this procedure is lacking. An application of this idea to logistic regression is given by Landwehr, Pregibon, and Shoemaker (1984).

## 4.4 General *F* testing

Thus far we have encountered several situations that lead to computation of a statistic that has a nominal *F* distribution when a null hypothesis and normality hold. The theory for the *F*-tests is quite general. In the basic structure, a smaller model (null hypothesis) is compared to a larger model (alternative hypothesis), and the smaller model can be obtained from the larger by setting some parameters in the larger model equal to zero, equal to each other, or equal to some specific value. One example previously encountered is testing to see if the last $q$ predictors in a multiple regression are needed after fitting the first $p' - q$. In matrix notation, partition $\mathbf{X} = (\mathbf{X}_1, \mathbf{X}_2)$, where $\mathbf{X}_1$ is $n \times (p' - q)$, $\mathbf{X}_2$ is $n \times q$, and partition $\boldsymbol{\beta}^T = (\boldsymbol{\beta}_1^T, \boldsymbol{\beta}_2^T)$, where $\boldsymbol{\beta}_1$ is $(p'-q) \times 1$, $\boldsymbol{\beta}_2$ is $q \times 1$, so the two hypotheses, NH and AH, are

$$\text{NH:} \quad \mathbf{Y} = \mathbf{X}_1 \boldsymbol{\beta}_1 + \mathbf{e}$$

$$\text{AH:} \quad \mathbf{Y} = \mathbf{X}_1 \boldsymbol{\beta}_1 + \mathbf{X}_2 \boldsymbol{\beta}_2 + \mathbf{e} \tag{4.20}$$

The smaller model is obtained from the larger by setting $\boldsymbol{\beta}_2 = \mathbf{0}$.

To compute the $F$-test, both of the models must be fit to observed data. Under NH, find the residual sum of squares and its degrees of freedom $RSS_{NH}$ and d.f.$_{NH}$. Similarly, under the alternative model, find $RSS_{AH}$ and d.f.$_{AH}$. Clearly d.f.$_{NH}$ > d.f.$_{AH}$, since the alternative fits more parameters. Also, $RSS_{NH} - RSS_{AH} \geqslant 0$, since the fit of the AH must be at least as good as the fit of the NH. The $F$-test then gives evidence against NH if

$$F = \frac{(RSS_{NH} - RSS_{AH})/(\text{d.f.}_{NH} - \text{d.f.}_{AH})}{RSS_{AH}/\text{d.f.}_{AH}} \tag{4.21}$$

is large when compared to the percentage points of $F(\text{d.f.}_{NH} - \text{d.f.}_{AH}, \text{d.f.}_{AH})$.

**Non-null distributions.**  The numerator and denominator of (4.21) are independently distributed. Under normality and AH, each is distributed as $\sigma^2$ times a (noncentral) chi-squared variable divided by its degrees of freedom. In particular, the expected value of the numerator of (4.21) will be

$$E(\text{numerator of (4.21)}) = \sigma^2 (1 + \text{noncentrality parameter}) \tag{4.22}$$

For the particular hypothesis of (4.20), the noncentrality parameter is given by the expression

$$\frac{\boldsymbol{\beta}_2^T (\mathbf{X}_2^T \mathbf{X}_2 - \mathbf{X}_2^T \mathbf{X}_1 (\mathbf{X}_1^T \mathbf{X}_1)^{-1} \mathbf{X}_1^T \mathbf{X}_2) \boldsymbol{\beta}_2}{\sigma^2} \tag{4.23}$$

To help understand this, consider the special case of $\mathbf{X}_2^T \mathbf{X}_2 = I$ and $\mathbf{X}_1^T \mathbf{X}_2 = 0$ so the variables in $\mathbf{X}_2$ are orthogonal to each other and to the variables in $\mathbf{X}_1$. Then (4.22) becomes

$$E(\text{numerator}) = \sigma^2 + \boldsymbol{\beta}_2^T \boldsymbol{\beta}_2 \tag{4.24}$$

For this special case the expected value of the numerator of (4.21), and the power of the $F$-test, will be large if $\boldsymbol{\beta}_2$ is large. In the general case where $\mathbf{X}_1^T \mathbf{X}_2 \neq 0$, the results are more complicated and the size of the noncentrality parameter, and power of the $F$-test depend not only on $\boldsymbol{\beta}_2$ but also on the *sample* correlations between the variables in $\mathbf{X}_1$ and those in $\mathbf{X}_2$. If these correlations are large then the power of $F$ may be small even if $\boldsymbol{\beta}_2$ is large.

More general results on $F$-tests are presented in advanced linear model texts such as Seber (1977).

**Additional comments.**   The $F$-tests derived in this section have many important properties when the error terms are in fact normally distributed. For example they are likelihood ratio tests and all of the properties of such tests apply to the $F$-test. Since the exact normality of the error terms generally does not hold, discussion of optimality from a practical point of view is unnecessary. Fortunately, these procedures are "robust" to departures from normality of the errors; that is, estimates, tests, and confidence procedures are only modestly affected by departures from normality.

## 4.5   Joint confidence regions

Just as confidence intervals for a single parameter are based on the $t$ distribution, confidence regions for several parameters will require use of an $F$ distribution. These regions will be elliptical.

The $(1 - \alpha) \times 100\%$ confidence region for $\boldsymbol{\beta}$ is the set of vectors $\boldsymbol{\beta}$ such that

$$\frac{(\boldsymbol{\beta} - \hat{\boldsymbol{\beta}})^T(\mathbf{X}^T\mathbf{X})(\boldsymbol{\beta} - \hat{\boldsymbol{\beta}})}{p'\hat{\sigma}^2} \leqslant F(\alpha; p', n - p') \tag{4.25}$$

Often, the confidence region for $\boldsymbol{\beta}^*$, the parameter vector excluding $\beta_0$, is of interest. The $(1 - \alpha) \times 100\%$ region for $\boldsymbol{\beta}^*$ is, using the notation of Chapter 2, the set of points $\boldsymbol{\beta}^*$ such that

$$\frac{(\boldsymbol{\beta}^* - \hat{\boldsymbol{\beta}}^*)^T(\mathscr{X}^T\mathscr{X})(\boldsymbol{\beta}^* - \hat{\boldsymbol{\beta}}^*)}{p\hat{\sigma}^2} \leqslant F(\alpha; p, n - p') \tag{4.26}$$

The region (4.25) is a $p'$-dimensional ellipsoid centered at $\hat{\boldsymbol{\beta}}$, while (4.26) is a $p$-dimensional ellipsoid centered at $\hat{\boldsymbol{\beta}}^*$.

For example, the 95% confidence region for $\beta_1$, $\beta_2$ in the regression of FUEL on $X_1 = $ TAX and $X_2 = $ DLIC (Example 2.1) is given in Figure 4.3. This ellipse is centered at $(-32.075, 12.515)$. The orientation of the ellipse (the directions of the major and minor axes) is determined by $\mathbf{X}^T\mathbf{X}$ or, equivalently, by the sample correlation between $X_1$ and $X_2$. If $X_1$ and $X_2$ are uncorrelated, the axes of the ellipse will be parallel to the $X_1$ and $X_2$ axes.

**Confidence ellipsoid for an arbitrary subset of $\boldsymbol{\beta}$.**   Rather than give the general results, only a special case will be given from which the general rule is to be derived. Suppose the 95% confidence region for $(\beta_1, \beta_2)^T$ is desired from a model with four variables in the fuel consumption data. Let $\mathbf{S}$ be the $2 \times 2$ submatrix of $(\mathbf{X}^T\mathbf{X})^{-1}$ corresponding to $X_1$ and $X_2$ (TAX and DLIC in the example). That is, from Table 2.3,

$$\mathbf{S} = \begin{bmatrix} 0.0382636 & 0.0022158 \\ 0.0022158 & 0.0008411 \end{bmatrix}$$

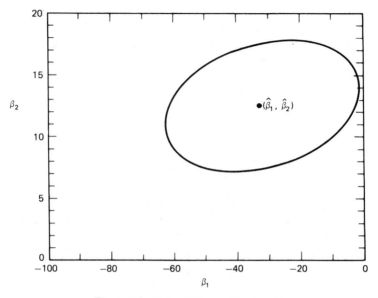

**Figure 4.3**  Joint 95% confidence region.

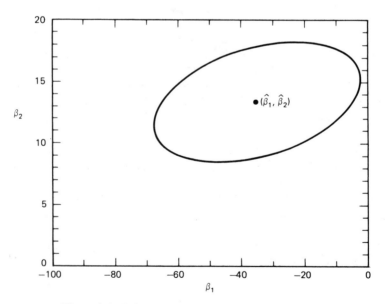

**Figure 4.4**  Joint conditional 95% confidence region.

Then the 95% confidence region is the set of points $\boldsymbol{\beta} = (\beta_1, \beta_2)^T$ such that

$$\frac{\left(\begin{pmatrix} \beta_1 \\ \beta_2 \end{pmatrix} - \begin{pmatrix} \hat{\beta}_1 \\ \hat{\beta}_2 \end{pmatrix}\right)^T \mathbf{S}^{-1} \left(\begin{pmatrix} \beta_1 \\ \beta_2 \end{pmatrix} - \begin{pmatrix} \hat{\beta}_1 \\ \hat{\beta}_2 \end{pmatrix}\right)}{2\hat{\sigma}^2} \leqslant F(\alpha; 2, n - p') \qquad (4.27)$$

where in (4.27), $(\hat{\beta}_1, \hat{\beta}_2)$ is computed from the four-variable model, $(\hat{\beta}_1, \hat{\beta}_2) = (-34.79, 13.36)$. This region is given in Figure 4.4, which for the example is not too different from the region in Figure 4.3. In other problems these regions may differ substantially, just as the parameter estimates may change depending on the model.

## Problems

**4.1  Galton's sweet peas.**    Many of the ideas of regression first appeared in the work of Sir Francis Galton on the inheritance of characteristics from one generation to the next. In a paper on "Typical Laws of Heredity," delivered to the Royal Institution on February 9, 1877, Galton discussed some experiments on sweet peas. By comparing the sweet peas produced by parent plants to those produced by offspring plants, he could observe inheritance from one generation to the next. Galton categorized parent plants according to the typical diameter of the peas they produced. For seven size classes from 0.15 to 0.21 inches he arranged for each of nine of his friends to grow ten plants from seed in each size class; however, two of the crops were total failures. A summary of Galton's data was later published by Karl Pearson (1930) (see Table 4.9). Only average diameter and standard deviation of the offspring peas are given by Pearson; sample sizes are unknown.

Table 4.9    Galton's data

| Diameter of Parent Peas (hundredths of an inch) | Mean Diameter of Offspring Peas (hundredths of an inch) | Standard Deviation |
|---|---|---|
| 21 | 17.26 | 1.988 |
| 20 | 17.07 | 1.938 |
| 19 | 16.37 | 1.896 |
| 18 | 16.40 | 2.037 |
| 17 | 16.13 | 1.654 |
| 16 | 16.17 | 1.594 |
| 15 | 15.98 | 1.763 |

**4.1.1.** Draw the scatter plot of $Y$ =average offspring diameter versus $X$ = parent diameter.

**4.1.2.** Assuming that the standard deviations given are population values, compute the weighted regression of $Y$ on $X$. Draw the fitted line on your scatter plot.

**4.1.3.** Galton wanted to know if characteristics of the parent plant such as size were passed on to the offspring plants. In fitting the regression, a parameter value of $\beta_1 = 1$ would correspond to perfect inheritance, while $\beta_1 < 1$ would suggest that the offspring are "reverting" toward "what may be roughly and perhaps fairly described as the average ancestral type." (The substitution of "regression" for "reversion" was probably due to Galton in 1885.) Test the hypothesis that $\beta_1 = 1$ versus the alternative that $\beta_1 < 1$.

**4.1.4.** Obtain a test for lack of fit of the straight-line model.

**4.1.5.** In his experiments, Galton took the average size of all peas produced by a plant to determine the size class of the parental plant. Yet for seeds to represent that plant and produce offspring, Galton chose seeds that were as close to the overall average size as possible. Thus for a small plant, the exceptional large seed was chosen as a representative, while larger more robust plants were represented by relatively smaller seeds. What effects would you expect these experimental biases to have on (1) estimation of the intercept and slope and (2) estimates of error?

**4.2   Physics data.**   Table 4.10 lists the results of experiments like those described in Example 4.1 with choices of $a$ and $c$ equal to $\pi^+$ and $\pi^-$ from Weisberg et al. (1978). The data with $a = c = \pi^-$ are also given in Table 4.1. In Table 4.10 the values to the right of the $\pm$ sign are the values of $\sigma/\sqrt{w_i}$. For one or more of these data sets, fit model (4.13) and test for lack of fit. If the model does not fit, try (4.17) and again test for lack of fit. Informally compare the fitted models for various combinations of $a$ and $c$.

**4.3   Apple shoots.**   Apply the analysis of Section 4.3 to the data on short shoots in Table 4.6. Informally compare the fitted regression for short and long shoots (be sure to draw a scatter plot of $\bar{y}$ versus DAY).

**4.4   Control charts.**   The data in Table 4.11 give the outside diameter of crankpins produced by an industrial process over several days (Jensen, 1977). All of the crankpins produced should be between 0.7425 and 0.7430 inches. The numbers given in the table are in units of 0.00001 inches deviation from 0.742 inches. For example, the number 93 means $0.742 + 0.00093 = 0.74293$ inches.

When the manufacturing process is "under control," the average size of crankpins produced should (1) fall near the middle of the specified range and (2) should not depend on time. Fit the appropriate model to see if the process is under control, and test for lack of fit of the model.

## Table 4.10  Strong interaction data

| $p_a^{lab}$ (GeV/$c$) | $s^{-1/2}$ GeV/$c^{-1}$ | $y(ap \to \pi^-)$ ($\mu$b) | $y(ap \to \pi^+)$ ($\mu$b) |
|---|---|---|---|
| | | $a = \pi^-$ | |
| 4 | 0.345 | $367 \pm 17$ | $284 \pm 13$ |
| 6 | 0.287 | $311 \pm 9$ | $288 \pm 9$ |
| 8 | 0.251 | $295 \pm 9$ | $304 \pm 9$ |
| 10 | 0.225 | $268 \pm 7$ | $284 \pm 8$ |
| 12 | 0.207 | $253 \pm 7$ | $281 \pm 7$ |
| 15 | 0.186 | $239 \pm 6$ | $276 \pm 7$ |
| 20 | 0.161 | $220 \pm 6$ | $275 \pm 7$ |
| 24 | 0.148 | | $257 \pm 10$ |
| 30 | 0.132 | $213 \pm 6$ | $275 \pm 7$ |
| 75 | 0.084 | $193 \pm 5$ | $277 \pm 7$ |
| 150 | 0.060 | $192 \pm 5$ | $281 \pm 7$ |
| | | $a = \pi^+$ | |
| 4 | 0.345 | $133 \pm 6$ | $499 \pm 18$ |
| 6 | 0.287 | $157 \pm 6$ | $464 \pm 13$ |
| 8 | 0.251 | $157 \pm 5$ | $442 \pm 12$ |
| 10 | 0.225 | $154 \pm 6$ | $389 \pm 11$ |
| 12 | 0.207 | $149 \pm 6$ | $388 \pm 12$ |
| 15 | 0.186 | $156 \pm 10$ | $341 \pm 14$ |
| 20 | 0.161 | $173 \pm 22$ | $375 \pm 22$ |
| 30 | 0.132 | $155 \pm 5$ | $331 \pm 9$ |
| 75 | 0.084 | $166 \pm 5$ | $303 \pm 8$ |
| 150 | 0.060 | $160 \pm 6$ | $311 \pm 9$ |
| 250 | 0.046 | $171 \pm 11$ | $306 \pm 18$ |

## Table 4.11  Crankpin data

| Day | Diameter of Crankpins |
|---|---|
| 1 | 93, 98, 90, 94, 94 |
| 4 | 93, 100, 88, 85, 89 |
| 7 | 89, 90, 92, 95, 100 |
| 10 | 93, 88, 87, 87, 87 |
| 13 | 88, 86, 91, 89, 86 |
| 16 | 82, 72, 80, 72, 89 |
| 19 | 81, 80, 78, 94, 90 |
| 22 | 90, 92, 82, 77, 89 |

**4.5   An F-test.**   In simple regression derive an explicit formula for the F-test of

$$\text{NH:} \quad y_i = x_i + e_i \qquad (\beta_0 = 0, \beta_1 = 1)$$
$$\text{AH:} \quad y_i = \beta_0 + \beta_1 x_i + e_i \qquad i = 1, 2, \ldots, n$$

**4.6   Snow geese.**   Aerial survey methods are regularly used to estimate the number of snow geese in their summer range areas west of Hudson Bay in Canada. To obtain estimates, small aircraft fly over the range and, when a flock of geese is spotted, an experienced person estimates the number of geese in the flock. To investigate the reliability of this method of counting, an experiment was conducted in which an airplane carrying two observers flew over $n = 45$ flocks, and each observer made an independent estimate of the number of birds in each flock. Also, a photograph of the flock was

**Table 4.12   Estimates of flock size**

| Photo | Observer Number 1 | Observer Number 2 | Photo | Observer Number 1 | Observer Number 2 |
|-------|-------------------|-------------------|-------|-------------------|-------------------|
| 56 | 50 | 40 | 119 | 75 | 200 |
| 38 | 25 | 30 | 165 | 100 | 200 |
| 25 | 30 | 40 | 152 | 150 | 150 |
| 48 | 35 | 45 | 205 | 120 | 200 |
| 38 | 25 | 30 | 409 | 250 | 300 |
| 22 | 20 | 20 | 342 | 500 | 500 |
| 22 | 12 | 20 | 200 | 200 | 300 |
| 42 | 34 | 35 | 73 | 50 | 40 |
| 34 | 20 | 30 | 123 | 75 | 80 |
| 14 | 10 | 12 | 150 | 150 | 120 |
| 30 | 25 | 30 | 70 | 50 | 60 |
| 9 | 10 | 10 | 90 | 60 | 100 |
| 18 | 15 | 18 | 110 | 75 | 120 |
| 25 | 20 | 30 | 95 | 150 | 150 |
| 62 | 40 | 50 | 57 | 40 | 40 |
| 26 | 30 | 20 | 43 | 25 | 35 |
| 88 | 75 | 120 | 55 | 100 | 110 |
| 56 | 35 | 60 | 325 | 200 | 400 |
| 11 | 9 | 10 | 114 | 60 | 120 |
| 66 | 55 | 80 | 83 | 40 | 40 |
| 42 | 30 | 35 | 91 | 35 | 60 |
| 30 | 25 | 30 | 56 | 20 | 40 |
| 90 | 40 | 120 | | | |

taken so that an exact count of the number of birds in the flock could be made. The resulting data are given in Table 4.12 (Cook and Jacobsen, 1978).

**4.6.1.** Draw scatter plots of $Y =$ photo count versus $X_1 =$ count by observer 1 and versus $X_2 =$ count by observer 2. Do these graphs suggest that a simple regression model might be appropriate? Why or why not? For the simple regression model of $Y$ on $X_1$ or on $X_2$, what do the error terms measure? Why is it appropriate to fit the regression of $Y$ on $X_1$ or $X_2$ rather than the regression of $X_1$ or $X_2$ on $Y$?

**4.6.2.** Compute the regression of $Y$ on $X_1$ and $Y$ on $X_2$ via ordinary least squares, and test the hypothesis of Problem 4.5 for each observer. State in words the meaning of this hypothesis, and the result of the test. Are either of the observers reliable (you must define reliable)? Summarize your results.

**4.6.3.** Repeat 4.6.2, except fit the regression of $Y^{1/2}$ on $X_1^{1/2}$ and $Y^{1/2}$ on $X_2^{1/2}$. The square-root scale is used to stabilize the error variance.

**4.6.4.** Repeat 4.6.2, except assume that $\text{var}(e_i) = x_i \sigma^2$.

As a result of this experiment, the practice of using visual counts of flock size to determine population estimates was discontinued in favor of using photographs.

**4.7   Jevons' gold coins.**   The data in this example are deduced from a diagram in a paper written by W. Stanley Jevons (1868), and provided by Stephen M. Stigler. In a study of coinage, Jevons weighed 274 gold sovereigns that he had collected from circulation in Manchester, England. For each coin, he recorded the weight after cleaning to the nearest .001 gram, and the date of issue. Table 4.13 lists the average, minimum and maximum weight for each age class. The age classes are coded 1 to 5, roughly corresponding to the age of the coin in decades. The standard weight of a gold sovereign was supposed to be 7.9876 grams; the minimum legal weight was 7.9379 grams.

**Table 4.13   Gold coinage data**

| Age, $x$, Decades | Sample Size $= n$ | Average Weight $= \bar{y}$ | SD | Minimum Weight | Maximum Weight |
|---|---|---|---|---|---|
| 1 | 123 | 7.9725 | 0.01409 | 7.900 | 7.999 |
| 2 | 78 | 7.9503 | 0.02272 | 7.892 | 7.993 |
| 3 | 32 | 7.9276 | 0.03426 | 7.848 | 7.984 |
| 4 | 17 | 7.8962 | 0.04057 | 7.827 | 7.965 |
| 5 | 24 | 7.8730 | 0.05353 | 7.757 | 7.961 |

*Source:* Stephen M. Stigler.

**4.7.1.** Let $x =$ coded age and $\bar{y} =$ average weight. Draw a scatter plot of $\bar{y}$ versus $x$, and comment on the applicability of the usual assumptions of the linear regression model. Also draw a scatter plot of the SD's versus $x$, and summarize the information in this plot.

**4.7.2.** Since the numbers of coins $n$ in each age class are all fairly large, it is reasonable to pretend that the variance of coin weight at $x$ is well approximated by $SD^2$, and hence var($\bar{y}$) is given by $SD^2/n$. Use the values given to find weights, and compute the weighted regression of $\bar{y}$ on $x$.

**4.7.3.** Compute a lack of fit test for the linear regression model, and summarize results.

**4.7.4.** Is the fitted regression consistent with the known standard weight for a new coin?

**4.7.5.** For a previously unsampled coin of age $x = 1, 2, 3, 4, 5$, estimate the probability that the weight of the coin is less than the legal minimum. (Hint: For these calculations, make use of the assumption that the residual variance for a coin of age $x$ decades is the known value $SD^2$. Hence, predicted values will have normal, not $t$, distributions.) Estimate the proportion of all $x = 4$ decade coins below the legal minimum. Also of interest would be determination of the age $x$ at which the predicted weight of coins is equal to the legal minimum. A point estimate for this value of $x$ can be obtained by setting $y = 7.9379$, and solving the fitted regression equation for $x$. This problem is called *inverse regression*, and is discussed by Williams (1959, Chapter 6).

**4.8  Computer program.** To adapt a Sweep algorithm program for weighted least squares, it is necessary only to modify the algorithm for the computation of the matrix of corrected sums of squares and cross products. Appropriate modifications to the method of Appendix 1A.5 are given by West (1979). Let $\bar{x}_m$, $\bar{y}_m$ be the averages of two variables $x$ and $y$ after $m$ cases are read, $\sum w_i$ is the sum of the first $m$ case weights, and $SSX_m = \sum w_i(x_i - \bar{x}_m)^2$, $SSY_m = \sum w_i(y_i - \bar{y}_m)^2$, $SXY_m = \sum w_i(x_i - \bar{x}_m)(y_i - \bar{y}_m)$, where all sums are over the first $m$ cases. Then, setting $\bar{x}_0 = \bar{y}_0 = SXX_0 = SYY_0 = SXY_0 = 0$, we can update estimates for the $(m + 1)$st case by

$$SXX_{m+1} = SXX_m + \frac{w_{m+1}}{\sum w_i + w_{m+1}} \left( \sum w_i \right)(x_{m+1} - \bar{x}_m)^2$$

$$SYY_{m+1} = SYY_m + \frac{w_{m+1}}{\sum w_i + w_{m+1}} \left( \sum w_i \right)(y_{m+1} - \bar{y}_m)^2$$

$$SXY_{m+1} = SXY_m + \frac{w_{m+1}}{\sum w_i + w_{m+1}} \left( \sum w_i \right)(x_{m+1} - \bar{x}_m)(y_{m+1} - \bar{y}_m)$$

$$\bar{x}_{m+1} = \bar{x}_m + \frac{w_{m+1}}{\sum w_i + w_{m+1}} (x_{m+1} - \bar{x}_m)$$

$$\bar{y}_{m+1} = \bar{y}_m + \frac{w_{m+1}}{\sum w_i + w_{m+1}} (y_{m+1} - \bar{y}_m)$$

These formulas give the updating for any pair of variables for sums of squares and cross products and averages. Modify the Sweep program written for problem 2.6 to do weighted least squares.

# 5

# DIAGNOSTICS I:
# RESIDUALS AND INFLUENCE

The methods for obtaining estimates, tests and other summaries developed so far tell only half the story of regression analysis. All of these methods are computed as if the model and the assumptions are correct, but in any practical problem assumptions are in doubt. A second phase of analysis designed to check assumptions and build a model is usually required. Whereas the initial phase of modeling produced combinations of the data as summary statistics, this latter phase requires examination of statistics that generally have values for each case. As a class, we call these *diagnostic statistics*, since they are designed to find problems with assumptions in an analysis.

The primary concerns of diagnostics are two interrelated questions. First, we ask how well the model used resembles the data actually observed. The basic statistic here is a useful transformation of the residuals. If the fitted model does not give a set of residuals that appear to be reasonable, then some aspect of the model may be called into doubt. The second question of interest is the effect of each case on estimation and other aspects of aggregate analysis. In some data sets, the observed aggregate statistics may change in important ways if one case is deleted from the data. Such a case is called *influential*, and we shall learn to detect such cases. We will be led to study and use two relatively unfamiliar diagnostic statistics, called distance measures and potential or leverage values.

### Example 5.1  The usefulness of plots (Anscombe, 1973)

The need for case analysis is well illustrated by the four artificial data sets given in Table 5.1. Each set consists of 11 pairs of points $(x_i, y_i)$,

**Table 5.1   Four hypothetical data sets**

| Case Number | 1–3 X | 1 Y | 2 Y | 3 Y | 4 X | 4 Y |
|---|---|---|---|---|---|---|
| 1 | 10.0 | 8.04 | 9.14 | 7.46 | 8.0 | 6.58 |
| 2 | 8.0 | 6.95 | 8.14 | 6.77 | 8.0 | 5.76 |
| 3 | 13.0 | 7.58 | 8.74 | 12.74 | 8.0 | 7.71 |
| 4 | 9.0 | 8.81 | 8.77 | 7.11 | 8.0 | 8.84 |
| 5 | 11.0 | 8.33 | 9.26 | 7.81 | 8.0 | 8.47 |
| 6 | 14.0 | 9.96 | 8.10 | 8.84 | 8.0 | 7.04 |
| 7 | 6.0 | 7.24 | 6.13 | 6.08 | 8.0 | 5.25 |
| 8 | 4.0 | 4.26 | 3.10 | 5.39 | 19.0 | 12.50 |
| 9 | 12.0 | 10.84 | 9.13 | 8.15 | 8.0 | 5.56 |
| 10 | 7.0 | 4.82 | 7.26 | 6.42 | 8.0 | 7.91 |
| 11 | 5.0 | 5.68 | 4.74 | 5.73 | 8.0 | 6.89 |

(Data Set Number; Variable)

to which the simple linear regression model $y_i = \beta_0 + \beta_1 x_i + e_i$ is fit. Each data set leads to an identical aggregate analysis, namely,

$$\hat{\beta}_0 = 3.0$$
$$\hat{\beta}_1 = 0.5$$
$$\hat{\sigma}^2 = 13.75$$
$$R^2 = 0.667$$

Since the aggregate statistics are the same for each data set, one might conclude that the linear regression model is equally appropriate for each of them. However, the simplest case analysis suggests that this is not true, as seen by examination of the scatter plots in Figures 5.1a through 5.1d.

The first data set, graphed in Figure 5.1a, is as one might expect to observe if the simple linear regression model were appropriate. The graph of the second data set given in Figure 5.1b suggests a different conclusion, that the analysis based on simple linear regression is incorrect, and that a smooth curve, perhaps a quadratic polynomial, could be fit to the data with little remaining variability.

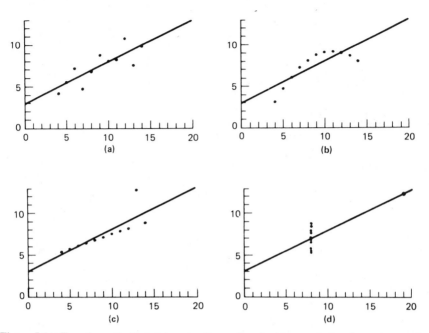

**Figure 5.1**    Four hypothetical data sets. Reproduced with permission from Anscombe (1973).

Figure 5.1$c$ suggests that the prescription of simple regression may be correct for most of the data, but one of the cases is too far away from the fitted regression line. This is called the outlier problem. Possibly the case that does not match the others should be deleted from the data set, and the regression should be refit from the remaining 10 cases. If this is done, the fitted equation is given by $\hat{y} = 4. + 0.346x$, which is quite different from that obtained from all 11 cases. Without a context for the data, we cannot judge one line "correct" and the other "incorrect." The differences between these two lines need to be understood and reported.

The final set, graphed in Figure 5.1$d$, is different from the other three in that there is really not enough information to make a judgment concerning the fitted model. The slope parameter estimate $\hat{\beta}_1$ is largely determined by the value of $y_8$. If the eighth case were deleted, we could not even estimate $\beta_1$. We must distrust an aggregate analysis that is so heavily dependent upon a single case.

For case statistics to be useful, we must understand their behavior when the model is correct, and, if possible, when the model is incorrect. Comparison of the case statistics to their expected behavior can diagnose the success or failure of the model with respect to the observed data. Definitions and properties of the important case statistics are the main topics of this chapter. Methods using the statistics developed here are presented in Chapter 6. A more complete treatment of the material in this chapter, at a higher mathematical level, is given by Cook and Weisberg (1982a, Chapters 2 and 3).

## 5.1   The residuals

To study the residuals, we use matrix notation as outlined in Chapter 2. The basic model is

$$\mathbf{Y} = \mathbf{X}\boldsymbol{\beta} + \mathbf{e} \qquad \text{var}(\mathbf{e}) = \sigma^2 \mathbf{I} \tag{5.1}$$

where $\mathbf{X}$ is a known full rank matrix with $n$ rows and $p'$ columns, with $p' =$ number of predictors $+ 1 = p + 1$ if a column of ones is included in the model, and $p' = p$ if the column of ones is not in the model. Similarly, $\boldsymbol{\beta}$ is a $p' \times 1$ unknown vector. The vector $\mathbf{e}$ consists of unknown errors that we assume throughout this chapter are equally variable and uncorrelated. Development for weighted least squares is similar, with details given by Cook and Weisberg (1982a, Appendix A.1).

In fitting model (5.1), we estimate $\boldsymbol{\beta}$ by $\hat{\boldsymbol{\beta}} = (\mathbf{X}^T\mathbf{X})^{-1}\mathbf{X}^T\mathbf{Y}$, and the <u>fitted values</u> $\hat{\mathbf{Y}}$ corresponding to the observed values $\mathbf{Y}$ are then given by

$$\begin{aligned}
\hat{\mathbf{Y}} &= \mathbf{X}\hat{\boldsymbol{\beta}} \\
&= \mathbf{X}[(\mathbf{X}^T\mathbf{X})^{-1}\mathbf{X}^T\mathbf{Y}] \\
&= \mathbf{X}(\mathbf{X}^T\mathbf{X})^{-1}\mathbf{X}^T\mathbf{Y} \\
&= \mathbf{H}\mathbf{Y}
\end{aligned} \tag{5.2}$$

where $\mathbf{H}$ is the $n \times n$ matrix defined by‡

$$\mathbf{H} = \mathbf{X}(\mathbf{X}^T\mathbf{X})^{-1}\mathbf{X}^T \tag{5.3}$$

$\mathbf{H}$ is called the <u>*hat matrix*</u> because it transforms the vector of observed responses $\mathbf{Y}$ into the vector of fitted responses $\hat{\mathbf{Y}}$, usually read as <u>y-hat</u>. The

---

‡In the first edition of this book, $\mathbf{V}$ was used to denote the hat matrix, and its elements were denoted by $v_{ij}$.

vector of residuals $\hat{\mathbf{e}}$ is defined by

$$\hat{\mathbf{e}} = \mathbf{Y} - \hat{\mathbf{Y}}$$
$$= \mathbf{Y} - \mathbf{X}(\mathbf{X}^T\mathbf{X})^{-1}\mathbf{X}^T\mathbf{Y}$$
$$= [\mathbf{I} - \mathbf{X}(\mathbf{X}^T\mathbf{X})^{-1}\mathbf{X}^T]\mathbf{Y}$$
$$= [\mathbf{I} - \mathbf{H}]\mathbf{Y} \qquad (5.4)$$

**Difference between e and ê.** The errors $\mathbf{e}$ are unobservable random variables, assumed to have zero mean and uncorrelated elements, each with common variance $\sigma^2$. The residuals $\hat{\mathbf{e}}$ are computed quantities that can be graphed or otherwise studied. Their mean and variance, using (5.4) and Appendix 2A.2, are

$$E(\hat{\mathbf{e}}) = \mathbf{0}$$
$$\text{var}(\hat{\mathbf{e}}) = \sigma^2(\mathbf{I} - \mathbf{H}) \qquad (5.5)$$

Like the errors, each of the residuals has zero mean, but each residual may have a different variance, and they are correlated. If the errors are normally distributed, then so are the residuals, since by (5.4) they are a linear combination of the errors. Also, if the intercept is included in the model, then the sum of the residuals is zero, $\hat{\mathbf{e}}^T\mathbf{1} = 0$. In scalar form, the variance of the $i$th residual is

$$\text{var}(\hat{e}_i) = \sigma^2(1 - h_{ii}) \qquad (5.6)$$

where $h_{ii}$ is the $i$th diagonal element of $\mathbf{H}$. Diagnostic procedures are based on the computed residuals, which we would like to assume behave as would the unobservable errors. The usefulness of this assumption depends on the hat matrix, since it is $\mathbf{H}$ that relates $\hat{\mathbf{e}}$ to $\mathbf{e}$ and also gives the variances and covariances of $\hat{\mathbf{e}}$.

**The hat matrix.** $\mathbf{H}$ is $n \times n$ and symmetric with many special properties that are easy to verify directly from definition (5.3). For example, multiplying $\mathbf{X}$ on the left by $\mathbf{H}$ gives $\mathbf{X}$ back, $\mathbf{HX} = \mathbf{X}$. Similarly, $(\mathbf{I} - \mathbf{H})\mathbf{X} = \mathbf{0}$. The property $\mathbf{HH} = \mathbf{H}^2 = \mathbf{H}$ also shows that $\mathbf{H}(\mathbf{I} - \mathbf{H}) = \mathbf{0}$, so the covariance between the fitted values $\mathbf{HY}$ and residuals $(\mathbf{I} - \mathbf{H})\mathbf{Y}$ is $\sigma^2\mathbf{H}(\mathbf{I} - \mathbf{H}) = \mathbf{0}$. $\mathbf{H}$ is called the *orthogonal projection operator on the column space of* $\mathbf{X}$. It is generally not invertible and has the same rank as $\mathbf{X}$, usually $p'$. The $(i, j)$th element of $\mathbf{H}$, denoted by $h_{ij}$, is given by the formula

$$h_{ij} = \mathbf{x}_i^T(\mathbf{X}^T\mathbf{X})^{-1}\mathbf{x}_j = \mathbf{x}_j^T(\mathbf{X}^T\mathbf{X})^{-1}\mathbf{x}_i = h_{ji} \qquad (5.7)$$

The diagonal elements, the $h_{ii}$, are given by

$$h_{ii} = \mathbf{x}_i^T(\mathbf{X}^T\mathbf{X})^{-1}\mathbf{x}_i \qquad (5.8)$$

Many relationships can be found between the $h_{ij}$. For example,

$$\sum_{i=1}^{n} h_{ii} = \text{rank}(\mathbf{X}) = p' \qquad (5.9)$$

In addition, each $h_{ii}$ is bounded below by $1/n$, for models with an intercept, and above by $1/r$, if $r$ is the number of rows of $\mathbf{X}$ that are identical to $\mathbf{x}_i$. Also, for models with an intercept, $\mathbf{H1} = \mathbf{1}$, or, in scalar form,

$$\sum_{i=1}^{n} h_{ij} = \sum_{j=1}^{n} h_{ij} = 1 \qquad (5.10)$$

As can be seen from (5.6), cases with large values of $h_{ii}$ will have small values for $\text{var}(\hat{e}_i)$; as $h_{ii}$ gets closer to one, this variance will approach zero. For such a case, no matter what value of $y_i$ is observed for the $i$th case, we are nearly certain to get a residual near zero. Hoaglin and Welsch (1978) pointed this out using a scalar version of (5.2),

$$\hat{y}_i = \sum_{j=1}^{n} h_{ij} y_j = h_{ii} y_i + \sum_{j \neq i} h_{ij} y_j \qquad (5.11)$$

In combination with (5.10), this shows that as $h_{ii}$ approaches 1, $\hat{y}_i$ approaches $y_i$. For this reason, they call $h_{ii}$ the *leverage* of the $i$th case. However, this particular definition ignores the role played by the random $y_j$ in determining $\hat{y}_i$, so a less picturesque name for the $h_{ii}$ might be in order. Cook and Weisberg (1982a) use the name *potential* for $h_{ii}$ to remind us that the effect of the $i$th case on the regression is more likely to be large if $h_{ii}$ is large, but its importance is uncertain depending on the $y_j$. Measures that combine the roles of the $h_{ii}$ and the $y_j$'s are presented in Section 5.3.

Characteristics of a case that make $h_{ii}$ large or small can be described geometrically. Assuming that the intercept is in the model, consider again the deviations from averages form of the cross products matrix discussed in Chapter 2. Again let $\mathscr{X}^T \mathscr{X}$ be the corrected cross products matrix (2.16), $\bar{\mathbf{x}}$ is the $p \times 1$ vector of sample averages of the $p$ predictors, and redefine $\mathbf{x}_i^T$ to be the $i$th row of $\mathbf{X}$ *without* the one for the intercept. Then, one can show that $h_{ii}$ can be written as

$$h_{ii} = \frac{1}{n} + (\mathbf{x}_i - \bar{\mathbf{x}})^T (\mathscr{X}^T \mathscr{X})^{-1} (\mathbf{x}_i - \bar{\mathbf{x}}) \qquad (5.12)$$

Geometrically, the second term on the right-hand side of (5.12) gives the equation of an ellipsoid centered at $\bar{\mathbf{x}}$.

For example, consider again the data on fuel consumption first discussed in Chapter 2. We will only use the model with $X_1 = \text{TAX}$ and $X_2 = \text{DLIC}$ as predictors so that a two-dimensional picture is possible. The data for $(X_1, X_2)$ are given in the scatter plot in Figure 5.2. The ellipses drawn on the

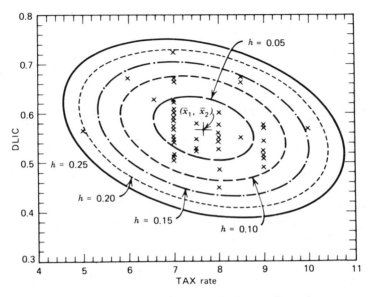

**Figure 5.2**    Contours of constant $h_{ii}$ in two dimensions.

graph correspond to elliptical contours of constant $h_{ii}$ for $h_{ii}=0.25$, 0.20, 0.15, 0.10, and 0.05. Thus, for example, any point that falls exactly on the outer contour would have $h_{ii}=0.25$, while points on the innermost contour have $h_{ii}=0.05$. This definition of distance depends on the data. Points near the long or major axis of the ellipsoid need to be much farther away from $\bar{x}$, in the usual Euclidean distance sense, than do points closer to the minor axis, to have the same values for $h_{ii}$.

In the example, the state with the lowest taxes—Texas, with 5¢ tax—has the largest $h_{ii}$, about 0.2. None of the $h_{ii}$ is very large, although in other data sets finding cases with $h_{ii}$ between 0.5 and 1.0 is not unusual. Cases with large $h_{ii}$ are potentially the most influential in fitting a model. Potential influence does not depend on $Y$, but only on the $X$'s. Comprehensive measures of influence must consider both $\mathbf{Y}$ and $\mathbf{X}$.

**The Mahalanobis distance.**    If the $1/n$ term in (5.12) is dropped from the right-hand side of the equation, and the remaining term is multiplied by $(n-1)$, the resulting quantity is called the Mahalanobis distance from $\mathbf{x}_i$ to the center of the data $\bar{\mathbf{x}}$. The Mahalanobis distance is widely used in multivariate analysis, most notably in discriminant analysis, as a measure for assigning units to different populations based on their relative location. The use of the Mahalanobis distance in discriminant analysis is founded on multivariate normality of the rows of $\mathbf{X}$, an assumption we neither need nor use.

*Example 5.2   Simple regression*

Consider now the simple regression model of Chapter 1, in which $\hat{y}_i = \hat{\beta}_0 + \hat{\beta}_1 x_i$. The matrix $\mathbf{H}$ is $n \times n$, and even for simple regression the number of entries in $\mathbf{H}$, namely, $n^2$, may be quite large, and it is rarely computed in full. However, using (2.25) to get $(\mathbf{X}^T\mathbf{X})^{-1}$, a formula may be obtained for an individual $h_{ij}$. We find

$$h_{ij} = \mathbf{x}_i^T(\mathbf{X}^T\mathbf{X})^{-1}\mathbf{x}_j$$

$$= [1 \ \ x_i] \begin{bmatrix} \dfrac{\sum x_i^2}{nSXX} & \dfrac{-\bar{x}}{SXX} \\[2ex] \dfrac{-\bar{x}}{SXX} & \dfrac{1}{SXX} \end{bmatrix} \begin{bmatrix} 1 \\[1ex] x_j \end{bmatrix}$$

$$= \frac{1}{n} + \frac{(x_i - \bar{x})(x_j - \bar{x})}{SXX} \tag{5.13}$$

By setting $j$ equal to $i$ in (5.13), we find the diagonal elements $h_{ii}$ to be

$$h_{ii} = \frac{1}{n} + \frac{(x_i - \bar{x})^2}{SXX} \tag{5.14}$$

Direct summation in (5.14) will verify the result of (5.9) that $\sum h_{ii} = p' = 2$. $h_{ii}$ will achieve its minimum value of $1/n$ when $x_i = \bar{x}$, and will increase as $x_i$ deviates from $\bar{x}$. The maximum value of 1 can be achieved only when deletion of the $i$th case from the data makes one of the parameters in the model not estimable. In data set 4 of Example 5.1, $h_{88} = 1$, since deletion of this case makes slope estimation impossible.

**Studentized residuals.**   As already pointed out, var($\hat{e}_i$) will be small whenever $h_{ii}$ is large, so cases with $\mathbf{x}_i$ near $\bar{\mathbf{x}}$ will have larger residuals, on the average, than cases far from $\bar{\mathbf{x}}$. This is particularly undesirable because models are most likely to break down far from $\bar{\mathbf{x}}$ where residuals tend to be smaller. This suggests that improved diagnostics may be obtained by scaling the $\hat{e}_i$ by dividing by an estimate of standard error. If the model is correct, these scaled or *Studentized* residuals will all have common variance. We consider two very closely related Studentizations that differ only by the choice of estimator for $\sigma^2$. The first uses $\hat{\sigma}^2$ to estimate $\sigma^2$, giving the formula

$$r_i = \frac{\hat{e}_i}{\hat{\sigma}\sqrt{1 - h_{ii}}} \tag{5.15}$$

The $r_i$ are called *internally Studentized residuals* because the estimate of $\sigma^2$

SAS calls this "Student"

uses all of the data including the $i$th case. The second scaling uses an estimate of $\sigma^2$ obtained when the $i$th case is excluded from the regression, and is discussed in Section 5.2.

Unlike the $\hat{e}_i$'s, the sum of the $r_i$'s is not zero, although $E(r_i)=0$, $i=1$, $2, \ldots, n$. The $r_i$'s are slightly correlated with the $\hat{y}_i$'s, but in practice this correlation is negligible. When the model is correct, the Studentized residuals all have constant variance equal to 1, independent of $\sigma^2$ and the $h_{ii}$. The covariance of $r_i$ and $r_j$ is equal to the correlation between $\hat{e}_i$ and $\hat{e}_j$,

$$\text{cov}(r_i, r_j) = \frac{-h_{ij}}{(1-h_{ii})^{1/2}(1-h_{jj})^{1/2}}$$

Under the assumptions of the model and normality, $r_i^2/(n-p')$ is distributed as a beta random variable with parameters $\frac{1}{2}$ and $(n-p'-1)/2$. Since beta variables are bounded between 0 and 1, $r_i$ is bounded between $-(n-p')^{1/2}$ and $+(n-p')^{1/2}$.

## 5.2   Outliers

One important assumption made in regression analysis is that the model used is appropriate for all of the data. In applications it is not unusual for one or more cases to have an observed response that does not seem to correspond to the model fitted to the bulk of the data. In a simple regression problem such as Figure 5.1c, this may be obvious from a plot of $y$ versus $x$ where most of the cases lie near a fitted line but a few do not. Cases that do not follow the same model as the rest of the data are called *outliers*, and one important function of case analysis is to identify such cases.

It is helpful to define carefully just what we will mean by the term *outlier*. For this we use an explicit formulation called a *mean shift outlier model*. Suppose that the $i$th case is a candidate for an outlier. We assume that the model for all other cases is

$$y_j = \mathbf{x}_j^T \boldsymbol{\beta} + e_j \qquad j \neq i$$

but for case $i$, the model is

$$y_i = \mathbf{x}_i^T \boldsymbol{\beta} + \delta + e_i$$

The $i$th response $y_i$ has expected value different from $\mathbf{x}_i^T\boldsymbol{\beta}$ by an amount $\delta$. Therefore, we can test the $i$th case to be an outlier if we have a test of $\delta=0$.

Before deriving the test explicitly, some comments on the nature of outliers may be helpful. As we shall see, candidates for outliers will be cases with large $|\hat{e}_i|$. Not all large residual cases are outliers, since according to the model large errors $e_i$ will occur with the frequency prescribed by the generating probability distribution. Whatever testing procedure we develop must offer protection against declaring too many cases to be outliers. This leads to the use of simultaneous testing procedures. Also, not all outliers are bad. We can imagine, for example, a geological model in which an outlier corresponds to cases with likely oil deposits or some other valuable characteristic that is absent from the majority of cases. Finding outliers might then be the goal of the analysis. Outlier identification is done relative to a specified model. If the form of the model is modified, the status of individual cases as outliers may change. Finally, some outliers will have greater effect on the regression estimates than will others, a point that is pursued in the next section.

**An outlier test.**   Suppose that the $i$th case is suspected to be an outlier. First, define a new predictor variable, say $U$, with the $j$th element of $U$, $u_j = 0$ for $j \neq i$, and the $i$th element $u_i = 1$. Then, simply compute the regression of $Y$ on $X$ *and* $U$. The estimated coefficient for $U$ is the estimate of the mean shift $\delta$. The $t$ statistic for testing $\delta = 0$ against a two-sided alternative is the appropriate test statistic. If errors are normally distributed, then this test will be nominally distributed as Student's $t$ with $n - p' - 1$ degrees of freedom.

We will now consider an alternative approach that will lead to the same test, but from a different point of view. The equivalence of the two approaches is left as an exercise.

Again suppose that the $i$th case is suspected to be an outlier. We can proceed as follows.

1. Delete the $i$th case from the data. The remaining $n - 1$ cases will be used to fit the linear model.

2. Using the reduced data set, estimate $\beta$ and $\sigma^2$. Call these estimates $\hat{\beta}_{(i)}$ and $\hat{\sigma}_{(i)}^2$ to remind us that case $i$ was not used in estimation. $\hat{\sigma}_{(i)}^2$ has $n - p' - 1$ degrees of freedom.

3. For the deleted case, compute the fitted value $\tilde{y}_i = x_i^T \hat{\beta}_{(i)}$. Since the $i$th case was not used in estimation, $y_i$ and $\tilde{y}_i$ are independent. The variance of $y_i - \tilde{y}_i$ is given by

$$\operatorname{var}(y_i - \tilde{y}_i) = \sigma^2 + \sigma^2 x_i^T (X_{(i)}^T X_{(i)})^{-1} x_i \tag{5.16}$$

where $X_{(i)}$ is the matrix $X$ with the $i$th row deleted. This variance is estimated by replacing $\sigma^2$ with $\hat{\sigma}_{(i)}^2$ in (5.16).

4. Now, if $y_i$ is not an outlier, $E(y_i - \tilde{y}_i) = 0$. Assuming normal errors, a Student's $t$-test of the hypothesis $E(y_i - \tilde{y}_i) = 0$ is given by

$$t_i = \frac{y_i - \tilde{y}_i}{\hat{\sigma}_{(i)}[1 + x_i^T (X_{(i)}^T X_{(i)})^{-1} x_i]^{1/2}} \tag{5.17}$$

This test has $n - p' - 1$ degrees of freedom.

It is a remarkable fact that, not only do the two preceding approaches lead to the same test, but the test is closely related to the internally Studentized residuals discussed in the last section. With the aid of Appendix 5A.1, one can show that $t_i$ can be computed as

SAS calls this

RSTUDENT

$$t_i = r_i \left( \frac{n - p' - 1}{n - p' - r_i^2} \right)^{1/2} = \frac{\hat{e}_i}{\hat{\sigma}_{(i)}\sqrt{1 - h_{ii}}} \tag{5.18}$$

$t_i$ is called an *externally Studentized residual* since case $i$ is not used in computing the estimate of $\sigma^2$. We see that $r_i$ and $t_i$ are monotonically related.

Either the $r_i$ or the $t_i$ are available in many computer packages. Unfortunately, the names given to these two statistics are not standard, and the only way to be certain of the one actually available is to see the formula used.

**Significance levels for the outlier test.**    If the investigator suspects in advance that the $i$th case is an outlier, then $t_i$ should be compared to the central $t$ distribution with the appropriate number of degrees of freedom. Usually, the experimenter has no a priori choice for the outlier. If we test the case with the largest value of $t_i$ to be an outlier, we are in reality performing $n$ significance tests, one for each of $n$ cases. Suppose, for example, that there were no outlier, and that $n = 65$, $p' = 4$. The probability that a $t$ statistic with 60 degrees of freedom exceeds 2.000 in absolute value is .05; however, the probability that the largest of 65 independent $t$-tests exceeds 2.000 is .964, suggesting quite clearly the need for a different critical value. (Of course, the tests in our problem are correlated, so this computation is only a guide.‡) The technique we use to find critical values is based on the *Bonferroni inequality*, which states that for $n$ tests each of size $\alpha$, the probability of falsely labeling at least one point an outlier is no greater than $n\alpha$. This procedure is conservative, since the Bonferroni inequality would specify only that the probability of the maximum of 65 tests exceeding 2.00 is no greater than 65(0.05), which is larger than 1. However, choosing the critical value to be the $(\alpha/n) \times 100\%$ point of $t$ will give a significance level of no more than $n(\alpha/n) = \alpha$. We would choose a level of $.05/65 = .00077$ for each test to give an overall level of no more than 65(.00077) = .05.

To facilitate these computations of critical values, either special tables or charts of the extreme tails of the $t$ distribution or special tables of critical values, like Table E at the end of the book, are needed. To use Table D, select an $\alpha$ level (both .01 or .05 are available) and enter the table in the row corresponding to the sample size $n$ and the column corresponding to the number of parameters $p'$. The corresponding entry in the table is $\alpha/n \times$

‡Excellent discussions of this and other multiple-test problems are presented by Miller (1981).

100% of $t$ with $n - p' - 1$ degrees of freedom. For example, if $n = 39$, $p' = 5$, we would compare $|t_i|$ to 3.52 for a test at level $\alpha = 0.05$.

In Forbes' data, Example 1.1, case 12 was suspected to be an outlier because of its large residual. To perform the outlier test, we first need the Studentized residual, which is computed using (5.15) from $\hat{e}_i = 1.36$ (Table 1.5), $\hat{\sigma} = 0.379$ (Table 1.6), and $h_{12,12}$, which can be computed from (5.14) to be 0.0639. Thus

$$r_{12} = \frac{1.3592}{0.379\sqrt{1 - 0.0639}} = 3.7078$$

and the outlier test is

$$t_{12} = 3.7078 \left( \frac{17 - 2 - 1}{17 - 2 - 3.7078^2} \right)^{1/2} = 12.40$$

This statistic has 14 degrees of freedom. Entering Table D with $p' = 2$ and $n = 17$, we find the critical value for a 0.01 test to be 4.41. Since $t_i$ clearly exceeds this value, evidence is provided that this case is an outlier.

**Additional comments.** There is a vast literature on methods for handling outliers including two books, the first by Barnett and Lewis (1978) and the second by Hawkins (1981) and a recent review article by Beckman and Cook (1983). Outliers can be modeled in ways other than by a shift in mean of the response, for example, via a shift in variance. In addition, the method described here is for single outliers. If a set of data has more than one outlier, the cases may mask each other, making finding outliers difficult. Cook and Weisberg (1982a, p. 28) provide the generalization of the mean shift model given here to multiple cases. Hawkins, Bradu, and Kass (1984) provide a promising method for searching all subsets of cases for outlying subsets. Bonferroni bounds for outlier tests are discussed by Cook and Prescott (1981). They find that for one-case-at-a-time methods the bound is very accurate, but it is much less accurate for multiple case methods. Butler (1984) presents a method of finding more precise estimates of $p$-values using second-order bounds.

The approach to outliers taken here is that of *identification*: the goal is to find outliers to make them available for further study. Alternatively, we can think of using statistical methods that can tolerate or *accommodate* some proportion of bad or outlying data. This is the justification for the development of "robust" statistical methods, briefly discussed in Chapter 11.

Little has been said about what should be done once outliers are identified, as their disposition is context dependent. In some problems, just finding the outliers is the goal, and these cases may be the only ones of further interest. In other problems, outlying cases can be discarded as not representative of the process under study. In still other problems, outlying cases can be corrected. Sometimes, repeating an analysis with and without suspected outliers is necessary.

## 5.3    Influence of cases

Another aspect of case analysis is the attempt to understand the influence or importance of each case in fitting models. The general idea is to study changes in a specific part of the analysis when the data are slightly perturbed. Whereas statistics such as residuals are used to find problems with a model, influence analysis is done as if the model were correct, and we study the robustness of conclusions, given a particular model, to the perturbations. The most useful and important method of perturbing the data is deleting the cases from the data one at a time. We then study the effects or influence of each individual case by comparing the full data analysis to the analysis obtained with a case removed. Cases whose removal causes major changes in the analysis are called *influential*.

We will now formalize some notation used in the last section. A subscript $(i)$ shall mean "with the $i$th case deleted," so, for example, $\hat{\boldsymbol{\beta}}_{(i)}$ is the estimate of $\boldsymbol{\beta}$ computed without case $i$, $\mathbf{X}_{(i)}$ is the $(n-1) \times p'$ matrix obtained from $\mathbf{X}$ by deleting the $i$th row, and so on. In particular, then,

$$\hat{\boldsymbol{\beta}}_{(i)} = (\mathbf{X}_{(i)}^T \mathbf{X}_{(i)})^{-1} \mathbf{X}_{(i)}^T \mathbf{Y}_{(i)} \qquad (5.19)$$

As an example, Figure 5.3 is a graph of $\hat{\beta}_{1,(i)}$ versus $\hat{\beta}_{2,(i)}$ for the fuel consumption data, with only $X_1 = \text{TAX}$ and $X_2 = \text{DLIC}$ as predictors. The point labeled Wyoming corresponds to the estimate of the slopes for TAX and DLIC that would be obtained if Wyoming were deleted from the data and coefficients were estimated from the remaining 47 cases. To display completely the effect of deleting a case, a three-dimensional graph is required, as values for the intercept are not shown.

As is evident from Figure 5.3, deleting a case will result in changing the estimates, but for most cases, the change will be small. The full data estimates are $(-32.075, 12.515)$; if Texas were deleted, the estimates are $(-33.933, 12.436)$, and if only Wyoming were deleted, the estimates are $(-30.933, 10.691)$. From the graph, this latter point appears to be relatively far from the original estimate. A method of measuring the distance between these points is needed to judge if the change in the estimate is large enough to alter conclusions substantially.

**Cook's distance.**    We can measure influence by comparing $\hat{\boldsymbol{\beta}}$ to $\hat{\boldsymbol{\beta}}_{(i)}$. Since each of these is a $p'$ vector, the comparison requires a method of com-

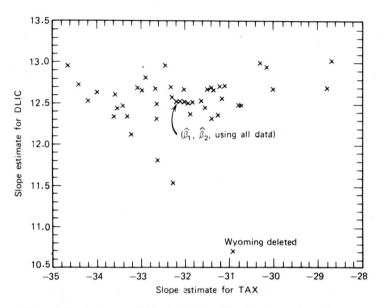

**Figure 5.3**  Estimates of $(\beta_1, \beta_2)$ obtained by deleting each case in turn.

bining information from each of the $p'$ components into a single number. Several ways of doing this have been proposed in the literature, but most of them will result in roughly the same information, at least for multiple linear regression. The method we use is due to Cook (1977). We define _Cook's distance_ $D_i$ to be

$$D_i = \frac{(\hat{\boldsymbol{\beta}}_{(i)} - \hat{\boldsymbol{\beta}})^T (\mathbf{X}^T\mathbf{X})(\hat{\boldsymbol{\beta}}_{(i)} - \hat{\boldsymbol{\beta}})}{p'\hat{\sigma}^2} \qquad (5.20)$$

This statistic has several desirable properties. First, contours of constant $D_i$ are ellipsoids, with the same shape as confidence ellipsoids. Second, the contours can be thought of as defining the distance _from_ $\hat{\boldsymbol{\beta}}_{(i)}$ _to_ $\hat{\boldsymbol{\beta}}$. Third, $D_i$ does not depend on parameterization, so if the columns of $\mathbf{X}$ are modified by linear transformation, the $D_i$ are not effected. Finally, if we define as usual $\hat{\mathbf{Y}} = \mathbf{X}\hat{\boldsymbol{\beta}}$, and $\hat{\mathbf{Y}}_{(i)} = \mathbf{X}\hat{\boldsymbol{\beta}}_{(i)}$, then (5.20) can be rewritten as

$$D_i = \frac{(\hat{\mathbf{Y}}_{(i)} - \hat{\mathbf{Y}})^T(\hat{\mathbf{Y}}_{(i)} - \hat{\mathbf{Y}})}{p'\hat{\sigma}^2} \qquad (5.21)$$

so $D_i$ is the ordinary Euclidean distance between $\hat{\mathbf{Y}}$ and $\hat{\mathbf{Y}}_{(i)}$. Cases for which $D_i$ is large have substantial influence on $\hat{\boldsymbol{\beta}}$ and on fitted values, and deletion of them may result in important changes in conclusions.

**Magnitude of $D_i$.**    Cases with large values of $D_i$ are ones whose deletion will result in substantial changes in the analysis. Typically, the case with the largest $D_i$, or in large data sets the cases with the largest few $D_i$, will be of interest. One method of calibrating $D_i$ is obtained by analogy to confidence regions. If $D_i$ were exactly equal to the $a \times 100\%$ point of the $F$ distribution with $p'$ and $n - p'$ degrees of freedom, then deletion of the $i$th case would move the estimate of $\beta$ to the edge of a $(1 - a) \times 100\%$ confidence region based on the complete data. Since for most $F$ distributions the 50% point is near 1, a value of $D_i = 1$ will move the estimate to the edge of about a 50% confidence region, a potentially important change. If the largest $D_i$ is substantially less than 1, deletion of a case will not change the estimate of $\beta$ by much. To investigate the influence of a case more closely, the analyst should delete the large $D_i$ case and recompute the analysis to see exactly what aspects of it have changed.

**Computing $D_i$.**    From the derivation of Cook's distance it is not clear that using these statistics is computationally convenient. However, the results sketched in Appendix 5A.1 can be used to write $D_i$ using more familiar quantities. The simplest form for $D_i$ is

$$D_i = \frac{1}{p'} r_i^2 \left( \frac{h_{ii}}{1 - h_{ii}} \right) \qquad (5.22)$$

$D_i$ is a product of the square of the $i$th Studentized residual $r_i$ and a monotonic function of $h_{ii}$. If $p'$ is fixed, the size of $D_i$ will be determined by two different sources: the size of $r_i$, a random variable reflecting lack of fit of the model at the $i$th case, and the potential $h_{ii}$, reflecting the location of $x_i$ relative to $\bar{x}$. A large value of $D_i$ may be due to large $r_i$, large $h_{ii}$, or both.

Table 5.2 gives the case statistics for three of the states in the fuel consumption data, again using only TAX and DLIC as predictors. To compute $r_i$ and $D_i$ from $\hat{e}_i$ and $h_{ii}$, we also need to know $\hat{\sigma}^2 = 5796.31$. For Wyoming, we can compute

$$r_{40} = \frac{242.6}{\sqrt{5796.31(1 - 0.0930)}} = 3.3453$$

$$D_{40} = \frac{(3.3453)^2}{3} \left( \frac{0.0930}{1 - 0.0930} \right) = 0.3826$$

Texas, the state with the largest potential has relatively little observed influence because its Studentized residual $r_i$ is very small. Similarly, Wyoming is relatively influential because of its large value of $r_i$; it appears that people in Wyoming consume more fuel than would be predicted from DLIC and TAX. However, in this example none of the cases alone will exert substantial influence on the estimated coefficients.

Table 5.2 Selected case statistics for the
fuel consumption data

| State | $\hat{e}_i$ | $h_{ii}$ | $r_i$ | $D_i$ |
|-------|-------------|----------|-------|-------|
| 18 ND | 153.8 | .0444 | 2.0659 | .0661 |
| 37 TX | − 16.94 | .2067 | − 0.2497 | .0054 |
| 40 WY | 242.6 | .0930 | 3.3453 | .3826 |

A complete analysis requires consideration of the $D_i$, $r_i$, and the $h_{ii}$, or equivalent functions of them, for each case.

## Example 5.3 Rat data (Dennis Cook)

An experiment was conducted to investigate the amount of a particular drug present in the liver of a rat. Nineteen rats were randomly selected, weighed, placed under light ether anesthesia and given an oral dose of the drug. Because it was felt that large livers would absorb more of a given dose than smaller livers, the actual dose an animal received was approximately determined as 40 mg of the drug per kilogram of body weight. (Liver weight is known to be strongly related to body weight.) After a fixed length of time each rat was sacrificed, the liver weighed, and the percent of the dose in the liver determined.

The experimental hypothesis was that, for the method of determining the dose, there is no relationship between the percentage of the dose in the liver ($Y$) and the body weight ($X_1$), liver weight ($X_2$), and relative dose ($X_3$).

The data and sample correlations are given in Tables 5.3 and 5.4. As had been expected, the sample correlations between the response and the independent variables are all small, and none of the simple regressions of dose on any of the independent variables is significant, all having $t$-values less than 1, as shown in Table 5.5. However, the regression of $Y$ on $X_1$, $X_2$, and $X_3$ gives a different and contradictory result: two of the independent variables, $X_1$ and $X_3$, have significant $t$-tests, with $p < 0.05$ in both cases, indicating that the two measurements combined are a useful indicator of $Y$; if $X_2$ is dropped from the model, the same phenomenon appears. The analysis so far, based only on aggregate statistics, might lead to the conclusion that a combination of dose and rat weight is associated with the response; however, dose was

Table 5.3    Rat data

| $X_1 =$ Body Weight | $X_2 =$ Liver Weight | $X_3 =$ Dose | $Y$ |
|---|---|---|---|
| 176 | 6.5 | .88 | .42 |
| 176 | 9.5 | .88 | .25 |
| 190 | 9.0 | 1.00 | .56 |
| 176 | 8.9 | .88 | .23 |
| 200 | 7.2 | 1.00 | .23 |
| 167 | 8.9 | .83 | .32 |
| 188 | 8.0 | .94 | .37 |
| 195 | 10.0 | .98 | .41 |
| 176 | 8.0 | .88 | .33 |
| 165 | 7.9 | .84 | .38 |
| 158 | 6.9 | .80 | .27 |
| 148 | 7.3 | .74 | .36 |
| 149 | 5.2 | .75 | .21 |
| 163 | 8.4 | .81 | .28 |
| 170 | 7.2 | .85 | .34 |
| 186 | 6.8 | .94 | .28 |
| 146 | 7.3 | .73 | .30 |
| 181 | 9.0 | .90 | .37 |
| 149 | 6.4 | .75 | .46 |

Table 5.4    Sample correlations—rat data

| $X_1 =$ Body weight (g) | 1.000 | | | |
|---|---|---|---|---|
| $X_2 =$ Liver weight (g) | 0.500 | 1.000 | | |
| $X_3 =$ Relative dose | 0.990 | 0.490 | 1.000 | |
| $Y$ | 0.151 | 0.203 | 0.228 | 1.000 |
| | Body weight | Liver weight | Dose | $Y$ |

(approximately) determined to be a multiple of rat weight so, at least to a first approximation, rat weight and dose were measuring the same thing!

We turn to case analysis to attempt to resolve this paradox. In Table 5.6 the residuals and related statistics are listed for the model $Y$ on $X_1, X_2, X_3$. The residuals ($\hat{e}_i$ or $r_i$) do not display any unusual features or reasons for the paradox: the $|r_i|$, for example, are all less than 2, without obvious trends or patterns. However, the $D_i$ immediately locate a possible cause: case 3 has $D_3 = .93$; no other case has $D_i$ bigger than .27, suggesting that case number 3 alone may have large

**Table 5.5   Regression of $Y$ on various predictors ($t$-values in parentheses)**

| Coefficient | Model Including | | | |
|---|---|---|---|---|
| | $X_1$ | $X_2$ | $X_3$ | $(X_1, X_2, X_3)$ |
| Intercept | 0.196 | 0.220 | 0.133 | 0.266 |
| | (0.89) | (1.64) | (0.63) | (1.37) |
| $\beta_1$ (rat weight) | 0.0008 | | | −0.0212 |
| | (0.63) | | | (−2.66) |
| $\beta_2$ (liver weight) | | 0.0147 | | 0.0143 |
| | | (0.86) | | (0.83) |
| $\beta_3$ (dose) | | | 0.235 | 4.178 |
| | | | (0.96) | (2.74) |

**Table 5.6   Residuals from $Y$ on $X_1, X_2, X_3$, all $n = 19$ cases**

| Case Number | $y_i$ | $\hat{e}_i$ | $r_i$ | $h_{ii}$ | $D_i$ |
|---|---|---|---|---|---|
| 1 | .42 | .124 | 1.77 | .178 | .17 |
| 2 | .25 | −.089 | −1.27 | .179 | .09 |
| 3 | .56 | .024 | .81 | .851 | .93 |
| 4 | .23 | −.101 | −1.38 | .108 | .06 |
| 5 | .23 | −.068 | −1.12 | .392 | .20 |
| 6 | .32 | .007 | .10 | .161 | .00 |
| 7 | .37 | .057 | .79 | .137 | .02 |
| 8 | .41 | .050 | .74 | .254 | .05 |
| 9 | .33 | .012 | .16 | .067 | .00 |
| 10 | .38 | −.003 | −.04 | .120 | .00 |
| 11 | .27 | −.080 | −.11 | .120 | .04 |
| 12 | .36 | .042 | .60 | .172 | .02 |
| 13 | .21 | −.098 | −1.54 | .316 | .27 |
| 14 | .28 | −.027 | −.38 | .131 | .01 |
| 15 | .34 | .032 | .43 | .076 | .00 |
| 16 | .28 | −.059 | −.86 | .217 | .05 |
| 17 | .30 | −.018 | −.26 | .195 | .00 |
| 18 | .37 | .061 | .85 | .149 | .03 |
| 19 | .46 | .135 | 1.92 | .178 | .20 |

Table 5.7    Regression with case 3 deleted

| Variable | Estimate | Standard Error | t-Value |
|----------|----------|----------------|---------|
| Intercept | 0.3114 | 0.2051 | 1.52 |
| X1 | −0.0078 | 0.0187 | −.42 |
| X2 | 0.0090 | 0.0187 | .48 |
| X3 | 1.4849 | 3.7131 | .40 |

$\hat{\sigma}^2 = 0.00612$,    $R^2 = 0.0211$,    d.f. $= 14$

enough influence on the fit to induce the anomaly. The value of $h_{33} = .85$ indicates that the problem with this case is that the vector $\mathbf{x}_3$ is different from the others.

One suggestion at this point is to delete the third case and recompute the regression. These computations are given in Table 5.7. Here, the paradox dissolves, and the apparent relationship found in the first analysis can thus be ascribed to the third case alone.

The careful analyst must now try to understand exactly why the third case is so influential. Inspection of the data indicates that this rat, with weight 190 g, was reported to have received a full dose of 1.000, which was a larger dose than it should have received according to the rule for assigning doses (for example, rat 8 with weight of 195 g got a lower dose of 0.98). A number of causes for the result found in the first analysis are possible: (1) the dose or weight recorded for case 3 was in error, so the case should probably be deleted from the study, or (2) the regression fit in the second analysis is not appropriate except in the region defined by the 18 points excluding case 3. This has many implications concerning the experiment. It is possible that the combination of dose and rat weight chosen was fortuitous, and that the lack of relationship found would not persist for any other combinations of them, since inclusion of a data point apparently taken under different conditions leads to a different conclusion. This suggests the need for collection of additional data, with dose determined by some rule other than a constant proportion of weight.

**Other measures of influence.**    Cook's distance was derived by analogy to confidence ellipsoids for the estimated vector $\hat{\boldsymbol{\beta}}$. A more general approach to influence requires use of more fundamental principles. Cook and Weisberg

(1982a, Sec. 5.2) define a class of influence measures by examination of the change in height of a log-likelihood function when a case is deleted. When interest centers on estimation of $\boldsymbol{\beta}$, this turns out to be equivalent to Cook's distance for the linear model. If interest centers on both estimation of $\boldsymbol{\beta}$ and of $\sigma^2$, a different measure results. The likelihood approach is also useful for applying the ideas of influence to other problems.

Other ad hoc measures of influence in the spirit of Cook's original derivation of $D_i$ have also been proposed. For example, Belsley, Kuh, and Welsch (1980) suggested a very similar statistic they called $\text{DFFITS}_i$, defined by

$$(\text{DFFITS}_i)^2 = \frac{(\hat{\boldsymbol{\beta}}_{(i)} - \hat{\boldsymbol{\beta}})^T (\mathbf{X}^T \mathbf{X})(\hat{\boldsymbol{\beta}}_{(i)} - \hat{\boldsymbol{\beta}})}{\hat{\sigma}^2_{(i)}} \qquad (5.23)$$

This statistic differs from $D_i$ by a scale factor and replacement of $\hat{\sigma}^2$ by $\hat{\sigma}^2_{(i)}$. Atkinson (1982) has suggested using a multiple of (5.23) that he calls a "modified Cook Statistic," in graphical procedures. Comparisons (Cook and Weisberg, 1982a, Chapter 4) suggest that all these influence measures will usually give essentially the same information.

As with the outlier problem, influential *groups* of cases may serve to mask each other and may not be found by examination of cases one at a time. In some problems, multiple-case methods may be desirable. See Cook and Weisberg (1982a), Sec. 3.6) as well as Gray and Ling (1984) for further discussion.

## Problems

**5.1**   In a regression problem with $n = 54$, $p' = 5$, the results included $\hat{\sigma} = 2.0$, and the following statistics for four of the cases.

| $\hat{e}_i$ | $h_{ii}$ |
|---|---|
| 0.6325 | .9000 |
| 1.732 | .7500 |
| 9.000 | .2500 |
| 10.295 | .0185 |

For each of these four cases, compute $r_i$, $D_i$, and $t_i$. Test each of the four cases to be an outlier. Make a qualitative statement about the influence of each case on the analysis.

**5.2**   In the two predictor fuel consumption example, the state with the largest $r_i$ was Wyoming.

  **5.2.1.** Test Wyoming to be an outlier (see Table 5.2).

  **5.2.2.** The following table gives data on all 4 predictors and the response for Alaska and Hawaii, the two states excluded from earlier analysis.

|        | TAX | DLIC | INC   | ROAD  | FUEL |
|--------|-----|------|-------|-------|------|
| Alaska | 8.0 | 45.2 | 5.162 | 3.246 | 551  |
| Hawaii | 5.0 | 64.8 | 4.995 | 0.602 | 345  |

Is the model fit in Chapter 2 to the other 48 states applicable to Alaska and Hawaii? If either or both of these states were included, would the fitted model change in important ways?

**5.3**   Using the QR factorization defined in Appendix 2A.3, and Problem 2.4, show that

$$H = QQ^T$$

Hence, if $q_i^T$ is the $i$th row of $Q$,

$$h_{ii} = q_i^T q_i \qquad h_{ij} = q_i^T q_j$$

Thus if the QR factorization of $X$ is computed, the $h_{ii}$ and the $h_{ij}$ are easily obtained.

**5.4**   Let $U$ be an $n \times 1$ vector with one as its first element and zeros elsewhere. Consider computing the regression of $U$ on an $n \times p'$ full rank matrix $X$. As usual, let $H = X(X^T X)^{-1} X^T$ be the orthogonal projection operator on the columns of $X$, with elements $h_{ij}$.

  **5.4.1.** Show that the elements of the vector of fitted values from the regression of $U$ on $X$ are the $h_{1j}, j = 1, 2, \ldots, n$.

  **5.4.2.** Show that the vector of residuals have $1 - h_{11}$ as the first element, and the other elements are $-h_{1j}, j > 1$.

  **5.4.3.** What are the internally Studentized residuals in terms of the $h_{ij}$? Using this method, the $h_{ij}$ can be found using any regression program that computes residuals.

**5.5**   Two $n \times n$ matrices $A$ and $B$ are orthogonal if $AB = BA = 0$. Show that $I - H$ and $H$ are orthogonal. Use this result to show that as long as the intercept is in a regression model, the sample correlation between the residuals

$\hat{\mathbf{e}} = (\mathbf{I} - \mathbf{H})\mathbf{Y}$ and the fitted values $\hat{\mathbf{Y}} = \mathbf{H}\mathbf{Y}$ is exactly zero or, equivalently, the slope of the regression of $\hat{\mathbf{e}}$ on $\hat{\mathbf{Y}}$ is zero.

**5.6**   The matrix $(\mathbf{X}_{(i)}^T \mathbf{X}_{(i)})$ can be written as $(\mathbf{X}_{(i)}^T \mathbf{X}_{(i)}) = \mathbf{X}^T \mathbf{X} - \mathbf{x}_i \mathbf{x}_i^T$, where $\mathbf{x}_i^T$ is the $i$th row of $\mathbf{X}$. Use this definition to prove that (5A.1) holds.

**5.7**   The quantity $y_i - \mathbf{x}_i^T \hat{\boldsymbol{\beta}}_{(i)}$ is the residual for the $i$th case when $\boldsymbol{\beta}$ is estimated without the $i$th case. Use (5A.1) to show that

$$y_i - \mathbf{x}_i^T \hat{\boldsymbol{\beta}}_{(i)} = \frac{\hat{e}_i}{1 - h_{ii}}$$

This quantity is called the predicted or PRESS residual and will be used in Chapters 8 and 9.

**5.8**   Use (5A.1) to verify (5.22).

**5.9**   Suppose that interest centered on $\boldsymbol{\beta}^*$ rather than $\boldsymbol{\beta}$, where $\boldsymbol{\beta}^*$ is the parameter vector excluding the intercept. Using (4.26) as a basis, define a distance measure $D_i^*$ like Cook's $D_i$ and show that (Cook, 1979)

$$D_i^* = \frac{r_i^2}{p} \left( \frac{h_{ii} - 1/n}{1 - h_{ii}} \right)$$

**5.10**   Use (5A.1) to prove (5.18), the relationship between $r_i$ and $t_i$. Also, show that the two approaches to testing for outliers outlined in Section 5.2 lead to the same statistic, $t_i$. *Hint:* You will need to prove that

$$\mathbf{x}_i^T (\mathbf{X}_{(i)}^T \mathbf{X}_{(i)})^{-1} \mathbf{x}_i = \frac{h_{ii}}{1 - h_{ii}}$$

**5.11**   Prove (5A.4).

# 6

# DIAGNOSTICS II:
# SYMPTOMS AND REMEDIES

The statistics defined in Chapter 5, residuals $\hat{e}_i$, Studentized residuals $r_i$ and $t_i$, potential values $h_{ii}$, and Cook's distance $D_i$ provide the building blocks for the study of the assumptions made in a linear model. Diagnostic methods using them are the main topic of this chapter.

Box (1980) has provided a useful paradigm for the fitting of statistical models, as outlined in modified form in Figure 6.1. The left-hand square represents the formulation of an interesting problem to study, choosing a model, making assumptions, and collecting data. The upper arrow in the diagram corresponds to the more traditional role of statistics in obtaining estimates. Estimation is done as if the contents of the formulation square were true, often using a general method like maximum likelihood. The result is the fitted model, tests, inferences, and the like, in the right-hand square.

Diagnostic methods correspond to the bottom arrow, labeled *criticism*. In this stage of analysis, we *condition on the fitted model* in the right-hand square to give information concerning the assumptions in the left square. Methods will naturally be based on residual-like quantities since these contain the information concerning lack of fit. Frequently, diagnostics will suggest modification of the model or the assumptions, and iterating again through the diagram.

Several criteria for designing useful diagnostic methods can be suggested (Weisberg, 1983). First, the behavior of a diagnostic procedure should be known, at least approximately, both under the correct model and under the model with one assumption changed. This suggests that diagnostic methods must be designed for specific purposes, and general omnibus methods such

**128**

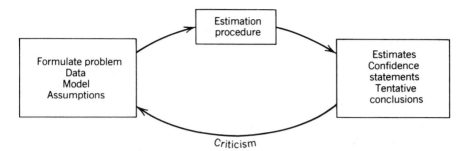

**Figure 6.1**   Modeling paradigm (after Box, 1980).

as a plot of residuals versus fitted values are likely to be less useful. An important class of diagnostic methods are based on *model expansion* (Cox, 1977), in which we fit a larger linear model than the one originally contemplated, with additional parameters that have specified values when an assumption concerning the model is true. A diagnostic method can then be obtained from a test concerning the additional parameters. We make use of model expansion many times in this chapter.

Second, diagnostics should be easy to compute. For linear regression, it turns out that many diagnostics use only the basic statistics and usual regression calculations applied to residuals. Third, diagnostics should be graphical or have graphical equivalents. The added variable plots of Section 2.4 are often useful. Finally, the diagnostics should suggest remedial action to the analyst. This last requirement is often possible to satisfy only approximately, as we shall see when considering specific methods.

## 6.1   Scatter plots

The most common diagnostic is the scatter plot, either of data or of derived statistics like residuals. A less traditional plot is the *matrix plot* (Chambers, Cleveland, Kleiner, and Tukey, 1983), shown in Figure 6.2 for the fuel consumption data. This plot is used to examine all two-dimensional plots that can be obtained by taking pairs of variables in a multidimensional data set. In the fuel data with 5 variables, there are 10 pairs of variables, and, if we consider the plot of FUEL versus INC different from the plot of INC versus FUEL, 20 possible plots. These are systematically arranged in Figure 6.2, with all plots in a row having the same variable plotted on the y-axis, and all plots in a column having the same variable plotted on the x-axis. For example, the last plot in the first row is of TAX versus FUEL, the second plot in the last row is of FUEL versus TAX, and the last plot in the bottom row is of

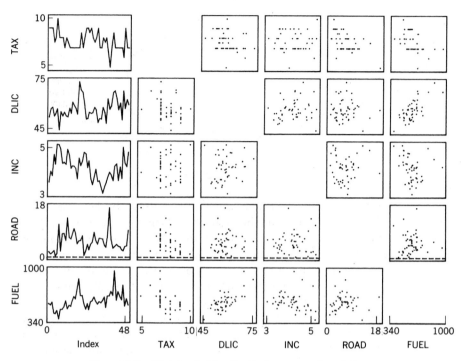

**Figure 6.2**   Matrix plot for the fuel consumption data.

FUEL versus ROAD. In Figure 6.2 these plots have been augmented on the left by a column of *index plots*, in which the quantity on the y-axis is plotted against sequence number on the x-axis. Index plots can be used as a substitute for lists of statistics, and in small problems they can be easy to use. Although each individual plot in the matrix plot is tiny, enough resolution is kept to display the main features of the two-dimensional relationships. As a whole, the plot may be taken as a graphical equivalent of the correlation matrix. Figure 6.2 shows that FUEL and DLIC are fairly closely related, while the relationship between FUEL and INC seems to have three anomalous points.

**Plots of residuals.**   The single diagnostic method with the longest tradition is the plot of residuals or Studentized residuals versus fitted values or one of the predictors. This is intended as a "one plot fits all" type of diagnostic, since many problems can potentially be diagnosed, depending on the shape of this plot. Since the $r_i$ or the $t_i$ and the $\hat{y}_i$ are nearly uncorrelated (the $e_i$ and the $y_i$ are *exactly* uncorrelated), when residuals are plotted against fitted values, we will always have a plot with slope near zero and as long as the intercept is in the model with scatter about the value 0. Eight idealized residual versus fitted value plots are given in Figure 6.3. The first plot is as one would expect if the specified model were correct; it is a *null plot*. This figure is just a swarm of points without pattern. Nonconstant variance might be indicated by any of Figures 6.3b to 6.3d. The right opening megaphone, Figure 6.3b, suggests variance increasing with the quantity plotted on the x-axis. This will often occur if an intrinsically positive response varies over a wide range, say from near zero into the thousands, since large values generally have more "room" to vary than do small values. The left opening megaphone of Figure 6.3c suggests that variance decreases with the quantity on the x-axis. The double bow of Figure 6.3d can occur if the response is constrained to lie between a minimum and a maximum value, for example, a percentage between 0 and 100. Large and small percentages are often less variable than are percentages near 50%.

Figures 6.3e and 6.3f suggest that the response is a *nonlinear* function of the quantity plotted on the x-axis since after removing the linear trend, only a nonlinear curved trend remains. This will often call for transformation of the data, either the response or the predictors, or use of nonlinear models. Figures 6.3g and 6.3h are supposed to be combinations of the two symptoms of nonconstant variance and nonlinearity.

Isolated points in these plots far from the zero point on the y-axis will be indicative of possible outliers. If the $r_i$ or the $t_i$ are plotted, only about 5 in 100 of the plotted values should exceed $\pm 2$, while about 1 in 100 will exceed $\pm 3$.

Unfortunately, these idealized plots cover up one very important point; in real data sets, the true state of affairs is rarely this clear. Consider the residual plot given as Figure 6.4. The data used to generate this plot is from Example 2.3.4 of Cook and Weisberg (1982a); see also Cook and Weisberg (1982b). This is a plot of $r_i$ versus $\hat{y}_i$. Depending on point of view, any or all of the symptoms described previously can be found in this plot. The point in the upper right corner may either be a possible outlier or indicate a right-opening megaphone. The shape is also reminiscent of Figure 6.3h, suggesting nonlinearity or possibly nonconstant variance. In all, this plot is not very helpful.

**Plots against predictors or other quantities.**   As with the fitted values, the $r_i$'s and each of the predictors are nearly uncorrelated, so plots against predictors have the same value, and limitation, as plots against fitted values. Plots against other quantities such as case number or time may be of use if the model failures are dependent on these other quantities. Index plots of residuals or other diagnostic statistics are often useful in place of simple lists of statistics, since large values may be found more quickly from a graph than from a list of numbers.

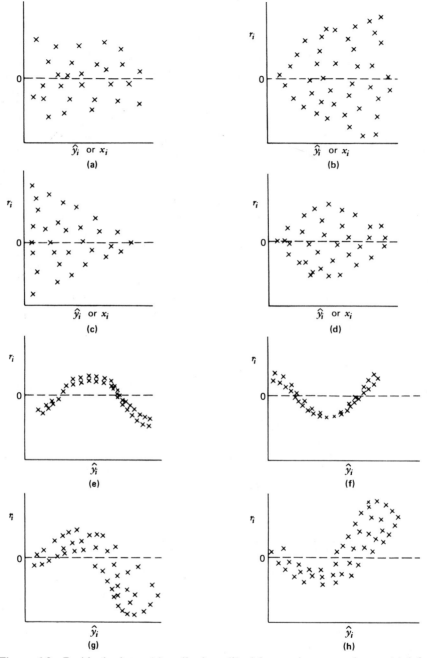

**Figure 6.3** Residual plots: (*a*) null plot; (*b*) right-opening megaphone; (*c*) left-opening megaphone; (*d*) double outward bow; (*e*) nonlinearity; (*f*) nonlinearity; (*g*) nonlinearity and nonconstant variance; (*h*) nonlinearity and nonconstant variance.

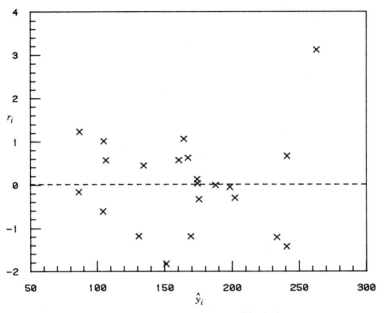

**Figure 6.4**  An ambiguous residual plot.

## 6.2  Nonconstant variance

The first assumption studied is that $\mathrm{var}(e_i) = \sigma^2$ for all cases in the data. This assumption is in doubt in many problems, as variances can depend on the response, on one or more of the predictors, or possibly on other factors, such as time or physical ordering.

If nonconstant variance is diagnosed, but exact variances are unknown, we could contemplate two remedies. First, weighted least squares could be used, with empirically chosen weights. Weights that are simple functions of single predictors, such as $\mathrm{var}(e_i) = \sigma^2 x_{1i}$, with $x_{1i} > 0$, can sometimes be justified theoretically. If large samples with replication are available, then within group variances may be used to provide approximate weights. Generally, however, empirical weights that are functions of the $\hat{y}_i$ or $\hat{e}_i$ from ordinary least squares cannot be recommended unless nonstandard methods are used to estimate variances; see, for example, Holt and Scott (1981).

The second alternative is to transform the response $Y$ via a *variance stabilizing transformation*. For almost any relationship between $\mathrm{var}(e_i)$ and an intrinsically positive response, an appropriate variance stabilizing transformation can be found (see Scheffé, 1959, Chapter 9 for technical details). In Table 6.1, the common variance stabilizing transformations are listed. The

**Table 6.1    Common variance stabilizers**

| Transformation | Situation | Comments |
|---|---|---|
| $\sqrt{Y}$ | $\mathrm{var}(e_i) \propto E(Y_i)$ | The theoretical basis is for counts from the Poisson distribution |
| $\sqrt{Y} + \sqrt{Y+1}$ | As above | For use when some $Y_i$'s are zero or very small; this is called the Freeman–Tukey (1950) transformation |
| $\log Y$ | $\mathrm{var}(e_i) \propto [E(Y_i)]^2$ | This transformation is very common; it is a good candidate if the range of $Y$ is very broad, say from 1 to several thousand; all $Y_i$ must be strictly positive |
| $\log(Y+1)$ | As above | Used if $Y_i = 0$ for some cases |
| $1/Y$ | $\mathrm{var}(e_i) \propto [E(Y_i)]^4$ | Appropriate when responses are "bunched" near zero, but, in markedly decreasing numbers, large responses do occur; e.g., if the response is a latency or response time for a treatment or a drug, some subjects may respond quickly while a few take much longer; the reciprocal transformation changes the scale of time per response to the rate of response, response per unit time; all $Y_i$ must be positive |
| $1/(Y+1)$ | As above | Used if $Y_i = 0$ for some cases |
| $\sin^{-1}(\sqrt{Y})$ | $\mathrm{var}(e_i) \propto E(Y_i)(1 - E(Y_i))$ | For binomial proportions $(0 \leqslant Y_i \leqslant 1)$ |

$Y^{1/2}$, $\log(Y)$ and $1/Y$, are appropriate when variance increases or decreases with the response, but each is more severe than the one before it. The square root transformation is relatively mild and is most appropriate when the $y_i$'s follow a Poisson distribution, usually the first model considered for errors in counts. The logarithm is the most commonly used transformation; the base of the logarithms is irrelevant. It is appropriate when the error standard deviation is a percent of the response, such as $\pm 10\%$ of the response, not $\pm 10$ units, so $[\mathrm{var}(e_i)]^{1/2} \propto E(y_i)$.

The reciprocal or inverse transformation is often applied when the response is a time until an event, such as time to complete a task, or until healing. This converts times per event to a rate per unit time; often the transformed measurements may be multiplied by a constant to avoid very small numbers. Rates can provide a natural measurement scale.

A technical problem may occur if the response has zero or negative values, since transformations such as the log are not defined. One solution is to transform [Y+(a small constant)], as shown in the table, with the constant set to 1. This may not be a good procedure if the Y's in the data are all small, but it may be a reasonable expedient if the values of Y are occasionally large.

**A diagnostic for nonconstant variance.** Cook and Weisberg (1983) have provided a diagnostic procedure for examining the assumption of nonconstant variance. The basic idea is to convert the constant variance assumption into a testable parametric hypothesis. We will specify the form that the variance will take if it is not constant.

Suppose now that var($e_i$) depends on an *unknown* vector parameter $\lambda$ and a *known* vector $z_i$ that may be different for each $i$. For example, if $z_i = \hat{y}_i$, the variance depends on the response. Similarly, $z_i$ may be the vector of predictors $x_i$, and the residual variance depends on all the predictors. $z_i$ may also be a subset of the predictors or of other quantities, such as time or spatial ordering of the units in a study and so on. Given $z_i$, we assume that

$$\text{var}(e_i) = \sigma^2 [\exp(\lambda^T z_i)] \tag{6.1}$$

This complicated form says that (1) var($e_i$) > 0 for all $z_i$; (2) variance depends on $z_i$ and $\lambda$ but only through $\lambda^T z_i$; (3) var($e_i$) is monotonic, either increasing or decreasing, in each component of $z_i$, (4) if $\lambda = 0$, then var($e_i$) = $\sigma^2$ for all $i$. The results of Chen (1983) suggest that the tests described here are not very sensitive to the exact functional form used in (6.1) as long as the four conditions are satisfied.

Given that the $e_i$ are independent and normally distributed, a *score test* of $\lambda = 0$ is particularly simple to compute using standard regression software. The test is carried out using the following steps:

**1.** Compute the regression of Y on all the X's in the model and save the residuals $\hat{e}_i$.

**2.** Compute *scaled squared residuals* $u_i$, defined by $u_i = \hat{e}_i^2/\tilde{\sigma}^2$, where $\tilde{\sigma}^2 = \sum \hat{e}_j^2/n$ is the maximum likelihood estimate of $\sigma^2$, and differs from the usual estimate of $\sigma^2$ only by the divisor of $n$ rather than $n - p'$.

**3.** Compute the regression of the $u_i$ on $z_i$, including an intercept. Obtain *SSreg* for this regression. If each $z_i$ has $q$ components not including the intercept, then *SSreg* has $q$ degrees of freedom. If variance is thought to be a function of the responses $y_i$, then compute the regression of $u_i$ on the fitted values $\hat{y}_i$ from the regression of $Y$ on the $X$'s. *SSreg* for this regression will have 1 d.f.

**4.** Compute the score test, $S = SSreg/2$. *P*-values for the test can be obtained by comparing $S$ to its asymptotic distribution, which, under the hypothesis $\lambda = \mathbf{0}$, is $\chi^2(q)$. If $\lambda \neq \mathbf{0}$, then $S$ will be too large, so large values of $S$ provide evidence against the hypothesis of constant variance.

If $q = 1$, variance depends on a single variable $z_i$, and a graphical equivalent of the test is obtained by plotting $r_i^2$ from the regression of $Y$ on the $X$'s versus $(1 - h_{ii})z_i$ ; for many problems a plot versus $z_i$ is simpler and will not lose much information. A wedge shape in this plot is indicative of nonconstant variance. However, the statistic should also be computed because a wedge shape in the plot may be obscured if the density of points on the $x$-axis is uneven. If $q > 1$, a graph can be obtained by plotting $r_i^2$ versus $(1 - h_{ii})\hat{\lambda}^T z_i$, where the values of $\hat{\lambda}^T z_i$ are simply the fitted values obtained in the regression in step 3.

### *Example 6.1    Snow geese*

In problem 4.6 of Chapter 4, the relationship between $Y$ = photo count and $X$ = observer count for flocks of snow geese in the Hudson Bay area of Canada is studied. Using the first observer, we can now examine the question of constant residual variance for the simple regression model $Y$ on $X$.

**Table 6.2**

*(a)    Regression summary of snow geese data from observer 1*

| Variable | Estimate | Standard Error | *t*-value |
|---|---|---|---|
| Intercept | 26.65 | 8.61 | 3.09 |
| X | 0.88 | 0.08 | 11.37 |

$\hat{\sigma}^2 = 1971.87$,    $R^2 = .750$,    d.f. $= 43$

*(b)    Anova of $u_i$ on $X_1$ for snow geese data*

| Source | d.f. | SS | MS |
|---|---|---|---|
| Regression | 1 | 162.8264 | 162.8264 |
| Residual | 43 | 137.8813 | 3.206541 |

Table 6.2(a) gives the regression summary for $Y$ on $X_1$. This summary suggests a reasonably strong relationship between $Y$ and $X_1$, although no information concerning the lack of fit of the model is contained in the summary. Table 6.2(b) is the analysis of variance table that results from the regression of the $u_i$ on $X_1$. The score test for nonconstant variance is $S = \frac{1}{2}SSreg = (\frac{1}{2})162.83 = 81.41$, which, when compared to the $\chi^2$ distribution with 1 d.f., gives an extremely small $p$-value. The hypothesis of constant residual variance is not tenable. The analyst must now cope with the almost certain nonconstant variance evident in the data. Two courses of action are outlined in problems 4.6.3 and 4.6.4.

---

*Example 6.2   Sniffer data (John Rice)*

---

When gasoline is pumped into a tank, hydrocarbon vapors are forced out of the tank and into the atmosphere. To reduce this significant source of air pollution, devices are installed to capture the vapor. In testing these vapor recovery systems, the amount that escapes cannot be measured, but a "sniffer" can determine if some vapor is escaping. Also, the amount that is recovered can be measured. To estimate the efficiency of the system, some method of estimating the total amount given off must be used. To this end, a laboratory experiment was conducted in which the amount of vapor given off was measured under carefully controlled conditions. Four variables are relevant for modeling:

$X_1$ = initial tank temperature (°F)
$X_2$ = temperature of the dispensed gasoline (°F)
$X_3$ = initial vapor pressure in the tank (psi)
$X_4$ = vapor pressure of the dispensed gasoline (psi)

In an experiment, these conditions were varied and the quantity of emitted hydrocarbons $Y$ was measured in grams. The data for 32 runs are displayed in Table 6.3.

We shall investigate the possibility of nonconstant variance. Figure 6.5a is a usual residual plot of $r_i$ versus $\hat{y}_i$ for the regression of $Y$ on $X_1 X_2 X_3 X_4$. While this plot is far from perfect, it does not suggest the need to worry about the assumption of nonconstant variance. Table

**Table 6.3   Gas vapor data**

| $X_1$ | $X_2$ | $X_3$ | $X_4$ | $Y$ | $X_1$ | $X_2$ | $X_3$ | $X_4$ | $Y$ |
|------|------|------|------|-----|------|------|------|------|-----|
| 33. | 53. | 3.32 | 3.42 | 29. | 90. | 64. | 7.32 | 6.70 | 40. |
| 31. | 36. | 3.10 | 3.26 | 24. | 90. | 60. | 7.32 | 7.20 | 46. |
| 33. | 51. | 3.18 | 3.18 | 26. | 92. | 92. | 7.45 | 7.45 | 55. |
| 37. | 51. | 3.39 | 3.08 | 22. | 91. | 92. | 7.27 | 7.26 | 52. |
| 36. | 54. | 3.20 | 3.41 | 27. | 61. | 62. | 3.91 | 4.08 | 29. |
| 35. | 35. | 3.03 | 3.03 | 21. | 59. | 42. | 3.75 | 3.45 | 22. |
| 59. | 56. | 4.78 | 4.57 | 33. | 88. | 65. | 6.48 | 5.80 | 31. |
| 60. | 60. | 4.72 | 4.72 | 34. | 91. | 89. | 6.70 | 6.60 | 45. |
| 59. | 60. | 4.60 | 4.41 | 32. | 63. | 62. | 4.30 | 4.30 | 37. |
| 60. | 60. | 4.53 | 4.53 | 34. | 60. | 61. | 4.02 | 4.10 | 37. |
| 34. | 35. | 2.90 | 2.95 | 20. | 60. | 62. | 4.02 | 3.89 | 33. |
| 60. | 59. | 4.40 | 4.36 | 36. | 59. | 62. | 3.98 | 4.02 | 27. |
| 60. | 62. | 4.31 | 4.42 | 34. | 59. | 62. | 4.39 | 4.53 | 34. |
| 60. | 36. | 4.27 | 3.94 | 23. | 37. | 35. | 2.75 | 2.64 | 19. |
| 62. | 38. | 4.41 | 3.49 | 24. | 35. | 35. | 2.59 | 2.59 | 16. |
| 62. | 61. | 4.39 | 4.39 | 32. | 37. | 37. | 2.73 | 2.59 | 22. |

6.4 gives the results of several nonconstant variance score tests, each computed using a different choice for $z_i$. Each of these tests is just half the sum of squares for regression for $u_i$ on $z_i$ for the various choices of $z_i$ shown. Remarkably, nonconstant variance is seen to be evident perhaps as a function of $X_1$ and $X_4$ jointly, even though it apparently is not a function of $X_1$ or $X_4$ separately; for $z = (X_1, X_4)$, $S = 9.28$, which compared to $\chi^2(2)$ gives $p$-value $= .01$. Figures 6.5b and 6.5c are the nonconstant variance plots for $z = \hat{y}_i$ and for $z = X_1$. The characteristic wedge shape is absent from both of these plots. Figure 6.5d gives the plot for $z = (X_1, X_4)$; the values on the x-axis are $(1 - h_{ii})$ times the fitted values from the regression of $u_i$ on $X_1$, $X_4$. For this plot, the wedge shape is evident.

The experiment that generated these data has 125 cases; the 32 in Table 6.3 were chosen as a homework problem for the first edition of this book. We might wish to study further the finding that variance seems to depend on $X_1$ and $X_4$ jointly by fitting the nonconstant variance model to the complete data. The tests are shown in the last two columns of Table 6.4: With more data, and therefore more powerful tests, nonconstant variance seems to be apparent for $X_1$ and $X_4$ alone, jointly, or as a function of the fitted values.

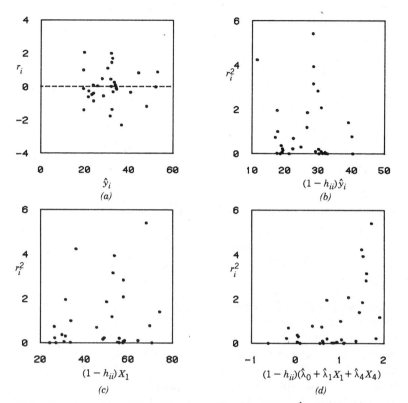

**Figure 6.5** Gas vapor data: (a) residual plot; (b) $r_i^2$ vs $(1-h_{ii})\hat{y}_i$; (c) $r_i^2$ vs $(1-h_{ii})X_1$; (d) $r_i^2$ vs $(1-h_{ii})(\hat{\lambda}_0 + \hat{\lambda}_1 X_1 + \hat{\lambda}_4 X_4)$.

**Table 6.4** Score tests, gas vapor data

| Choice for $\mathbf{z}_i$ | d.f. | $n=32$ S | $p$-value | $n=125$ S | $p$-value |
|---|---|---|---|---|---|
| $X_1$ | 1 | 1.40 | .24 | 9.46 | .002 |
| $X_4$ | 1 | 0.01 | .92 | 5.00 | .019 |
| $X_1, X_4$ | 2 | 9.28 | .01 | 11.76 | .003 |
| $X_1, X_2, X_3, X_4$ | 4 | 10.30 | .04 | 13.76 | .008 |
| Fitted values | 1 | .00 | .99 | 9.70 | .002 |

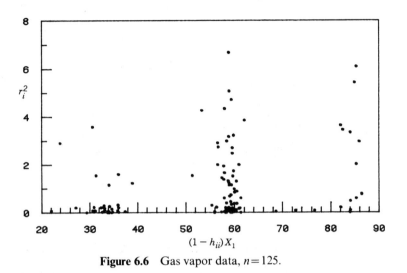

**Figure 6.6** Gas vapor data, $n = 125$.

Figure 6.6 is a plot of $r_i^2$ versus $(1 - h_{ii})X_1$ for all 125 cases. The wedge shape appears to be absent in the plot as the middle cluster of points has more large values than the outer clusters. This is an illusion, because there simply are more points in the middle cluster and this alone will result in more apparent variation in the middle. The test statistic $S = 9.46$, with 1 d.f. calibrates the plot. We see that to a good approximation, $X_1$ takes only three values, and the variance is different for each of these values. The larger sample size seems to have allowed simplification of the apparent nonconstant variance: The residual variance increases with $X_1$.

**Additional comments.** With some programs, it may be more convenient to compute the score test as follows: (1) Compute the residuals $\hat{e}_i$ from the regression of $Y$ on $\mathbf{X}$; let $\hat{\sigma}^2$ be the usual residual mean square from this regression. (2) Compute the regression of $\hat{e}_i^2$ on $\mathbf{z}$ for the choice of $\mathbf{z}$ of interest, and let $SSreg(\mathbf{z})$ be the resulting sum of squares for regression. (3) Compute $S = \frac{1}{2}SSreg(\mathbf{z})/[(n - p')\hat{\sigma}^2/n]^2$.

## 6.3 Nonlinearity

Although not all relationships between a response and a set of predictors are linear, the linear model is more generally useful than might be apparent at first. While a functional relationship may be nonlinear over the entire range of the predictors, a linear approximation over a restricted range may be an adequate model; different restricted ranges may require different approximating linear models. Such models can be useful for inference for limited ranges of the predictors; outside the relevant range, they may not be meaningful.

Also, suitable transformations of data can sometimes be found that will permit a nonlinear model to be approximated by a linear one. For example, suppose that the true relationship between the response and a single predictor is given by the power curve,

$$Y = \alpha X^\beta$$

Some of the members of this family for fixed $\alpha = 1$ and varying $\beta$ are shown in Figure 6.7. This form is linearized by taking logarithms,

$$\log Y = \log \alpha + \beta \log X \tag{6.2}$$

and the regression of $\log Y$ on $\log X$ is linear. Some attention, however, should be paid to errors. Multiplicative errors of the form "$k$ percent of the response" can be incorporated into (6.2) by writing

$$Y = (\alpha X^\beta)(e)$$

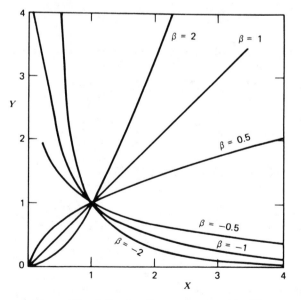

**Figure 6.7** A family of power curves, $Y = X^\beta$.

**Table 6.5    Linearizing transformations**

| Transformation | | Simple Regression Form | Multiple Regression Form |
|---|---|---|---|
| $\log Y$ | $\log X$ | $Y = \alpha X^{\beta}$ | $Y = \alpha X_1^{\beta_1} X_2^{\beta_2} \cdots X_p^{\beta_p}$ |
| $\log Y$ | $X$ | $Y = \alpha e^{\beta X}$ | $Y = \alpha e^{\Sigma \beta_j X_j}$ |
| $Y$ | $\log X$ | $Y = \alpha + \beta (\log X)$ | $Y = \alpha + \sum \beta_j \log(X_j)$ |
| $\dfrac{1}{Y}$ | $\dfrac{1}{X}$ | $Y = \dfrac{X}{\alpha X + \beta}$ | $Y = \dfrac{1}{\alpha + \sum (\beta_j / X_j)}$ |
| $\dfrac{1}{Y}$ | $X$ | $Y = \dfrac{1}{\alpha + \beta X}$ | $Y = \dfrac{1}{\alpha + \sum \beta_j x_j}$ |
| $Y$ | $\dfrac{1}{X}$ | $Y = \alpha + \beta \left(\dfrac{1}{X}\right)$ | $Y = \alpha + \sum \beta_j \left(\dfrac{1}{X_j}\right)$ |

Taking logarithms will reduce this model to a linear model with additive errors,

$$\log Y = \log(\alpha) + \beta \log(X) + \log(e)$$

If the $\log(e)$ have zero expectation and constant variance, usual methods of analysis can be used.

Table 6.5 lists common linearizable forms and the transformations needed to achieve linearity. If any of these forms are reasonable on theoretical grounds, the given transformations should be used. Of course transformations are not always chosen from a theoretical model, but the data are used to suggest a reasonable form.

Not all functions are linearizable, nor in some cases is it desirable to linearize. For example, a sum of two exponentials,

$$Y = \alpha_1 e^{\beta_1 X_1} + \alpha_2 e^{\beta_2 X_2}$$

is not linearizable. The logistic function, for $0 \leqslant Y \leqslant 1$

$$Y = \frac{e^{-(\alpha + \beta X)}}{1 + e^{-(\alpha + \beta x)}}$$

can be linearized by transforming $Y$ to $\ln[Y/(1 - Y)]$, but the resulting errors may not have zero mean and constant variance. These models are best handled using nonlinear or generalized linear models, as discussed in Chapter 12.

**Symptoms of nonlinearity.**    Graphical methods may help find nonlinearity. Curved trends in plots of residuals versus fitted values or predictors can indicate the need to transform. In multiple regression, the added variable plot or its close relative, the partial residual plot, both described in Section 2.4, may be more useful for diagnosing nonlinearity than the usual residual plot. Again, a curved trend is indicative of the need to transform.

If repeated measurements are made at each value of the predictors, the $F$-test for lack of fit described in Section 4.3 can be used to provide a symptom for nonlinearity. However, the $F$ statistic must be viewed as a guide, not as a test, as the exploratory procedure may invalidate the distributional assumptions of the $F$-test.

**Remedies.**  Transformations or addition of polynomial or, possibly, cross product terms of the original predictors are the remedies for nonlinearity. The symptoms that call for polynomials are often similar to those that call for transformations, and only experience with similar problems can help decide between the two.

### Example 6.3  Brain and body weights

Table 6.6 gives the average brain and body weights for 62 species of mammals. We shall consider the problem of modeling brain weight as a function of body weight. These data were taken from a larger study and were collected for another purpose (Allison and Cicchetti, 1976).

An initial attempt to graph brain weight (in grams) versus body weight (in kilograms), as given in Figure 6.8a, indicates immediately that some transformation is required. Most of the points in the plot are jammed into the lower left corner with only a few stragglers elsewhere. Because of the wide variation of both variables, log transformations are obvious candidates. This is like assuming that the correct functional relationship is of the form brain weight $= \alpha_0$(body weight)$^{\alpha_1}$. Models like this seem to work very well for relationships between parts of objects; see Sprent (1972) or Gould (1966) for discussion. The base 10 logarithms of the weights are given in Table 6.6. The scatter plot in the log scale is given in Figure 6.8b, suggesting that there is a strong linear relationship in the log scale. The scatter plot of the residuals from the fitted regression is given in Figure 6.8c. The fitted line corresponds to the observed data remarkably well. The regression summaries are given in Table 6.7.

In this example, the transformation used seems to achieve two important goals: linearity and constant variance. In many problems, the

**Table 6.6  Average brain weights and body weights of 62 species of mammals**

| | | Body Weight (kg) | Brain Weight (g) | Log(Body Weight) | Log(Brain Weight) |
|---|---|---|---|---|---|
| 1 | Arctic fox | 3.385 | 44.500 | 0.530 | 1.648 |
| 2 | Owl monkey | 0.480 | 15.500 | —0.319 | 1.190 |
| 3 | Mountain beaver | 1.350 | 8.100 | 0.130 | 0.908 |
| 4 | Cow | 465.000 | 423.000 | 2.667 | 2.626 |
| 5 | Gray wolf | 36.330 | 119.500 | 1.560 | 2.077 |
| 6 | Goat | 27.660 | 115.000 | 1.442 | 2.061 |
| 7 | Roe deer | 14.830 | 98.200 | 1.171 | 1.992 |
| 8 | Guinea pig | 1.040 | 5.500 | 0.017 | 0.740 |
| 9 | Vervet | 4.190 | 58.000 | 0.622 | 1.763 |
| 10 | Chinchilla | 0.425 | 6.400 | − 0.372 | 0.806 |
| 11 | Ground squirrel | 0.101 | 4.000 | − 0.996 | 0.602 |
| 12 | Arctic ground squirrel | 0.920 | 5.700 | − 0.036 | 0.756 |
| 13 | African giant pouched rat | 1.000 | 6.600 | − 0.000 | 0.820 |
| 14 | Lesser short-tailed shrew | 0.005 | 0.140 | 2.301 | − 0.854 |
| 15 | Star-nosed mole | 0.060 | 1.000 | − 1.222 | − 0.000 |
| 16 | Nine-banded armadillo | 3.500 | 10.800 | 0.544 | 1.033 |
| 17 | Tree hyrax | 2.000 | 12.300 | 0.301 | 1.090 |
| 18 | N. American opossum | 1.700 | 6.300 | 0.230 | 0.799 |
| 19 | Asian elephant | 2547.000 | 4603.000 | 3.406 | 3.663 |
| 20 | Big brown bat | 0.023 | 0.300 | − 1.638 | − 0.523 |
| 21 | Donkey | 187.100 | 419.000 | 2.272 | 2.622 |
| 22 | Horse | 521.000 | 655.000 | 2.717 | 2.816 |
| 23 | European hedgehog | 0.785 | 3.500 | − 0.105 | 0.544 |
| 24 | Patas monkey | 10.000 | 115.000 | 1.000 | 2.061 |
| 25 | Cat | 3.300 | 25.600 | 0.519 | 1.408 |
| 26 | Galago | 0.200 | 5.000 | − 0.699 | 0.699 |
| 27 | Genet | 1.410 | 17.500 | 0.149 | 1.243 |
| 28 | Giraffe | 529.000 | 680.000 | 2.723 | 2.833 |
| 29 | Gorilla | 207.000 | 406.000 | 2.316 | 2.609 |
| 30 | Gray seal | 85.000 | 325.000 | 1.929 | 2.512 |
| 31 | Rock hyrax[a] | 0.750 | 12.300 | − 0.125 | 1.090 |
| 32 | Human | 62.000 | 1320.000 | 1.792 | 3.121 |
| 33 | African elephant | 6654.000 | 5712.000 | 3.823 | 3.757 |
| 34 | Water opossum | 3.500 | 3.900 | 0.544 | 0.591 |
| 35 | Rhesus monkey | 6.800 | 179.000 | 0.833 | 2.253 |
| 36 | Kangaroo | 35.000 | 56.000 | 1.544 | 1.748 |
| 37 | Yellow-bellied marmot | 4.050 | 17.000 | 0.607 | 1.230 |
| 38 | Golden hamster | 0.120 | 1.000 | − 0.921 | 0.000 |
| 39 | Mouse | 0.023 | 0.400 | − 1.638 | − 0.398 |
| 40 | Little brown bat | 0.010 | 0.250 | − 2.000 | − 0.602 |

**Table 6.6    (continued)**

|    |                      | Body Weight (kg) | Brain Weight (g) | Log(Body Weight) | Log(Brain Weight) |
|----|----------------------|-----------------:|-----------------:|-----------------:|------------------:|
| 41 | Slow loris           | 1.400            | 12.500           | 0.146            | 1.097             |
| 42 | Okapi                | 250.000          | 490.000          | 2.398            | 2.690             |
| 43 | Rabbit               | 2.500            | 12.100           | 0.398            | 1.083             |
| 44 | Sheep                | 55.500           | 175.000          | 1.744            | 2.243             |
| 45 | Jaguar               | 100.000          | 157.000          | 2.000            | 2.196             |
| 46 | Chimpanzee           | 52.160           | 440.000          | 1.717            | 2.643             |
| 47 | Baboon               | 10.550           | 179.500          | 1.023            | 2.254             |
| 48 | Desert hedgehog      | 0.550            | 2.400            | −0.260           | 0.380             |
| 49 | Giant armadillo      | 60.000           | 81.000           | 1.778            | 1.908             |
| 50 | Rock hyrax[b]        | 3.600            | 21.000           | 0.556            | 1.322             |
| 51 | Raccoon              | 4.288            | 39.200           | 0.632            | 1.593             |
| 52 | Rat                  | 0.280            | 1.900            | −0.553           | 0.279             |
| 53 | Eastern American mole| 0.075            | 1.200            | −1.125           | 0.079             |
| 54 | Mole rat             | 0.122            | 3.000            | −0.914           | 0.477             |
| 55 | Musk shrew           | 0.048            | 0.330            | −1.319           | −0.481            |
| 56 | Pig                  | 192.000          | 180.000          | 2.283            | 2.255             |
| 57 | Echidna              | 3.000            | 25.000           | 0.477            | 1.398             |
| 58 | Brazilian tapir      | 160.000          | 169.000          | 2.204            | 2.228             |
| 59 | Tenrec               | 0.900            | 2.600            | −0.046           | 0.415             |
| 60 | Phalanger            | 1.620            | 11.400           | 0.210            | 1.057             |
| 61 | Tree shrew           | 0.104            | 2.500            | −0.983           | 0.398             |
| 62 | Red fox              | 4.235            | 50.400           | 0.627            | 1.702             |

[a] *Heterohyrax brucci.*
[b] *Procavia habessinica.*


**Table 6.7    Regression summary: brain weight/ body weight data**

|           | Estimate | Standard Error | $t$-Value |
|-----------|---------:|---------------:|----------:|
| Intercept | 0.927    | 0.0417         | 22.23     |
| Slope     | 0.752    | 0.0285         | 26.41     |

$R^2 = 0.92$, $\hat{\sigma} = 0.0909$, d.f. $= 60$

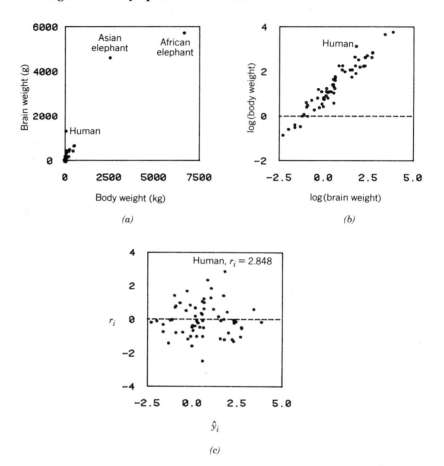

**Figure 6.8** Brain weight data. (*a*) Scatter plot; (*b*) Scatter plot, log scale; (*c*) residual plot.

transformation appropriate for one of these will not be appropriate for the other. From Figure 6.8c, the largest residual is for humans, with $r_i = 2.848$, and the $t_i$, computed from (5.18), is 3.04. This statistic can be used to test humans as an outlier. If interest in humans as an outlying species existed before examination of the data, then regular $t$-tables, rather than Bonferroni $t$-tables, can be used to obtain significance levels. From Table A, the significance level is about .005, so humans would be declared to have brain weight that is too large to be consistent with the model for the data.

## 6.4   Transforming the response

The methods for choosing transformations discussed so far use either specific knowledge about the relationship between variables or informal graphical aids. In this section and the next, the idea of model expansion is used to find objective methods for choosing transformations. Although these methods generally require several strong assumptions, they can be useful in many problems. We have already seen that transforming can be a remedy for either nonconstant variance or for nonlinearity. Also, errors in the transformed regression may be more nearly normally distributed than are the errors in the untransformed scale. A transformation of scale may also allow use of a simpler model to describe a set of data. These four goals for transformation will not always be met by the same transformation, and compromises may be required.

Box and Cox (1964) present a systematic approach to the problem of choosing a transformation. They expand the model, and turn the problem of finding a transformation into one of estimating a parameter. We assume for now that each $y_i > 0$, $i = 1, 2, \ldots, n$.

Suppose we let $\mathbf{Y}^\lambda$ be an $n \times 1$ vector whose $i$th element is $y_i^\lambda$; if $\lambda = 0$, we agree to set $y_i^\lambda = \ln(y_i)$. If $\lambda = 1$, the data are untransformed; otherwise $\lambda$ determines a *power transformation*; the usual values like the inverse, $\lambda = -1$, logarithmic, $\lambda = 0$, and the square and cube root, $\lambda = \frac{1}{2}$ and $\frac{1}{3}$, respectively, are included in this family. Box and Cox suggest examination of the model

$$\mathbf{Y}^\lambda = \mathbf{X}\boldsymbol{\beta} + e \qquad \text{var}(\mathbf{e}) = \sigma^2 \mathbf{I} \qquad (6.3)$$

Given data and this model, one could estimate $(\boldsymbol{\beta}, \sigma^2, \lambda)$ simultaneously to determine an estimated transformation. Unlike most of the models previously encountered in this book, (6.3) is not a linear model, so estimation of $\lambda$ requires estimation techniques other than least squares. We shall consider two methods of estimating $\lambda$. The first is a maximum likelihood type estimate suggested by Box and Cox. The second method, due to Atkinson (1973, 1981), replaces (6.3) by an approximating linear model, so least squares calculations can be used to estimate $\lambda$.

Box and Cox formulate the problem of choosing $\lambda$ to make the errors as nearly like a normal sample as possible. For this reason, their method gives a transformation toward normality. Their approach has technical difficulties that need not concern us here; see Hernandez and Johnson (1980) or Cook and Weisberg (1982a, Sec. 2.4) for discussion.

Suppose for a moment that we know $\lambda$. Given this knowledge, estimation of $\boldsymbol{\beta}$ and computation of the residual sum of squares is immediate:

$$\hat{\boldsymbol{\beta}} = (\mathbf{X}^T \mathbf{X})^{-1} \mathbf{X}^T \mathbf{Y}^\lambda \qquad (6.4)$$

$$RSS_\lambda = (\mathbf{Y}^\lambda)^T(\mathbf{I} - \mathbf{H})\mathbf{Y}^\lambda \tag{6.5}$$

These can be computed in an ordinary linear regression program by regressing $\mathbf{Y}^\lambda$ on the predictors.

Since $\lambda$ is unknown, (6.5) can be computed for a range of reasonable values of $\lambda$, say values in the range from about $-2$ to $+2$; if $\lambda$ is outside this range, the usefulness of this method is doubtful. To compare the various values of $\lambda$, we cannot compare the residual sums of squares directly because for each $\lambda$ the residual sum of squares is in different units. This can be handled in two equivalent ways. Both methods require conversion to a scale that gives comparable values for each $\lambda$; one such scale is the *log-likelihood function*. Using (6.5), the log-likelihood function for $\lambda \neq 0$ is

$$L(\lambda) = n\ln(|\lambda|) - \frac{n}{2}\ln(RSS_\lambda) + n(\lambda - 1)\ln(GM(y)) \tag{6.6a}$$

where $GM(y)$ is the *geometric mean* of the $y$'s, $GM(y) = (\prod y_i)^{1/n}$. If $\lambda = 0$, then compute $L(0)$ as

$$L(0) = -\frac{n}{2}\ln(RSS_0) - n\ln(GM(y)) \tag{6.6b}$$

The value of $\lambda$ that *maximizes* $L(\lambda)$ is the estimator, say $\hat{\lambda}$. Box and Cox suggest drawing a graph of $L(\lambda)$ versus $\lambda$, and reading the value of the maximum off the graph. This is adequate precision for determining a transformation, since it is usual to round $\hat{\lambda}$ to a nearby common value like $-1$, $-\frac{1}{2}$, 0, and so on.

There is a perhaps simpler way to do these computations if we use a more complicated version of the power family of transformations. Let $\mathbf{Z}^\lambda$ be an $n \times 1$ vector with $i$th element $z_i^\lambda$ defined by

$$z_i^\lambda = \begin{cases} \dfrac{y_i^\lambda - 1}{\lambda[GM(y)]^{\lambda-1}} & \lambda \neq 0 \\ GM(y)\ln(y_i) & \lambda = 0 \end{cases} \tag{6.7}$$

where, as before, $GM(y)$ is the geometric mean. Unlike $y_i^\lambda$, $z_i^\lambda$ is continuous at $\lambda = 0$. If we fit the model

$$\mathbf{Z}^\lambda = \mathbf{X}\boldsymbol{\beta} + \mathbf{e} \tag{6.8}$$

and compute the regression of $\mathbf{Z}^\lambda$ on the predictors, the residual sum of squares, say $RSS_\lambda(\mathbf{Z})$, is in the same scale for each $\lambda$, and therefore these values can be compared for different values of $\lambda$. Although this result is not obvious, the required proof is not hard and is given, for example, by Cook and Weisberg (1982a, Section 2.4). $\lambda$ can be chosen to minimize $RSS_\lambda(\mathbf{Z})$.

Alternatively, the log-likelihood function can be computed for any $\lambda$ by

$$L(\lambda) = -\frac{n}{2}\ln[RSS_\lambda(\mathbf{Z})] \qquad \text{\Large ✳} \qquad (6.9)$$

This will give the same values as will (6.6a) and (6.6b).

### Example 6.4   Size of Romanesque churches

We have seen in Example 6.3 the remarkably strong relationship between brain weight and body weight of mammals. Similar relationships are found for many other body parts. Gould (1973) has speculated on the applicability of biological "laws" of shape to other objects. To study this, he chose a "simple minded" example: medieval churches. These were built in a very wide range of sizes and shapes, to serve fairly similar purposes. Because of limitations due to the use of stone as a building material, we might speculate that the relationship between various measurements of the churches will be very strong. Table 6.8 lists the perimeter in hundreds of meters and area in hundreds of square meters for 25 post-Conquest Romanesque churches in Britain. The data were measured from ground plans given by Clapham (1934) and kindly provided by S. J. Gould. Since few churches are rectangular,

**Table 6.8   Perimeter (in hundreds of meters) and area (in hundreds of square meters) for 25 Romanesque churches**

| Church | Perimeter | Area | Church | Perimeter | Area |
|--------|-----------|------|--------|-----------|------|
| St. Albans | 3.48 | 38.83 | Byland | 3.14 | 34.27 |
| Durham | 3.69 | 43.92 | Roche | 2.04 | 17.61 |
| Blyth | 1.43 | 9.14 | Carmel | 1.77 | 13.37 |
| Binham | 2.05 | 16.66 | Bengeo | 0.59 | 2.04 |
| Gloucester | 3.05 | 36.16 | Copford | 0.69 | 2.22 |
| Norwich | 4.19 | 38.66 | Kempley | 0.50 | 1.46 |
| Leominster | 2.43 | 17.74 | Birkin | 0.69 | 1.92 |
| Southwell | 2.40 | 19.46 | Hales | 0.63 | 1.86 |
| Chertsey | 2.72 | 23.00 | Moccas | 0.58 | 1.69 |
| Hereford | 2.99 | 29.75 | Peterchurch | 0.86 | 3.31 |
| Canterbury | 4.78 | 51.19 | Little Tey | 0.41 | 1.13 |
| Lindesfarne | 1.33 | 6.60 | Melbourne | 1.23 | 6.74 |
| Tintern | 1.67 | 9.04 | | | |

*Source:* S. J. Gould.

the relationship between these two measurements is not obvious (see also Example 3.1). The scatter plot, Figure 6.9a, indicates a possibly nonlinear relationship with variance increasing with perimeter. Gould argues transformations for both quantities based on the need for light from windows to penetrate the relatively thick walls. We will ignore his theoretical arguments and use the Box and Cox procedure to choose a transformation for the response, which we will take to be area.

To find $\hat{\lambda}$, we compute (6.7) for a range of values in the interval $-2$ to $+2$; the values $-1$, $-.5$, $0$, $.5$, and $1$ are usually adequate for this purpose. For each value of $\lambda$, we compute the residual sum of squares $RSS_\lambda(\mathbf{Z})$. For example, at $\lambda = 0$, we find $RSS_0(\mathbf{Z}) = 377.3043$, while at $\lambda = .5$, we compute $RSS_{.5}(\mathbf{Z}) = 116.2636$, a much smaller value. Using (6.9), we find $L(0) = -74.16$, while $L(.5) = -59.45$; the complete curve for the interval $\lambda \in (0, 1)$ is shown in Figure 6.9b. Since $L(\lambda)$ is unimodal,

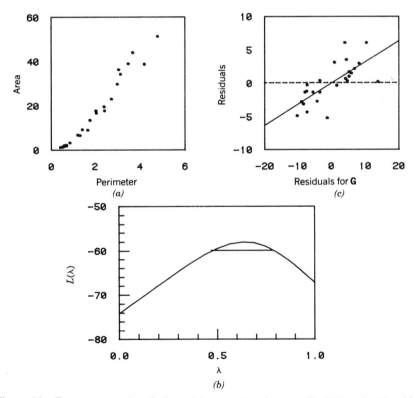

**Figure 6.9**   Romanesque church data: (a) area vs perimeter; (b) $L(\lambda)$ vs $\lambda$; (c) added variable plot for Atkinson's score.

the curve decreases outside this interval. The maximum of $L(\lambda)$ occurs at about $\hat{\lambda}=.63$, where $L(.63)=-58.05$. Clearly, $L(\lambda)$ is nearly the same at $\lambda=.5$ and $\lambda=.63$, so either of these values for $\lambda$ is likely to work about as well. We can obtain an approximate $(1-\alpha)\times 100\%$ confidence interval for $\lambda$ as the set of all $\lambda$ such that $L(\lambda)>L(\hat{\lambda})-(1/2)\chi^2(\alpha;1)$. For the example, a 95% interval consists of all $\lambda$ for which $L(\lambda)>-58.05$ $-\frac{1}{2}(3.84)=-59.97$. This interval corresponds to the set $.45\leqslant\lambda\leqslant.80$ marked by the horizontal line in the plot. We would surely choose the square root transformation for area, both because it is close to $\hat{\lambda}$ and because the square root of an area is in the same units as perimeter.

---

**Atkinson's score method.**   Although the Box and Cox method of choosing a transformation has gained great popularity in the last few years, it does suffer from the difficulty of requiring a fairly large amount of computing, as well as some specialized software for drawing the log-likelihood function. Atkinson (1973) proposed a quick test of the need to choose $\lambda$ to have a value other than 1; later (1981), he proposed the use of this method to give a quick estimate and also a graphical diagnostic.

Atkinson's method is to replace (6.8) by another model that can be analyzed using linear calculations. The method may be derived by approximating $\mathbf{Z}^\lambda$ by the first two terms of its Taylor expansion about the value of $\lambda=1$:

$$\mathbf{Z}^\lambda\cong\mathbf{Z}^1+(\lambda-1)\left.\frac{\partial\mathbf{Z}^\lambda}{\partial\lambda}\right|_{\lambda=1} \tag{6.10}$$

$$=\mathbf{Z}^1+(\lambda-1)\mathbf{G}$$

where the $i$th element $g_i$ of $\mathbf{G}$ is obtained by differentiating the $i$th element of $\mathbf{Z}^\lambda$ with respect to $\lambda$ and evaluating at $\lambda=1$,

$$g_i=y_i\{\ln[\,y_i/GM(y)]-1\}+\{\ln[GM(y)]+1\} \tag{6.11}$$

The second bracketed term on the right-hand side of (6.11) that does not depend on $i$ can be ignored. Substituting (6.10) into (6.8), we get, approximately,

$$\mathbf{Z}^1\cong\mathbf{X}\boldsymbol{\beta}+(1-\lambda)\mathbf{G}+\mathbf{e} \tag{6.12}$$

Since $\mathbf{Z}^1=\mathbf{Y}-\mathbf{1}$, we can replace $\mathbf{Z}^1$ by $\mathbf{Y}$. Setting $\phi=1-\lambda$,

$$\mathbf{Y}=\mathbf{X}\boldsymbol{\beta}+\phi\mathbf{G}+\mathbf{e} \tag{6.13}$$

Thus, a test of $\phi=0$ is approximately the same as a test of $\lambda=1$. The test of $\phi=0$ is the usual $t$ statistic from the regression of $\mathbf{Y}$ on $\mathbf{X}$ and $\mathbf{G}$. This statistic, which Atkinson has labeled $t_p$, can be compared to the standard normal distribution to get approximate $p$-values; a two-tailed test is appropriate. A quick estimate of $\lambda$ is also obtained from this regression; if $\tilde{\phi}$ is the estimate of $\phi$ obtained from the regression, then the estimate of $\lambda$, say $\tilde{\lambda}$ is $1 - \tilde{\phi}$. A diagnostic plot for this test is the added variable plot for adding **G** to the model for the regression of **Y** on **X**.

For the churches example, the regression of area on perimeter and **G** gives $\tilde{\phi} = 0.32$, and $t_D = 4.87$; the corresponding $p$-value is very small, so we have evidence of the need to transform the response. The quick estimate of the transformation is $1 - .32 = .68$, giving very good agreement with the likelihood-type estimate. In most problems, such close agreement cannot be expected, although conditions under which these will disagree are unknown. Figure 6.9c is the added variable plot for **G** after perimeter. The overall linear trend in this plot supports the usefulness of transforming area. The square root transformation is again suggested. We return to this example in the next section.

**Additional comments.**    The Box and Cox method is applicable only if the response is strictly positive. If zero or negative values occur, the usual method is to add a constant to the response before applying the method; unfortunately, very little information is available in the data to help choose the added constant; see Atkinson (1983). Carroll (1980) has suggested a robust method of determining $\lambda$; see also Cook and Wang (1983).

An interesting literature has developed recently on the interpretation of inferences that are made after $\lambda$ is estimated from the data. Bickel and Doksum (1981) argue that the method may not be useful because, without conditioning on the scale $\lambda$, confidence statements will be imprecise and difficult to use. Others have suggested that this point of view is not practically important, and that once $\hat{\lambda}$ is calculated, all further analysis is made conditionally, given $\hat{\lambda}$; see Hinkley and Runger (1984) and the discussion following this paper for a summary of the state of this controversy.

## 6.5    Transforming the predictors

We will distinguish two cases in transforming the predictors. In the first, the response achieves a maximum or minimum within the range of the predictors; often the point at which the maximum or minimum occurs is of interest. It is then usual to fit powers of the predictors, like $X_1^2$, $X_1^3$, $X_1 X_2$, and so on, to model the response. This topic of *polynomial regression* is taken up in the next chapter. Alternatively, the response may be monotonically increasing or decreasing with the predictors, but not at a constant rate. If power transformations are to be used, powers in the range $(-2, +2)$ are likely values. This type of transformation is the topic of this section.

Through model expansion, it is possible to obtain tests of the need to transform the predictors, quick estimates of the needed transformation and a diagnostic plot. The method presented here is a simplification of that given in the first edition of this book.

Suppose that we have a multiple regression problem with $p$ predictors and we contemplate transforming one of them, say $X_1$. If we limit ourselves to power transformations, the linear model

$$Y = \beta_0 + \sum_{j=1}^{p} \beta_j X_j + e \tag{6.14}$$

can be expanded to

$$Y = \beta_0 + \beta_1 X_1^{\alpha_1} + \sum_{j=2}^{p} \beta_j X_j + e \tag{6.15}$$

where, as before, we take $X_1^{\alpha_1} = \ln(X_1)$ if $\alpha_1 = 0$. While the parameters of this nonlinear model can be estimated using nonlinear least squares, we approximate (6.15) via a linear model by expanding $X_1^{\alpha_1}$ in a Taylor series about $\alpha_1 = 1$:

$$X_1^{\alpha_1} \cong X_1 + (\alpha_1 - 1) \frac{\partial}{\partial \alpha_1} X_1^{\alpha_1} \Big|_{\alpha_1 = 1}$$

$$\cong X_1 + (\alpha_1 - 1) X_1 \ln(X_1) \tag{6.16}$$

Substituting (6.16) into (6.15) gives

$$Y \cong \beta_0 + \beta_1 [X_1 + (\alpha_1 - 1) X_1 \ln(X_1)] + \sum_{j=2}^{p} \beta_j X_j + e$$

$$= \beta_0 + \sum_{j=1}^{p} \beta_j X_j + \beta_1 (\alpha_1 - 1) X_1 \ln(X_1) + e$$

$$= \beta_0 + \sum_{j=1}^{p} \beta_j X_j + \eta X_1 \ln(X_1) + e \tag{6.17}$$

where $\eta = \beta_1(\alpha_1 - 1)$. $\eta = 0$ corresponds to either $\beta_1 = 0$ or to $\alpha_1 = 1$, so, roughly speaking, this is a test of the need to transform. $\eta$ can be estimated as usual from the regression of $Y$ on the predictors and $X_1 \ln(X_1)$. The usual $t$-test of $\eta = 0$, which has a $t(n - p' - 1)$ distribution, is appropriate for testing the need to transform. A diagnostic graph is the added variable plot for $X_1 \ln(X_1)$ after the other predictors in the model. A quick estimate of $\alpha_1$ is obtained by solving $\eta = \beta_1(\alpha_1 - 1)$ for $\alpha_1$:

$$\hat{\alpha}_1 = \frac{\hat{\eta}}{\hat{\beta}_1} + 1 \tag{6.18}$$

where $\hat{\eta}$ is estimated from (6.17), but $\hat{\beta}_1$ is estimated from (6.14), not from (6.17).

We apply this method to the church data of Example 6.4 and examine the need to transform perimeter by fitting

$$\sqrt{(\text{area})} = \beta_0 + \beta_1(\text{perimeter}) + e \qquad (6.19)$$

The fitted model is given in Table 6.9(a). The $t$-value for $\beta_1$ is 26.20; clearly, a zero slope is inappropriate. This method of determining a transformation will not work if the slope before transforming is not distinguishable from zero. The fitted model for

$$\sqrt{(\text{area})} = \beta_0 + \beta_1(\text{perimeter}) + \eta(\text{perimeter})[\ln(\text{perimeter})] + e \qquad (6.20)$$

is given in Table 6.9(b); only the estimate of $\eta$, its standard error, and $t$-value are of interest in this regression. The test of $\eta = 0$ has value $t = -4.45$, which is compared to $t(22)$, providing evidence of the need to transform perimeter. The suggested transformation is

$$\hat{\alpha}_1 = \frac{-.6726}{1.544} + 1 = 0.56 \qquad (6.21)$$

*[handwritten annotation: $n - p' - 1$   d.f.]*

Again rounding to a convenient multiple, we choose to take the square root of perimeter. The added variable plot for perimeter $\times \ln(\text{perimeter})$, Figure 6.10$a$, supports the need for transformation because of the general linear trend.

The scatter plot of $(\text{area})^{1/2}$ versus $(\text{perimeter})^{1/2}$ is given in Figure 6.10$b$.

Table 6.9   Transforming perimeter, church data

| (a) | Estimate | Standard Error | $t$-Value |
|---|---|---|---|
| Intercept | 0.5998 | 0.1374 | 4.36 |
| Perimeter | 1.5437 | 0.0589 | 26.20 |

$\hat{\sigma}^2 = 0.1344$, $R^2 = .9676$, d.f. $= 23$

| (b) | Estimate | Standard Error | $t$-Value |
|---|---|---|---|
| Intercept | −0.5036 | 0.2679 | −1.88 |
| Perimeter | 2.6966 | 0.2625 | 10.27 |
| Added variable | −0.6727 | 0.1510 | −4.45 |

$\hat{\sigma}^2 = 0.0739$, $R^2 = .9830$, d.f. $= 22$

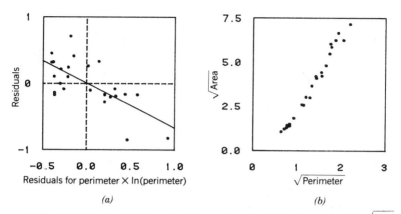

*(a)*                                          *(b)*

**Figure  6.10**   (*a*) Added variable plot for perimeter × ln(perimeter). (*b*) $\sqrt{\text{area}}$ vs $\sqrt{\text{perimeter}}$.

Although this plot is reasonably straight, the smaller churches seem to deviate systematically from the overall trend. The important lesson here is that automatic methods for choosing a transformation can fail to give completely satisfactory results. For further discussion of these data, see Gould (1968).

**Additional remarks.**   The method derived here was suggested in part by Box and Tidwell (1962). They proposed an iterative procedure for estimating $\alpha_1$ obtained by repeating the one-step process. The one-step estimator is equivalent to a *score test* and is in the same spirit as the tests for constant variance and Atkinson's test, described previously. In a multiple regression problem, transforming several of the predictors may be of interest. We suggest a sequential approach, studying each predictor in turn.

In many problems, this method of determining a transformation for a predictor may fail because little information is available in the data to distinguish between, for example, the regression of $Y$ on $X$ and the regression of $Y$ on $X^{1/2}$. In addition, absurd answers, such as $\hat{\alpha} = -74$ or $+18$, will occasionally result, when only powers in the range of about $(-2, 2)$ are suggested.

Consider the choice of a transformation for one predictor, say $X_1$. If the $t$ statistic for $\beta_1$ in (6.14) is not large, then $\hat{\alpha}_1$ from (6.18) will be poorly determined. Similarly, if the ratio of the largest value of $X_1$ to the smallest value of $X_1$ is less than about 10, the two quantities $X_1$ and $X_1 \ln(X_1)$ will be very highly correlated, so $\hat{\eta}$, and hence $\hat{\alpha}_1$ will again be very poorly determined. If this ratio is larger than 10, $X_1$ should probably be transformed to logarithms before any analysis begins, getting us back to collinear predictors.

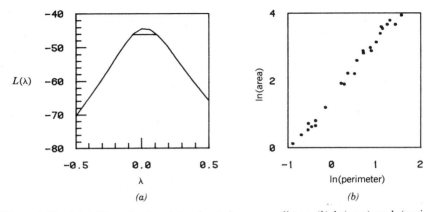

**Figure 6.11**   (a) $L(\lambda)$ vs $\lambda$ using ln(perimeter) as a predictor. (b) ln(area) vs ln(perimeter).

There is an obvious interplay between transforming the response and the predictors. While simultaneous methods exist (see Cook and Weisberg, 1982a, Example 2.4.5), for most problems their use will not be necessary. The following procedure can be recommended, assuming all the data are strictly positive. First, for any predictors with the ratio maximum/minimum bigger than about 10, transform to logarithms. Then, use the Box and Cox or Atkinson method to transform the response. Finally, for any predictors with large $t$-values, consider the Box–Tidwell method outlined in this section to examine the predictors.

If we apply these ideas to the church data, since maximum (perimeter)/minimum(perimeter) is about 10, we are led to consider ln(perimeter) as a predictor rather than perimeter. If we then repeat the Box and Cox procedure to choose a scale for area, we get the likelihood curve shown in Figure 6.11a, with $\hat{\lambda} = 0.02$ and $RSS(\mathbf{Z}) = 34.81$, clearly suggesting a log transform for area. The plot of ln(area) versus ln(perimeter), Figure 6.11b, is satisfactory.

## 6.6   Normality assumption

The usual distributional assumption in regression analysis is that the errors are normally distributed. This assumption is used to justify $F$- and $t$-tests and confidence procedures. Nonnormality of errors is very difficult to diagnose in small samples by examination of residuals. Suppose we have a linear model $\mathbf{Y} = \mathbf{X}\boldsymbol{\beta} + \mathbf{e}$, where $\text{var}(\mathbf{e}) = \sigma^2\mathbf{I}$, and we wish to examine the normality assumption. Of course, $\mathbf{e}$ is never observed, so the test of normality must be based on the residuals $\hat{\mathbf{e}}$. From (5.3), and substituting $\mathbf{X}\boldsymbol{\beta}+\mathbf{e}$ for $\mathbf{Y}$,

$$\hat{\mathbf{e}} = (\mathbf{I} - \mathbf{H})\mathbf{Y}$$
$$= (\mathbf{I} - \mathbf{H})(\mathbf{X}\boldsymbol{\beta} + \mathbf{e})$$
$$= (\mathbf{I} - \mathbf{H})\mathbf{X}\boldsymbol{\beta} + (\mathbf{I} - \mathbf{H})\mathbf{e}$$
$$= (\mathbf{I} - \mathbf{H})\mathbf{e}$$

since $(\mathbf{I} - \mathbf{H})\mathbf{X} = \mathbf{0}$. Writing $h_{ij}$ for the $(i, j)$th element of $\mathbf{H}$, this last equation in scalar form is

$$\hat{e}_i = e_i - \left( \sum_{j=1}^{n} h_{ij} e_j \right) \tag{6.22}$$

Thus $\hat{e}_i$ is equal to $e_i$ adjusted by subtracting off a weighted sum of all the $e_j$'s, including $e_i$. If the number of degrees of freedom for error, $n - p'$, is small, and some of the $h_{ij}$ are large, the term in parentheses in (6.22) may be more important than $e_i$ in determining the distribution of $\hat{e}_i$. By the central limit theorem, this sum will behave normally even if the $e_i$'s are not normal. Thus any test for nonnormality applied to the residuals cannot be expected to be very good, at least in small samples. Gnanadesikan (1977) refers to this as the *supernormality* of residuals.

As $n$ increases for fixed $p'$, the $h_{ii}$ will tend to zero and the $e_i$ term in (6.22) will tend to dominate, because the sum will have relatively small variance. For large samples, then, usual methods applied to residuals can be expected to give about as much information as would the same methods applied to the errors themselves.

**Probability plots.**   Our technique of choice for studying nonnormality is the normal probability or rankit plot (a general treatment of probability plotting is given by Wilk and Gnanadesikan, 1968 and Gnanadesikan 1977). Suppose we have a sample of $n$ numbers $z_1, z_2, \ldots, z_n$, and we wish to examine the hypothesis that the $z$'s are a homogeneous sample from a normal distribution with unknown mean $\mu$ and variance $\sigma^2$. A useful way to proceed is as follows:

**1.** Order the $z$'s to get $z_{(1)} \leqslant z_{(2)} \leqslant \cdots \leqslant z_{(n)}$. The ordered $z$'s are called the sample order statistics.

**2.** Now, consider a normal sample of size $n$ with zero mean and unit variance. Let $u_{(1)} \leqslant u_{(2)} \leqslant \cdots \leqslant u_{(n)}$ be the mean values of the order statistics that would be obtained if we repeatedly took samples of size $n$ from the standard normal. The $u_{(i)}$'s are called the expected values of normal order statistics, or *rankits*. The rankits are frequently tabled, as in Table E, or can be easily approximated using a computer.‡

**3.** If the $z$'s are normal, then

$$E(z_{(i)}) = \mu + \sigma u_{(i)}$$

so that the regression of $z_{(i)}$ on $u_{(i)}$ will be a straight line. If the sample is not normal, then the *rankit plot* of $z_{(i)}$ versus $u_{(i)}$ should not approximate a straight line.

As an example of a rankit plot, a pseudorandom sample was generated on a computer with $n = 17$ as if the sample were drawn from $N(3, 4)$. Using the rankits from Table D, the rankit plot in Figure 6.12a was drawn. This plot is nearly, though not exactly, a straight line. In contrast, Figure 6.12b is a rankit plot of $n = 17$ pseudorandom numbers drawn as if they were from a uniform distribution on the interval $(0, 1)$. This latter plot exhibits a flattening at both ends of the plot—suggesting that too few values relatively far from the mean are in the sample for it to be considered a sample from a normal distribution. Besides the $\int$ observed in this plot, the other common shapes for nonnormal distributions include $\frown$, $\frown$, and $\underline{\phantom{x}}\diagup$. The first of these indicates too many extreme values, while the remaining two shapes indicate negative and positive skewness, respectively.

In general, judging whether or not a rankit plot indicates that a sample behaves as if it were normal requires an experienced observer. Daniel and

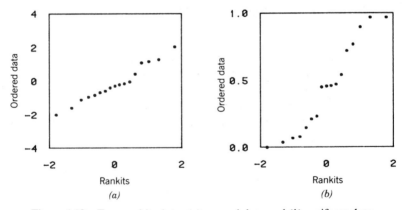

**Figure 6.12**   Two rankit plots: (*a*) normal data and (*b*) uniform data.

‡Using a subroutine that computes the cumulative normal distribution function $\Phi(t) = \int_{-\infty}^{t} (2\pi)^{-1/2} \exp(-x^2/2)\,dx$ and its inverse $\Phi^{-1}(p)$, a simple approximation to $u_{(i)}$ for a sample of size $n$ is $u_{(i)} \doteq \Phi^{-1}[(i - \tfrac{3}{8})/(n - \tfrac{1}{4})]$; see Blom (1958) or Weisberg and Bingham (1975), or Royston (1982b).

Wood (1981) and Daniel (1976) provide many pages of "training plots" that may help the analyst learn to interpret rankit plots.

The Box–Cox method outlined in Section 6.4 is often described as a transformation to normality; $\hat{\lambda}$ is that value of $\lambda$ that makes the resulting residuals as close to normal as possible. Hernandez and Johnson (1980) point out that even "as close as possible" may not be very close, and diagnostic checks, such as rankit plots, should be made after transformation.

**Additional comments.** Many statistics have been proposed for testing a sample for normality. One of these that works extremely well is the Shapiro and Wilk (1965) $W$ statistic, which is essentially the square of the correlation between the $z_{(i)}$ and the $u_{(i)}$ in the rankit plot; normality is rejected if $W$ is too small. Royston (1982abc) provides details and computer routines for the calculation of the test and for finding $p$-values.

For residuals from small samples, use of the ordered $r_{(i)}$ or $t_{(i)}$ in place of the $\hat{e}_{(i)}$ has been advocated, but there is little reason to believe that any one of these three choices will be better than the others. Atkinson (1981) has presented a computationally intensive method for approximating a significance test that can be applied even in small samples. The method is as follows: (1) Fix the sample size $n$ and the matrix of predictors $\mathbf{X}$. (2) Generate $m = 19$ $n \times 1$ vectors, say $\mathbf{A}_1, \ldots, \mathbf{A}_{19}$, such that all the elements of each vector are standard normal pseudorandom numbers. (3) Compute the regression of each $\mathbf{A}_k$ on $\mathbf{X}$, and save the residuals, calling the $k$th set $\mathbf{E}_k$. (4) Order the elements within each vector $\mathbf{E}_k$, from smallest to largest. (5) Among the 19 smallest values from the $\mathbf{E}_k$ vectors, keep only the smallest and largest. Among the 19 second smallest values, keep the smallest and largest, . . . , among the 19 largest values, keep the smallest and largest. These estimate a $19/20 \times 100\% = 95\%$ confidence interval for each of the order statistics. (6) Plot these smallest and largest values, and the observed values of $\hat{e}_i/\hat{\sigma}$ versus the $u_{(i)}$. The observed residuals should fall mostly within the simulated "envelope." The method can be modified by using Studentized residuals, changing $m$, or using a different standard distribution; Atkinson used the absolute values of the Studentized residuals, and the half-normal distribution.

Figure 6.13 gives the rankit plot of the $r_i$ for the final log model of the church data, along with the simulated envelope. We have no reason to doubt normality for these data.

**Correlated errors.** Another distributional assumption concerning the errors is that they are uncorrelated. This would mean that the value of the error for one case does not depend on the value of the error for any other case. This assumption will be violated in some problems, especially if the cases are ordered in time or in space, and adjacent cases influence each other.

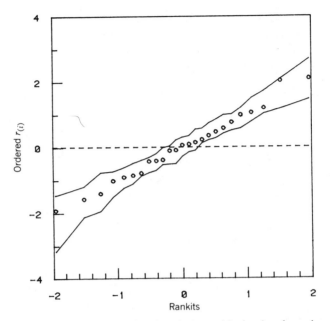

**Figure 6.13.**   Rankit plot for church data with simulated envelope.

Diagnostics for finding correlated errors are usually applicable only in special circumstances. For example, if cases are equally spaced in time, then the Durbin–Watson statistic (Durbin and Watson, 1950, 1951, 1971) is appropriate for testing for correlation between adjacent cases. In general, diagnosing correlated errors is very difficult, the best diagnostics coming from careful consideration of the process that generates the data.

## Problems

**6.1**   For Hooker's data, Problem 1.2, use the Box and Cox and Atkinson procedures to determine an appropriate transformation of PRES in the regression of PRES on TEMP. Find $\hat{\lambda}, \tilde{\lambda}$, the score test, and the added variable plot for the score. Summarize the results.

**6.2**   Examine the Longley data, Problem 3.3, for applicability of assumptions of the linear model.

**6.3**   The following data were collected in a study of the effect of dissolved sulfur on the surface tension of liquid copper (Baes and Killogg, 1953).

| $X =$ Weight % Sulfur | $Y =$ Decrease in Surface Tension (dynes/cm), two Replicates | |
|:---:|:---:|:---:|
| 0.034 | 301 | 316 |
| 0.093 | 430 | 422 |
| 0.30 | 593 | 586 |
| 0.40 | 630 | 618 |
| 0.61 | 656 | 642 |
| 0.83 | 740 | 714 |

**6.3.1.** Find transformations of $X$ and $Y$ so that in the transformed scale the regression is linear.

**6.3.2.** Assuming that $X$ is transformed to $\ln(X)$, which choice of $Y$ gives better results, $Y$ or $\ln(Y)$? (Sclove, 1972).

**6.4** The following (hypothetical) data give stopping times for $n = 62$ trials of various automobiles traveling at speed $X$ (miles per hour) and the resulting stopping distances $Y$ (feet) (Ezekiel and Fox, 1959).

| $X$ | $Y$ | $X$ | $Y$ |
|:---:|:---|:---:|:---|
| 4 | 4 | 20 | 48 |
| 5 | 2, 4, 8, 8 | 21 | 39, 42, 55 |
| 7 | 7, 7 | 24 | 56 |
| 8 | 8, 9, 11, 13 | 25 | 33, 48, 56, 59 |
| 9 | 5, 5, 13 | 26 | 39, 41 |
| 10 | 8, 14, 17 | 27 | 57, 78 |
| 12 | 11, 19, 21 | 28 | 64, 84 |
| 13 | 15, 18, 27 | 29 | 54, 68 |
| 14 | 14, 16 | 30 | 60, 67, 101 |
| 15 | 16 | 31 | 77 |
| 16 | 14, 19, 34 | 35 | 85, 107 |
| 17 | 22, 29 | 36 | 79 |
| 18 | 29, 34, 47 | 39 | 138 |
| 19 | 30 | 40 | 110, 134 |

**6.4.1.** Draw the scatter plot of $Y$ versus $X$. Fit the simple regression model and plot $r_i$ versus $\hat{y}_i$. What problems are apparent? Compute the $F$-test for lack of fit (Section 4.3) for the fitted model.

6.4.2. Use the method of Section 6.5 to find a transformation for $X$. Be sure to obtain the added variable plot for the need to transform.

6.4.3. Use Atkinson's score method to examine the need to transform the response $Y$ with $X$ not transformed. Also, obtain the estimated transformation using the Box and Cox likelihood-type method. Again, obtain appropriate graphical summaries.

**Table 6.10   Land rent data**

| $Y$ | $X_1$ | $X_2$ | $X_3$ | $X_4$ | $Y$ | $X_1$ | $X_2$ | $X_3$ | $X_4$ |
|------|-------|-------|-------|-------|------|-------|-------|-------|-------|
| 18.38 | 15.50 | 17.25 | 0.24 | 0 | 51.79 | 56.00 | 14.25 | 0.15 | 1 |
| 20.00 | 22.29 | 18.51 | 0.20 | 1 | 96.67 | 71.41 | 21.37 | 0.05 | 0 |
| 11.50 | 12.36 | 11.13 | 0.12 | 0 | 50.83 | 65.00 | 13.24 | 0.08 | 1 |
| 25.00 | 31.84 | 5.54 | 0.12 | 1 | 34.33 | 36.28 | 5.85 | 0.10 | 1 |
| 52.50 | 83.90 | 5.44 | 0.04 | 0 | 48.75 | 59.88 | 32.99 | 0.21 | 0 |
| 82.50 | 72.25 | 20.37 | 0.05 | 1 | 25.80 | 23.62 | 28.89 | 0.24 | 1 |
| 25.00 | 27.14 | 31.20 | 0.27 | 0 | 20.00 | 24.20 | 6.29 | 0.06 | 1 |
| 30.67 | 40.41 | 4.29 | 0.10 | 1 | 16.00 | 17.09 | 33.34 | 0.66 | 0 |
| 12.00 | 12.42 | 8.69 | 0.41 | 0 | 48.67 | 44.56 | 16.70 | 0.15 | 1 |
| 61.25 | 69.42 | 6.63 | 0.04 | 1 | 20.78 | 34.46 | 4.20 | 0.03 | 1 |
| 60.00 | 48.46 | 27.40 | 0.12 | 0 | 32.50 | 31.55 | 23.47 | 0.19 | 0 |
| 57.50 | 69.00 | 31.23 | 0.08 | 0 | 19.00 | 26.94 | 8.28 | 0.10 | 1 |
| 31.00 | 26.09 | 28.50 | 0.21 | 1 | 51.50 | 58.71 | 7.40 | 0.04 | 1 |
| 60.00 | 62.83 | 29.98 | 0.17 | 0 | 49.17 | 65.74 | 7.71 | 0.02 | 1 |
| 72.50 | 77.06 | 13.59 | 0.05 | 0 | 85.00 | 69.05 | 46.18 | 0.22 | 1 |
| 60.33 | 58.83 | 45.46 | 0.16 | 0 | 58.75 | 57.54 | 14.98 | 0.11 | 1 |
| 49.75 | 59.48 | 35.90 | 0.32 | 0 | 19.33 | 21.73 | 6.58 | 0.06 | 0 |
| 8.50 | 9.00 | 8.89 | 0.08 | 0 | 5.00 | 6.17 | 13.68 | 0.18 | 0 |
| 36.50 | 20.64 | 23.81 | 0.24 | 0 | 65.00 | 51.00 | 50.50 | 0.24 | 0 |
| 60.00 | 81.40 | 4.54 | 0.05 | 1 | 20.00 | 18.25 | 16.12 | 0.32 | 0 |
| 16.25 | 18.92 | 29.62 | 0.72 | 0 | 62.50 | 69.88 | 31.48 | 0.07 | 0 |
| 50.00 | 50.32 | 21.36 | 0.19 | 1 | 35.00 | 26.68 | 58.60 | 0.23 | 0 |
| 11.50 | 21.33 | 1.53 | 0.10 | 1 | 99.17 | 75.73 | 35.43 | 0.05 | 0 |
| 35.00 | 46.85 | 5.42 | 0.08 | 1 | 40.25 | 41.77 | 4.53 | 0.08 | 1 |
| 75.00 | 65.94 | 22.10 | 0.09 | 0 | 39.17 | 48.50 | 6.82 | 0.08 | 1 |
| 31.56 | 38.68 | 14.55 | 0.17 | 1 | 37.50 | 21.89 | 43.70 | 0.36 | 0 |
| 48.50 | 51.19 | 7.59 | 0.13 | 1 | 26.25 | 38.33 | 2.83 | 0.04 | 1 |
| 77.50 | 59.42 | 49.86 | 0.13 | 0 | 52.14 | 53.95 | 42.54 | 0.25 | 0 |
| 21.67 | 24.64 | 11.46 | 0.21 | 1 | 22.50 | 17.17 | 24.16 | 0.36 | 0 |
| 19.75 | 26.94 | 2.48 | 0.10 | 1 | 90.00 | 82.00 | 7.89 | 0.03 | 1 |
| 56.00 | 46.20 | 31.62 | 0.26 | 0 | 28.00 | 40.60 | 3.27 | 0.02 | 1 |
| 25.00 | 26.86 | 53.73 | 0.43 | 0 | 50.00 | 53.89 | 53.16 | 0.24 | 0 |
| 40.00 | 20.00 | 40.18 | 0.56 | 0 | 24.50 | 54.17 | 5.57 | 0.06 | 1 |
| 56.67 | 62.52 | 15.89 | 0.05 | 0 | | | | | |

*Source*: Douglas Tiffany.

**6.4.4.** Hald (1960) has suggested on the basis of a theoretical argument that the model $Y = \beta_1 X + \beta_2 X^2 + e$ with $\text{var}(e) = \sigma^2 X^2$ is appropriate for data of this type. Compare the fit of this model to the model found in problem 6.4.3. For $X$ in the range 0 to 40 mph, draw the curves that give the predicted $Y$ from each model, and qualitatively compare them. Also, for selected values of $X$, compute the variances of the predicted values. Then, for extrapolation to $X = 60$ mph, compute predicted values and standard errors of prediction for each of the two models. This comparison should demonstrate that an empirical relationship like that found in problem 6.4.2 may be perfectly adequate for some purposes such as interpolatory predictions, but may give misleading results for others, such as extrapolations (Draper and Hunter, 1969).

**6.4.5.** Fit Hald's model, given in problem 6.4.4, but with constant variance, $\text{var}(e) = \sigma^2$. Use the score test for nonconstant variance to test for constant variance against the alternative that variance increases with $X$ and $X^2$. Obtain a graphical summary.

**6.5**   The data in Table 6.10 were collected to study the variation in rent paid in 1977 for agricultural land planted to alfalfa. The data include:

$Y$ = average rent per acre planted to alfalfa

$X_1$ = average rent paid for all tillable land

$X_2$ = density of dairy cows (number per square mile)

$X_3$ = proportion of farmland used as pasture

$X_4$ = 1 if liming is required to grow alfalfa; 0, otherwise

The unit of analysis is a county in Minnesota; the 67 counties with appreciable rented farmland are included.

Alfalfa is a high protein crop that is suitable feed for dairy cows. It is thought that rent for land planted to alfalfa relative to rent for other agricultural purposes would be higher in areas with a high density of dairy cows and rents would be lower in counties where liming is required, since that would mean additional expense.

Use all the techniques learned so far to explore these data with regard to understanding rent structure. Summarize your results.

# 7

# MODEL BUILDING I: DEFINING NEW PREDICTORS

## 7.1 Polynomial regression

When the relationship between a predictor $X$ and a response $Y$ is smooth, but not a straight line, linear regression models can often be used if there are transformations of $X$ and $Y$ that give a straight line relationship in the transformed scale. Alternatively, a model can be expanded by addition of terms that are powers of the predictors, since any smooth function can be approximated by a polynomial of high enough degree. The resulting model is then given by

$$Y = \beta_0 + \beta_1 X + \beta_2 X^2 + \cdots + \beta_d X^d + e \qquad (7.1)$$

where $d$ is the degree of the polynomial; if $d = 2$, the model is quadratic, $d = 3$ is cubic, and so on. In (7.1) we assume that the errors are independent with constant variance $\sigma^2$, or else $\text{var}(e_i) = \sigma^2/w_i$, with $w_i > 0$ known. When variance is not constant, a variance stabilizing transformation may be required before polynomials are fit. Polynomial models are generally used as approximations and almost never represent a physical model.

The model (7.1) can be analyzed via least squares by defining $d$ new variables $Z_1, Z_2, \ldots Z_d$, by $Z_1 = X, Z_2 = X^2, \ldots, Z_d = X^d$, so (7.1) can be written as

$$Y = \beta_0 + \beta_1 Z_1 + \beta_2 Z_2 + \cdots + \beta_d Z_d + e \qquad (7.2)$$

Then, any least squares program can, in principle, compute estimates of the $\beta$'s and the other usual regression statistics for the regression of $Y$ on the $Z$'s. However, if $d$ is large, say 3 or more, serious numerical problems may arise

**164**

and direct fitting of (7.2) can be unreliable. Some numerical accuracy can be retained by centering, so that $Z_k = (X - \bar{X})^k, k = 1, \ldots, d$. Even better methods are surveyed by Seber (1977, Chapter 8).

An example of quadratic regression has already been given with the physics data of Example 4.2. There, a test for lack of fit indicated that a straight-line model was not adequate for the data, while the test for lack of fit after the quadratic model indicated that this model was adequate. When a test for lack of fit is not available, comparison of the quadratic model

$$Y = \beta_0 + \beta_1 X + \beta_2 X^2 + \text{error} \tag{7.3}$$

to the simple linear regression model

$$Y = \beta_0 + \beta_1 X + \text{error} \tag{7.4}$$

is usually based on a $t$-test of $\beta_2 = 0$ in (7.3). Indeed, a strategy for choosing $d$ is to continue adding terms to a model until the $t$-test for the highest order term is nonsignificant. Alternatively, an elimination scheme can be used, in which a maximum value of $d$ is fixed, and terms are deleted from the model one at a time, starting with the highest order term, until the highest order remaining term has a significant $t$-value. Kennedy and Bancroft (1971) suggest using a significance level of about 0.10 for this procedure. However, in the most common uses of polynomial regression, it is enough to consider only $d = 1$ or $d = 2$. For larger values of $d$, the fitted polynomial curves become wiggly, providing an increasingly better fit by matching the variation in the observed data more and more closely. The curve is then modeling the random variation rather than the overall shape of the relationship between variables.

**Polynomials with several predictors.** The extension to polynomials in several variables is straightforward: for each polynomial term added, a new predictor is created. This also raises the possibility of models with cross-product terms that depend on several predictors. For example, a model of the form

$$Y = \beta_0 + \beta_1 X_1 + \beta_2 X_2 + \beta_{11} X_1^2 + \beta_{22} X_2^2 + \beta_{12} X_1 X_2 + \text{error} \tag{7.5}$$

might be contemplated with two predictors. To help understand the term $\beta_{12} X_1 X_2$, suppose first that the $\beta$'s are known and $\beta_{12} = 0$. If $X_1$ is changed to $X_1 + \delta$, then the response $Y$ will change from (7.5) to $Y'$, where

$$Y' = \beta_0 + \beta_1(X_1 + \delta) + \beta_{11}(X_1 + \delta)^2 + \beta_2 X_2 + \beta_{22} X_2^2 \tag{7.6}$$

and the change in $Y$ is

$$\Delta Y = Y' - Y = \beta_1 \delta + \beta_{11}(2X_1 \delta + \delta^2) \tag{7.7}$$

so $\Delta Y$ does not depend on $X_2$, although it does depend on $X_1$. Now, if $\beta_{12} \neq 0$, and if $X_1$ is changed to $X_1 + \delta$, the change in $Y$ is

$$\Delta Y = \beta_1 \delta + \beta_{11}(2X_1 \delta + \delta^2) + \beta_{12} \delta X_2 \qquad (7.8)$$

and the effect on $Y$ due to altering $X_1$ will depend on the value of $X_2$ as well as the value of $X_1$. Thus if $\beta_{12} \neq 0$, an interaction effect between $X_1$ and $X_2$ is modeled.

**Response surfaces.** Experimental designs to estimate parameters of a polynomial, perhaps to find combinations of $X$'s to produce a maximum or minimum for $Y$, are called response surface designs. Discussions of them are given by Box and Wilson (1951), John (1971), Myers (1971), and Box, Hunter, and Hunter (1978).

*Example 7.1    An experiment with a lathe*

The data in Table 7.1 are the results of an experiment to characterize the performance of a cutting-tool material in cutting steel on a lathe. A completely randomized experiment in 20 runs was used, with two factors, cutting speed (in feet per minute) and feed rate (in thousandths of an inch per revolution). For convenience, the levels of the two factors are coded and centered to give predictors $S = (\text{speed} - 900)/300$ and $F = (\text{feed rate} - 13)/6$. The response was $Y = $ tool life (in minutes). Figure 7.1 is a scatter plot of $S$ versus $F$; the numbers on the plot correspond to the number of runs at each of the experimental settings.

Table 7.1    An experiment with a lathe

| Speed | Feed | Life | Speed | Feed | Life |
|-------|------|------|-------|------|------|
| $-1$ | $-1$ | 54.5 | $-\sqrt{2}$ | 0 | 20.1 |
| $-1$ | $-1$ | 66.0 | $\sqrt{2}$ | 0 | 2.9 |
| 1 | $-1$ | 11.8 | 0 | 0 | 3.8 |
| 1 | $-1$ | 14.0 | 0 | 0 | 2.2 |
| $-1$ | 1 | 5.2 | 0 | 0 | 3.2 |
| $-1$ | 1 | 3.0 | 0 | 0 | 4.0 |
| 1 | 1 | 0.8 | 0 | 0 | 2.8 |
| 1 | 1 | 0.5 | 0 | 0 | 3.2 |
| 0 | $-\sqrt{2}$ | 86.5 | 0 | 0 | 4.0 |
| 0 | $\sqrt{2}$ | 0.4 | 0 | 0 | 3.5 |

*Source*: M. R. Delozier.

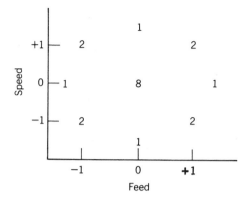

Figure 7.1  Scatter plot of $S$ versus $F$. Numbers in the plots give the number of runs at a given combination of speed and feed.

This layout of points is called a *central composite design* and is useful when the fitting of polynomials, and possibly interactions, is anticipated. Since several of the experimental conditions, especially the center point, have been replicated, pure error is available to estimate lack of fit of any model.

A glance at the data will verify that tool life is quite variable, less than 1 minute at very high speeds and up to an hour at low $(F, S)$ combinations. Additionally, the variability is much larger when tool life is larger. Both of these observations suggest a transformation of life, perhaps by taking logarithms. If the Box and Cox procedure is applied to the full second-order model given by

$$Y = \beta_0 + \beta_1 S + \beta_2 F + \beta_3 S^2 + \beta_4 F^2 + \beta_5 S \times F + e$$

the usefulness of taking logarithms of $Y$ is confirmed. We will use this model, with base 10 logarithms,

$$\log(Y) = \beta_0 + \beta_1 S + \beta_2 F + \beta_3 S^2 + \beta_4 F^2 + \beta_5 S \times F + e \qquad (7.9)$$

The summary of the regression for model (7.9) is given in Table 7.2. We see that all the coefficients have fairly large $t$-values, except for the $SF$ interaction. The $F$-test for lack of fit is $F = 4.09$ with (3, 11) d.f. for a $p$-value of about .04; if $SF$ is dropped from the model, we get $F = 3.26$ with (4, 11) d.f., and $p$-value = .05. Fitted and observed average log(life) in each of the nine experimental conditions are shown in Figure 7.2. We see that the averages and the fitted values seem to match very closely, except for the extreme combinations at $(\pm\sqrt{2}, 0)$. At $(+\sqrt{2}, 0)$, a single observation of 0.4 minutes was taken. The fitted value for this combination is $-0.210$, corresponding to tool life of only $10^{-0.210} = 0.6$

Table 7.2   Regression of log(life) on the full, second-order model

| Variable | Estimate | Standard Error | $t$-Value |
|---|---|---|---|
| Intercept | 0.5160 | 0.0456 | 11.31 |
| $S$ | $-0.6901$ | 0.0373 | $-18.52$ |
| $F$ | $-0.3432$ | 0.0373 | $-9.21$ |
| $S^2$ | 0.1251 | 0.0437 | 2.86 |
| $F^2$ | 0.1818 | 0.0437 | 4.16 |
| $S \times F$ | $-0.0316$ | 0.0456 | $-0.69$ |

$\hat{\sigma} = 0.1291$, d.f. $= 14$, $R^2 = .9706$

Analysis of Variance

| Source | d.f. | SS | MS | $F$ |
|---|---|---|---|---|
| Regression | 5 | 7.6003 | 1.5201 | |
| Lack of fit | 3 | 0.1228 | 0.0409 | 4.09 |
| Pure error | 11 | 0.1104 | 0.0100 | |

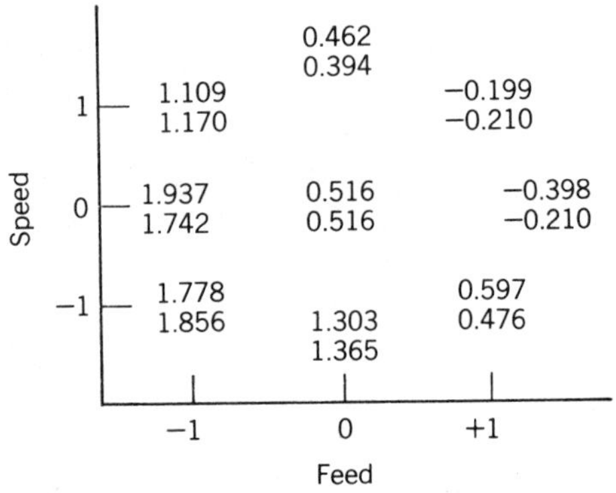

Figure 7.2   Average log(life), upper number, and fitted log(life), lower number, for the nine treatment combinations used in the experimental design.

minutes. If this one condition is removed from the data and the regression is refit, the $F$-test for lack of fit is $F = 2.59$, with $(2, 11)$ d.f., giving a $p$-value of about .12. If the other extreme point on the feed axis at $(-\sqrt{2}, 0)$ is also deleted before fitting the model, the $F$-test for lack of fit can be reduced even further to 1.25, with $(1, 11)$ degrees of freedom. Thus the lack of fit of the model may be due to the inability to account for lifetimes at relatively extreme feed rates. For the rest of the experiment, the fit may be considered adequate.

## 7.2 Dummy variables: Dichotomous

Dummy variables, also called indicator variables, are used to include categorical predictors in a regression analysis. Many of these variables have two categories, such as male or female, treatment or control, disease or no disease. Dummy variables usually assume only the two values 0 and 1 to indicate which category is the correct one for a given case; assignment of labels to the values is generally arbitrary.

*Example 7.2    Estimating the effect of cloud seeding*

Judging the success or failure of cloud seeding designed to increase rainfall is an important problem. Results from cloud seeding experiments have generally been mixed; sometimes the observed effect of seeding has been to increase rainfall, sometimes to decrease rainfall, and sometimes no effect is observed. Presumably, these mixed results are due at least in part to latent variation not directly controlled or measured in the experiments. Furthermore, the effects of seeding may be comparatively small, but small changes can be important.

The data given in Table 7.3 are taken from the Florida Area Cumulus Experiment (FACE) collected in 1975 (Woodley et al., 1977). A simplified version of the experimental protocol is as follows. A fixed target area of approximately 3000 square miles was established to the north and east of Coral Gables, Florida. During the summer of 1975, each day was judged on its suitability for seeding. The decision to use a particular day in the experiment was based primarily on a suitability criterion $S$, where $S$ is a computed number depending on a mathematical model for rainfall. Days with $S \geqslant 1.5$ were chosen as experimental days; there were 24 days chosen in 1975. On each day, the decision to seed was made by flipping a coin; as it turned out, 12 days were seeded, 12 unseeded. On seeded days, silver iodide was injected into the clouds from small aircraft. In all, for each experimental day, the following quantities were measured:

$A$ = action (1 = seed, 0 = do not seed).

$D$ = days after the first day of the experiment (June 16, 1975 = 0).

$S$ = suitability for seeding.

$C$ = percent cloud cover in the experimental area, measured using radar in Coral Gables, Florida.

$P$ = prewetness, amount of rainfall in the hour preceding seeding in $10^7$ cubic meters.

Table 7.3  Cloud seeding data

| Case | A | D | S | C | P | Log(P) | E | SA | CA | PA | Log(P)A | EA | Y | Log(Y) |
|---|---|---|---|---|---|---|---|---|---|---|---|---|---|---|
| 1 | 0 | 0 | 1.75 | 13.40 | 0.274 | −0.56225 | 2 | 0 | 0 | 0 | 0 | 0 | 12.85 | 1.10890 |
| 2 | 1 | 1 | 2.70 | 37.90 | 1.267 | 0.10278 | 1 | 2.70 | 37.90 | 1.267 | 0.10278 | 1 | 5.52 | 0.74194 |
| 3 | 1 | 3 | 4.10 | 3.90 | 0.198 | −0.70333 | 2 | 4.10 | 3.90 | 0.198 | −0.70333 | 2 | 6.29 | 0.79865 |
| 4 | 0 | 4 | 2.35 | 5.30 | 0.526 | −0.27901 | 1 | 0 | 0 | 0 | 0 | 0 | 6.11 | 0.78604 |
| 5 | 1 | 6 | 4.25 | 7.10 | 0.250 | −0.60206 | 1 | 4.25 | 7.10 | 0.250 | −0.60206 | 1 | 2.45 | 0.38917 |
| 6 | 0 | 9 | 1.60 | 6.90 | 0.018 | −1.74473 | 2 | 0 | 0 | 0 | 0 | 0 | 3.61 | 0.55751 |
| 7 | 0 | 18 | 1.30 | 4.60 | 0.307 | −0.51286 | 1 | 0 | 0 | 0 | 0 | 0 | 0.47 | −0.32790 |
| 8 | 0 | 25 | 3.35 | 4.90 | 0.194 | −0.71220 | 1 | 0 | 0 | 0 | 0 | 0 | 4.56 | 0.65896 |
| 9 | 0 | 27 | 2.85 | 12.10 | 0.751 | −0.12436 | 1 | 0 | 0 | 0 | 0 | 0 | 6.35 | 0.80277 |
| 10 | 1 | 28 | 2.20 | 5.20 | 0.084 | −1.07572 | 1 | 2.20 | 5.20 | 0.084 | −1.07572 | 1 | 5.06 | 0.70415 |
| 11 | 1 | 29 | 4.40 | 4.10 | 0.236 | −0.62709 | 1 | 4.40 | 4.10 | 0.236 | −0.62709 | 1 | 2.76 | 0.44091 |
| 12 | 1 | 32 | 3.10 | 2.80 | 0.214 | −0.66959 | 1 | 3.10 | 2.80 | 0.214 | −0.66959 | 1 | 4.05 | 0.60746 |
| 13 | 0 | 33 | 3.95 | 6.80 | 0.796 | −0.09909 | 1 | 0 | 0 | 0 | 0 | 0 | 5.74 | 0.75891 |
| 14 | 1 | 35 | 2.90 | 3.00 | 0.124 | −0.90658 | 1 | 2.90 | 3.00 | 0.124 | −0.90658 | 1 | 4.84 | 0.68485 |
| 15 | 1 | 38 | 2.05 | 7.00 | 0.144 | −0.84164 | 1 | 2.05 | 7.00 | 0.144 | −0.84164 | 1 | 11.86 | 1.07408 |
| 16 | 0 | 39 | 4.00 | 11.30 | 0.398 | −0.40012 | 1 | 0 | 0 | 0 | 0 | 0 | 4.45 | 0.64836 |
| 17 | 0 | 53 | 3.35 | 4.20 | 0.237 | −0.62525 | 2 | 0 | 0 | 0 | 0 | 0 | 3.66 | 0.56348 |
| 18 | 1 | 55 | 3.70 | 3.30 | 0.960 | −0.01773 | 1 | 3.70 | 3.30 | 0.960 | −0.01773 | 1 | 4.22 | 0.62531 |
| 19 | 0 | 56 | 3.80 | 2.20 | 0.230 | −0.63827 | 1 | 0 | 0 | 0 | 0 | 0 | 1.16 | 0.06446 |
| 20 | 1 | 59 | 3.40 | 6.50 | 0.142 | −0.84771 | 2 | 3.40 | 6.50 | 0.142 | −0.84771 | 2 | 5.45 | 0.73640 |
| 21 | 1 | 65 | 3.15 | 3.10 | 0.073 | −1.13668 | 1 | 3.15 | 3.10 | 0.073 | −1.13668 | 1 | 2.02 | 0.30535 |
| 22 | 0 | 68 | 3.15 | 2.60 | 0.136 | −0.86646 | 1 | 0 | 0 | 0 | 0 | 0 | 0.82 | −0.08619 |
| 23 | 1 | 82 | 4.01 | 8.30 | 0.123 | −0.91009 | 1 | 4.01 | 8.30 | 0.123 | −0.91009 | 1 | 1.09 | 0.03743 |
| 24 | 0 | 83 | 4.65 | 7.40 | 0.168 | −0.77469 | 1 | 0 | 0 | 0 | 0 | 0 | 0.28 | −0.55284 |

$E$ = echo motion category, either 1 or 2, a measure of the type of cloud.

$Y$ = rainfall following the action of seeding or not seeding in $10^7$ cubic meters.

The problem we shall study is estimating the effect on rainfall due to seeding. A secondary interest not considered here in any detail is modeling rainfall by the other measured variables. Randomization was used in this experiment because rainfall will be influenced by many other factors that have not been included in the study. One hopes that over the course of the experiment the effects of these other factors will balance, benefitting seeded and unseeded days equally. With this in mind, we are encouraged to consider a linear model for rainfall, even though the relationship between rainfall and the other predictors is probably much more complicated. As usual, we hope that a fitted linear model will provide a useful approximation to the true state of affairs.

Careful thought should be given to the possible nature of the effect due to the dummy variable $A$. This effect, if it exists, can manifest itself in at least two different ways. First, the effect on rainfall could be *additive*: For any set of combinations of other variables, the effect of seeding the clouds will be to increase or perhaps decrease rainfall by some fixed amount, without depending on any other variables. If this were so, then one would have a model of the form

$$Y = \beta_0 + \beta_1 A + (\text{a function of } D, S, C, E, \text{ and } P) + e$$

In the equation, the function of the other predictors is not specified exactly, since they may interact together or need to be transformed, polynomial terms may be required, or some of the predictors may be profitably dropped from the model. If this model were appropriate for seeding experiments, then the effect of seeding would be to change the expected rainfall by an amount $\beta_1$ cubic meters ($\times 10^7$), regardless of the other predictors.

The meaning of this model may be seen graphically in Figure 7.3 for a problem with a single additional predictor or covariate. In this graph, the $x$-axis is the value of the covariate and the $y$-axis is the value of the response. Separate regression lines are drawn for $A = 0$ and $A = 1$. Under the additive treatment effect model, the two lines are parallel.

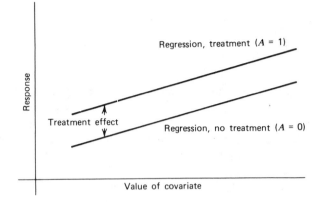

**Figure 7.3**   Additive treatment effect.

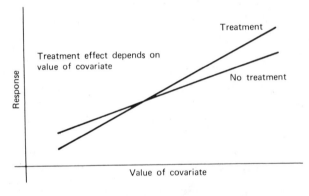

**Figure 7.4**   A nonadditive treatment effect.

A second more comprehensive class of models for seeding would allow the effect of seeding to be different for different combinations of the other predictors. For example, it may be true that if the seedability criterion $S$ is small, then seeding has a small effect, but if the seedability $S$ is large, then the effect of seeding is large. This is shown in Figure 7.4, with nonparallel regression lines for each treatment. When the two curves are parallel, the idea of a treatment effect is clearly defined as the distance between the two curves. In the nonparallel situation, the effect of the treatment depends on the values of the covariates, and may be positive for some values of the covariates and negative for others.

The model for nonparallel regressions is more complicated than the simple additive treatment effect model. We can write it down in two nearly equivalent ways. First, we could model a separate equation for seeded ($A = 1$) and unseeded ($A = 0$) days. Using this approach, one would also obtain two separate estimates of error variance, one from each group. If the error variance can be assumed to be the same in each group, an alternative and simpler approach is to write a single equation for all the data. This is done by defining a new set of predictors that are products of the indicator for seeding ($A$) and the other predictors ($S$, $C$, $E$, and $P$). These four new variables are $SA = S \times A$, $CA = C \times A$, $EA = E \times A$, and $PA = P \times A$; they are given in Table 7.3. For example, the values of $SA$ are those of $S$ if $A = 1$ and 0 if $A = 0$. The full model is given by

$$Y = \beta_0 + \beta_1 A + \beta_2 D + \beta_3 S + \beta_4 C + \beta_5 E + \beta_6 P + \beta_7 SA + \beta_8 CA$$
$$+ \beta_9 EA + \beta_{10} PA + e \qquad (7.10)$$

If the treatment effect does not depend on $S$, $C$, $E$, or $P$, then $\beta_7 = \beta_8 = \beta_9 = \beta_{10} = 0$, and we say that the treatment effect is additive. If there is no treatment effect at all, then $\beta_1 = \beta_7 = \beta_8 = \beta_9 = \beta_{10} = 0$. In an analysis where nonadditive treatment effects are suspected, (7.10) is a reasonable first model.

The next concern in analyzing these data is to find appropriate transformations for the variables. Since $P$ and $Y$ are volume measures, and both vary over more than a power of 10, transformation of these to perhaps logarithmic or cube root scale might be proposed. The use of $\log(Y)$ as a response has the added benefit of eliminating negative predicted rainfalls. This suggests use of the following model in place of (7.10):

$$\log(Y) = \beta_0 + \beta_1 A + \beta_2 D + \beta_3 S + \beta_4 C + \beta_5 E + \beta_6 (\log P)$$
$$+ \beta_7 SA + \beta_8 CA + \beta_9 EA + \beta_{10} (\log P)A + e \qquad (7.11)$$

In this model, the cloud seeding effect is measured in units of $\log(10^7$ cubic meters) rather than in cubic meters, which, for purposes of judging the existence of an effect, should be equally useful.

Index plots of $t_i$, $D_i$, and $h_{ii}$ are shown in Figure 7.5. Three cases are of interest: case 2 because of $D_2 = 1.51$, and cases 7 and 24, because $t_7 = -4.03$ and $t_{24} = -3.62$. From the large value of $D_2$, we know that deletion of case 2 may result in substantial changes in the fitted model; From Table 7.3 we see that on that day the cloud cover $C$ had a very large value of 37.9%; the authors of the original paper classify this as a

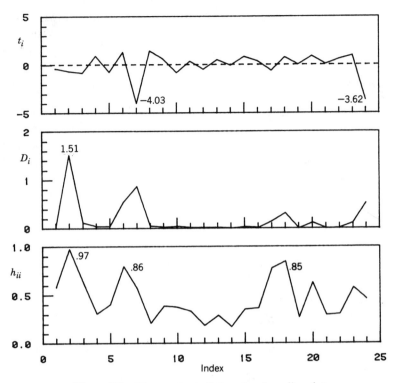

**Figure 7.5**    Diagnostic statistics, cloud seeding data.

"disturbed day." Case 2 is clearly different from the others in the data, and excluding it before analysis may be a reasonable procedure. However, this is not a general rule, and deletion of an influential case should only be done when there is reason to believe that the case is really different from the rest of those in the data. Both cases 7 and 24 have very large $t_i$, each with a $p$-value near 0.05 for the outlier test. Referring to the original data, we see that both these days are unseeded, and have very low observed rainfall. From their associated $h_{ii}$, neither case is particularly unusual, nor from their $D_i$ are they overly influential for estimating coefficients, at least for the model (7.11). Further information concerning cases 7 and 24 is available from the score test for non-constant variance. When all 24 cases are included in the data and model (7.11) is used, the test for nonconstant variance as a function of fitted values is $S = 5.38$, with 1 d.f., while if cases 2, 7, and 24 are deleted, we find $S = 0.66$. Because cases 7 and 24 are judged as possible outliers and contribute to an apparent problem with constant variance, we shall delete them before further analysis. A more thorough approach would require repeating the analysis with these points included; see Cook and Weisberg (1980, 1982a).

Before proceeding, we delete cases 2, 7, and 24, leaving $n = 21$ cases remaining. The residuals from the fit of (7.11) are plotted in Figure 7.6 and the regression summaries are given in Table 7.4. The plot and summaries indicate no clear violation of assumptions. The small $t$-values in Table 7.4 indicate that the fitted model may be improved by deleting some of the variables. Using methodology for variable selection that will be developed in the next chapter, we are led to the model summarized in Table 7.5. The fitted model is

$$\widehat{\log}(Y) = 0.492 + 1.294A - 0.007D + 0.022C + 0.399(\log(P))$$

$$+ 0.301E - 0.326SA \qquad (7.12)$$

The coefficient for $SA$ is nonzero and the effect of seeding depends on $S$. We are therefore in the situation of Figure 7.4. To explore this dependence, the change in $\widehat{\log}(Y)$ if an unseeded day were seeded is

$$\Delta[\widehat{\log}(Y)] = [(\widehat{\log}(Y)) \text{ if } A = 1] - [(\widehat{\log}(Y)) \text{ if } A = 0]$$

$$= \hat{\beta}_1 + \hat{\beta}_7(SA) = 1.294 - 0.326S \qquad (7.13)$$

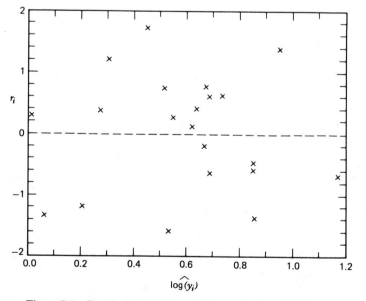

**Figure 7.6**   Residual plot, full model, cases 2, 7, and 24 deleted.

**Table 7.4   Regression summary, model (7.11), three cases deleted**

| Variable | Estimate | Standard Error | $t$-Value |
|----------|----------|----------------|-----------|
| Intercept | .4171 | .3995 | 1.04 |
| $A$ | 1.4263 | .5100 | 2.80 |
| $D$ | $-.0063$ | .0017 | $-3.72$ |
| $S$ | .0060 | .0085 | .07 |
| $C$ | .0301 | .0150 | 2.00 |
| $\log(P)$ | .3414 | .1458 | 2.34 |
| $E$ | .2652 | .1353 | 1.96 |
| $SA$ | $-.3334$ | .1073 | $-3.11$ |
| $CA$ | $-.0229$ | .0282 | $-.81$ |
| $\log(P)A$ | .0730 | .2236 | .33 |
| $EA$ | .0500 | .1783 | .28 |

$\hat{\sigma}^2 = 0.0194$, d.f. $= 10$, $R^2 = 0.90$

**Table 7.5   Final model for $\log(y)$, three cases deleted**

| Variable | Estimate | Standard Error | $t$-Value |
|----------|----------|----------------|-----------|
| Intercept | 0.4925 | 0.1291 | 3.81 |
| $A$ | 1.2944 | 0.1901 | 6.81 |
| $D$ | $-0.0066$ | 0.0013 | $-5.26$ |
| $C$ | 0.0219 | 0.0097 | 2.26 |
| $\log(P)$ | 0.3990 | 0.0830 | 4.80 |
| $E$ | 0.3010 | 0.0735 | 4.09 |
| $SA$ | $-0.3263$ | 0.0520 | $-6.28$ |

$\hat{\sigma}^2 = .0149$, d.f. $= 14$, $R^2 = .89$

Figure 7.7 is a graph of the change in fitted log(rainfall) as a function of the seedability criterion $S$. We have the curious result that as seeding suitability increases, the added rainfall appears to decrease and becomes negative if $S > 4$. The vertical distance between the two curved lines in Figure 7.7 gives the 95% confidence interval for $\Delta(\widehat{\log(Y)})$. To compute this, first find the variance of $\Delta(\widehat{\log(Y)})$,

$$\text{var}(\Delta(\widehat{\log(Y)})) = \text{var}(\hat{\beta}_1) + S^2\text{var}(\hat{\beta}_7) + 2S\,\text{cov}(\hat{\beta}_1, \hat{\beta}_7) \qquad (7.14)$$

The methodology for estimating each of the terms in (7.14) is discussed in Chapter 2. The 95% confidence bounds are then found in analogy to (1.38).

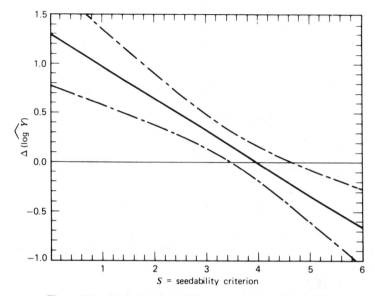

**Figure 7.7**   Added log(rainfall) as a function of seedability.

Kerr (1982) reports the result of a followup experiment called FACE-2, carried out in 1978. This latter experiment was a strictly randomized, double blind experiment in which none of the participants knew which days were seeded and which were not until well after the experiment had ended. When the analysis was finally done, conclusions concerning the success or failure of seeding in FACE-2 depended on one nonseeded day that had very heavy rainfall. If that day were excluded, seeding would be judged effective; if included, then the apparent effect of seeding disappears.

## 7.3   Dummy variables: Polytomous

If a categorical predictor has more than two categories, then several dummy variables may be required. For example, suppose the cloud seeding experiment consisted of three experimental conditions: unseeded, seeded with silver iodide, and seeded with dry ice. We could contemplate a single predictor variable with value 0 for unseeded, 1 for silver iodide, and 2 for dry ice, but this suggests that these three treatments are ordered and that the effect of changing from control to silver iodide is the same as changing from silver iodide to dry ice. In most problems, neither the assumptions of ordered categories nor of equal spacing are justifiable. Rather, we must define two dummy variables,

$$A_1 = 1 \text{ if unseeded,} \quad 0 \text{ otherwise}$$

$$A_2 = 1 \text{ if silver iodide,} \quad 0 \text{ otherwise}$$

Since a day in which dry ice was used is uniquely given by $A_1 = A_2 = 0$, all three treatment conditions can be specified by just these two dummy variables; the third variable,

$$A_3 = 1 \text{ if dry ice,} \quad 0 \text{ otherwise}$$

is redundant given the first two, as long as an intercept is in the model. If we consider only one other variable, say $S$, the additive, parallel regression model is

$$y = \beta_0 + \beta_1 A_1 + \beta_2 A_2 + \beta_s S + e \qquad (7.15)$$

Since the slope $\beta_s$ is assumed to be the same for all groups, the regressions are parallel for the three groups. The intercepts for the three groups are $\beta_0$ for dry ice, $\beta_0 + \beta_1$ for unseeded, and $\beta_0 + \beta_2$ for silver iodide. The $F$-test of $\beta_1 = \beta_2 = 0$ ($\beta_0, \beta_s$ arbitrary) tests for equality of intercepts for the three treatment conditions. Any two of $A_1$, $A_2$, and $A_3$ could have been used in the model, changing only the meaning of the resulting coefficients, but leading to the same $F$-test. As an alternative, we could consider leaving out the intercept, and fit the model

reg Y on $\times$ $A_1 A_2 A_3$ $\qquad y = \beta_1 A_1 + \beta_2 A_2 + \beta_3 A_3 + \beta_s S + e \qquad (7.16)$

The advantage of this latter parameterization is that each of the $\hat{\beta}_j$ from the fit of (7.16) is the estimated intercept for that group. Since standard errors of the $\hat{\beta}_j$ will be produced directly by most computer programs, use of formula (2.26) to find them can be avoided.

Both models (7.15) and (7.16) correspond to a parallel regression, or additive treatment effect, model. Both can be generalized by addition of $S \times A_j$ terms to the model to allow separate slopes for the covariate in each group. In problems with many dummy variables, addition of interactions of this type to allow for nonparallel regressions can quickly get out of hand because group sizes may be small. The analyst is forced to assume the parallel regression model for most categorical variables. The resulting regression analysis will only be as reliable as the assumption of parallel regressions.

# ⍟ 7.4   Comparing regression lines

For simplicity, consider simple regression with $m$ groups of cases, such that, in the $k$th group with $n_k$ cases, the correct model is

$$Y = \beta_{0k} + \beta_{1k}X + \text{error} \qquad k = 1, 2, \ldots, m$$

Comparison of these $m$ regression lines, based on the observed data, is often of interest. We distinguish four different situations:

**Model 1   Most general.**   If all the parameters are different, we have a circumstance like that in Figure 7.8*a*.

**Model 2   Parallel regressions.**   In this model, $\beta_{11} = \beta_{12} = \cdots = \beta_{1m}$, but the intercepts are arbitrary. These are the circumstances that lead to the additive model for dummy variables, as in Section 7.3 and shown in Figure 7.8*b*.

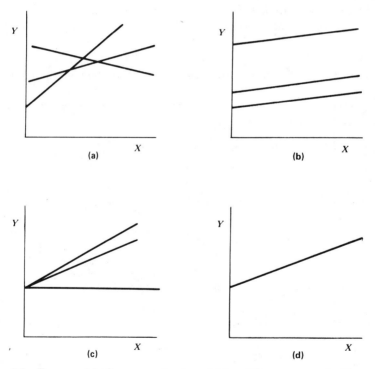

**Figure 7.8**   Four models for comparing regressions: (*a*) most general; (*b*) parallel; (*c*) concurrent; (*d*) coincident.

**Model 3   Concurrent regressions.**   In this model the intercepts are all equal, $\beta_{01} = \cdots = \beta_{0m}$, but slopes are arbitrary, as illustrated in Figure 7.8$c$.

**Model 4   Coincident regression lines.**   Here, all lines are the same, $\beta_{01} = \cdots = \beta_{0m}$ and $\beta_{11} = \cdots = \beta_{1m}$. This is the most stringent model, as illustrated in Figure 7.8$d$.

It is usually of interest to test the plausibility of models 4 or 2 against a different, less stringent model as an alternative. The form of these tests is immediate from the formulation of the general $F$-test given in Section 4.4. The methodology can be best illustrated by an example.

*Example 7.3   Twin data*

---

The data in Table 7.6 give the IQ scores of identical twins, one raised in a foster home ($Y$), and the other raised by natural parents ($X$). The data were originally used by Burt (1966). For the purposes of an example, we can divide cases into three groups according to the social class of the natural parents.

For the example, $m = 3$, $n_1 = 7$, $n_2 = 6$, and $n_3 = 14$, giving $n = \sum n_k = 27$. The data are graphed in Figure 7.9, suggesting that the regression lines

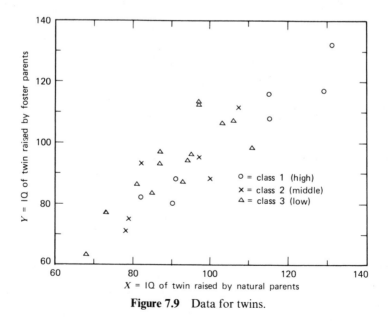

**Figure 7.9**   Data for twins.

Table 7.6 Twin data for comparing regression lines

| Case Number | Y | X | $G_1$ | $G_2$ | $G_3$ | $Z_1$ | $Z_2$ | $Z_3$ |
|---|---|---|---|---|---|---|---|---|
| 1 | 82 | 82 | 1 | 0 | 0 | 82 | 0 | 0 |
| 2 | 80 | 90 | 1 | 0 | 0 | 90 | 0 | 0 |
| 3 | 88 | 91 | 1 | 0 | 0 | 91 | 0 | 0 |
| 4 | 108 | 115 | 1 | 0 | 0 | 115 | 0 | 0 |
| 5 | 116 | 115 | 1 | 0 | 0 | 115 | 0 | 0 |
| 6 | 117 | 129 | 1 | 0 | 0 | 129 | 0 | 0 |
| 7 | 132 | 131 | 1 | 0 | 0 | 131 | 0 | 0 |
| 8 | 71 | 78 | 0 | 1 | 0 | 0 | 78 | 0 |
| 9 | 75 | 79 | 0 | 1 | 0 | 0 | 79 | 0 |
| 10 | 93 | 82 | 0 | 1 | 0 | 0 | 82 | 0 |
| 11 | 95 | 97 | 0 | 1 | 0 | 0 | 97 | 0 |
| 12 | 88 | 100 | 0 | 1 | 0 | 0 | 100 | 0 |
| 13 | 111 | 107 | 0 | 1 | 0 | 0 | 107 | 0 |
| 14 | 63 | 68 | 0 | 0 | 1 | 0 | 0 | 68 |
| 15 | 77 | 73 | 0 | 0 | 1 | 0 | 0 | 73 |
| 16 | 86 | 81 | 0 | 0 | 1 | 0 | 0 | 81 |
| 17 | 83 | 85 | 0 | 0 | 1 | 0 | 0 | 85 |
| 18 | 93 | 87 | 0 | 0 | 1 | 0 | 0 | 87 |
| 19 | 97 | 87 | 0 | 0 | 1 | 0 | 0 | 87 |
| 20 | 87 | 93 | 0 | 0 | 1 | 0 | 0 | 93 |
| 21 | 94 | 94 | 0 | 0 | 1 | 0 | 0 | 94 |
| 22 | 96 | 95 | 0 | 0 | 1 | 0 | 0 | 95 |
| 23 | 112 | 97 | 0 | 0 | 1 | 0 | 0 | 97 |
| 24 | 113 | 97 | 0 | 0 | 1 | 0 | 0 | 97 |
| 25 | 106 | 103 | 0 | 0 | 1 | 0 | 0 | 103 |
| 26 | 107 | 106 | 0 | 0 | 1 | 0 | 0 | 106 |
| 27 | 98 | 111 | 0 | 0 | 1 | 0 | 0 | 111 |

may well be the same. For the tests, we treat the IQ of the twin raised at home as a standard to which the twin from the foster home is to be compared, to study the effect of being raised apart. In addition to comparing the different lines for different social classes, one might also be interested in the actual values of the estimated slopes and whether they differ from 1. This point will not be pursued here.

In addition to $X$ and $Y$, six other variables are given, namely, three dummy variables $G_1$, $G_2$, and $G_3$ to indicate social class ($G_1 = 1$ for highest class, $G_2 = 1$ for middle, $G_3 = 1$ for lowest), and three additional variables $Z_1 = G_1 X$, $Z_2 = G_2 X$, and $Z_3 = G_3 X$. These variables will aid in fitting the four models suggested for comparing regression lines.

**Model 1.**   To fit Model 1, a separate regression calculation can be done for each group to obtain parameter estimates. The $RSS$, say $RSS_1$, is then obtained by adding the separate $RSS$ from each regression. This $RSS$ has $\mathrm{df}_1 = \sum (n_k - p')$ degrees of freedom (for simple regression, $p' = 2$). Alternatively and equivalently, one could fit the model

$$Y = \beta_{01}G_1 + \beta_{02}G_2 + \beta_{03}G_3 + \beta_{11}Z_1 + \beta_{12}Z_2 + \beta_{13}Z_3 + e$$

with the $G$'s and $Z$'s as given in Table 7.6 (no overall intercept is in this model). The results of this fit are given in Table 7.7.

**Model 2.**   To fit parallel regressions, use the model

$$Y = \beta_{01}G_1 + \beta_{02}G_2 + \beta_{03}G_3 + \beta_1 X + \text{error}$$

to get estimates of each intercept and of the common slope $\beta_1$. The $RSS$ from this fit, say $RSS_2$, has $\mathrm{df}_2 = n - m - p$ degrees of freedom (for a simple regression $p = 1$). This fit is summarized in Table 7.7.

**Model 3.**   This model requires fitting

$$Y = \beta_0 + \beta_{11}Z_1 + \beta_{12}Z_2 + \beta_{13}Z_3 + \text{error}$$

The $RSS$ for this model, $RSS_3$, has $\mathrm{df}_3 = n - mp - 1$ degrees of freedom, as shown in Table 7.7.

**Table 7.7   Computations for twin data**

| Variable | Estimates ($t$-Values) | | | |
| --- | --- | --- | --- | --- |
| | Model 1 | Model 2 | Model 3 | Model 4 |
| Intercept | — | — | 2.56 (0.24) | 9.21 (0.99) |
| $X$ | — | 0.97 (9.03) | | 0.90 (9.36) |
| $G_1$ | $-1.87$ ($-0.11$) | $-0.61$ ($-0.05$) | | |
| $G_2$ | 0.82 (0.03) | 1.43 (0.14) | | |
| $G_3$ | 7.20 (0.43) | 5.62 (0.56) | | |
| $Z_1$ | 0.98 (5.99) | | 0.94 (9.44) | |
| $Z_2$ | 0.97 (3.40) | | 0.95 (7.93) | |
| $Z_3$ | 0.95 (5.21) | | 1.00 (8.57) | |
| $RSS$ | 1317 | 1318 | 1326 | 1494 |
| d.f. | 21 | 23 | 23 | 25 |

✓ **Model 4.** This model assumes common regression lines, so the estimates are computed by pooling the data to fit

$$Y = \beta_0 + \beta_1 X + e$$

The RSS for this model, $RSS_4$, has $df_4 = n - p'$ degrees of freedom as shown in Table 7.7.

Most tests concerning the slopes and intercepts of different regression lines will use the general model (model 1) as the alternative model. The usual $F$-test, for testing models 2, 3, and 4, is then given, for $l = 2, 3, 4$ by

$$F_l = \frac{(RSS_l - RSS_1)/(df_l - df_1)}{RSS_1/df_1}$$

$$l = 2, 3, 4 \text{ with } (df_l - df_1, df_1) \text{ d.f.} \qquad (7.17)$$

⚠ If the hypothesis provides as good a model as does the alternative, then $F$ will be small. If the model is not adequate when compared to the general model, then $F$ will be large (when compared to the percentage points of the $F(df_l - df_1, df_1)$ distribution).

For the twin data, the $F$-statistics are

*Handwritten: $H_0: \beta_{11} = \beta_{12} = \beta_{13} = 0$ common slope arbitrary intercepts*

$$F_2 = \frac{(1318 - 1317)/2}{1317/21} = 0.01$$

$$F_3 = \frac{(1326 - 1317)/2}{1317/21} = 0.08$$

*Handwritten: $H_A:$ at least of $\beta_{1i}$ is not equal to other*

$$F_4 = \frac{(1494 - 1317)/4}{1317/21} = 0.71$$

Since all these $F$ values are much less than 1, comparison to percentage points of the corresponding $F$ distribution is unnecessary. The most restrictive model, model 4, is as good as the least restrictive model, model 1. The tiny value of $F_2 = 0.01$ may serve as a flag to the careful analyst, for, under the model assumed, observing an $F$-statistic this small is extremely unlikely (if all assumptions are true, the probability of $F_2$ being less than its observed value in this problem is less than 0.005). The observed data agree with the theory much more closely than chance theory would lead us to expect.

*coincident vs parallel*

**Analysis of covariance.**    Testing model 4 against model 2 is usually called the analysis of covariance. The predictor is called a covariate and is expected to have the same effect in all groups. The difference between groups is assumed to be an additive treatment effect. These two assumptions combine to give the assumption that the correct alternative hypothesis is equal slopes in groups, but possibly different intercepts (the difference between intercepts is the "treatment effect").

**Additional comments.**    Although the example given in this section uses simple regression, the identical methodology applies for comparing groups with many predictors. Computer programs are often designed to make comparison of groups easier. These programs seem to use three methods. The most primitive method allows the user to define weights in weighted least squares with values 0 and 1; cases given a weight of zero are not used in the computations. This, in combination with the ability to create dummy variables and interactions, allows all the computations needed for each of models 1 to 4. An older approach, used for example in the program BMDP1V (Dixon, 1983), computes some of the tests for comparing groups automatically, but without much user control. The most elegant method for comparing groups is that taken in the program GLIM, distributed by the Royal Statistical Society. This program includes a comprehensive language for specifying factors, essentially categorical variables, and covariates, essentially continuous predictors, that makes the fitting of models 1 to 4 especially easy; see McCullagh and Nelder (1983, Chapter 3) for details.

Probably the most common problem in comparing groups is testing for parallel slopes in simple regression with two groups. Since this $F$-test has 1 d.f. in the numerator, it is equivalent to a $t$-test. Let $\hat{\beta}_j$, $\hat{\sigma}_j^2$, $n_j$, and $SXX_j$ be, respectively, the estimated slope, residual mean square, sample size, and corrected sum of squares of the $X$'s in group $j$, $j = 1, 2$. Then a pooled estimate of $\sigma^2$ is

$$\hat{\sigma}^2 = \left( \frac{(n_1 - 2)\hat{\sigma}_1^2 + (n_2 - 2)\hat{\sigma}_2^2}{n_1 + n_2 - 4} \right) \tag{7.18}$$

and the $t$-test for equality of slopes is

$$t = \frac{\hat{\beta}_1 - \hat{\beta}_2}{\hat{\sigma}(1/SXX_1 + 1/SXX_2)^{1/2}} \tag{7.19}$$

with $n_1 + n_2 - 4$ d.f. The square of this $t$-statistic is numerically identical to the corresponding $F$ statistic.

The model for concurrent regressions, model 3 of Section 7.4, can be easily extended to the case where the regression lines are assumed concurrent at any fixed point, $X = c$. Suppose, in the twin data of Example 7.2, we wished to test for concurrence at $c = 100$, the average IQ. The concurrent regression model specifies that the lines agree at this one point but may differ elsewhere. This model may be fit by (1) replacing $X$ by $X - 100$; (2) defining the $G$'s and $Z$'s as in the example, but using the redefined $X$; and (3) fitting model 3 exactly as outlined in the text. Models 1, 2, and 4 will give the identical $RSS$ when using either the original $X$ or the redefined $X$. Another generalization of this method is to allow the regression lines to be concurrent at some arbitrary and unknown point, so the point of concurrency must be estimated from the data. This problem is more difficult, and is discussed by Saw (1966).

## 7.5   Scaling of variables

In models with an intercept, least squares regression has a very pleasant property called location and scale invariance: if any of the variables in the data is scaled by addition of a constant or by multiplication by a constant, then in the resulting regression, estimates will be changed in a predictable way and scale-free quantities, such as $R^2$ and $F$ and $t$ tests will be unaffected. In this section, the effects of scaling variables are summarized. Let $\hat{\beta}_0$, $\hat{\beta}_1, \ldots, \hat{\beta}_p$ and $\hat{\sigma}^2$ be the least squares estimates obtained before scaling. Only the effects of scaling on these and on $R^2$ and test statistics are given. The effects on estimated standard errors are easily derived from the results given here.

**Scaling.**   If a variable $X_j$ is replaced by $(X_j - c_j)/d_j$, then, in the resulting regression $\hat{\beta}_j$ becomes $d_j\hat{\beta}_j$, $\hat{\beta}_0$ becomes $\hat{\beta}_0 + (c_j/d_j)\hat{\beta}_j$, and $\hat{\sigma}^2$, $F$-tests, $t$-tests, and $R^2$ are unaffected.

**Scaling the response.**   If $Y$ is replaced by $(Y - f)/g$, then $\hat{\beta}_0$ is replaced by $(\hat{\beta}_0 - f)/g$, each $\hat{\beta}_j$ $(j = 1, \ldots, p)$ gets replaced by $\hat{\beta}_j/g$, and all sums of squares in the analysis of variance and $\hat{\sigma}^2$ are divided by $g^2$. However, $F$- and $t$-tests are unaffected.

A common use of scaling is to standardize the variables by subtracting sample averages and dividing by sample standard deviations. Thus for $j = 1, \ldots, p$, replace $X_j$ by

$$\frac{X_j - \bar{x}_j}{SD_j}$$

and replace $Y$ by

$$\frac{Y - \bar{y}}{SD_Y}$$

In the resulting scaled data, the estimate of the intercept is exactly zero, and the estimate of $\hat{\beta}_j$ becomes

$$\hat{\beta}_j(\text{standardized}) = \frac{SD_j}{SD_Y} \hat{\beta}_j \qquad j = 1, 2, \ldots, p \qquad (7.20)$$

Some investigators compare standardized coefficient estimates for different predictors. Under this logic, predictors with larger standardized coefficients are more important. Unfortunately, this reasoning is faulty because the scaling depends on the range of values for the variables in the data. For example, if two analysts collect data on the same predictors, one collecting data over a small range and the other over a larger range, they may come to completely different conclusions about the relative magnitudes of the standardized coefficients. Also, too much faith is placed in estimated coefficients when the model used is only an approximation to a more complicated relationship.

## 7.6    Linear transformations and principal components

The invariance of linear regression under location/scale changes is a special case of invariance of regression problems under linear transformations of the predictors. In a linear transformation, the $p$ predictors in a model are replaced by up to $p$ (linearly independent) linear combinations of them. We have already seen an example of this in the discussion of the Berkeley Guidance Study in Example 3.1. In that example, the three original variables WT2, WT9, and WT18 were replaced by WT2, WT9 − WT2, and WT18 − WT9, and regression on these three seems to allow a better interpretation of the information available than do the original data.

Another example of linear transformations we have encountered is in the $QR$ factorization in Appendix 2A.3, where a matrix $\mathbf{Q}$ with orthonormal columns that are linear combinations of the columns of $\mathbf{X}$ is found. Since $\mathbf{Q}$ has orthonormal columns, least squares computations are very simple, and translation from results using $\mathbf{Q}$ to the original variables can be done using numerically stable methods.

In general, a (nonsingular) linear transformation of an $n \times p'$ matrix $\mathbf{X}$ is specified by finding a $p' \times p'$ matrix $\mathbf{U}$ (of rank $p'$) so that the transformed variables $\mathbf{Z}$ are given by

$$\mathbf{Z} = \mathbf{XU} \qquad (7.21)$$

In the preceding guidance study example, $\mathbf{U}$ is $4 \times 4$ and is given by

$$U = \begin{bmatrix} 1 & 0 & 0 & 0 \\ 0 & 1 & -1 & 0 \\ 0 & 0 & 1 & -1 \\ 0 & 0 & 0 & 1 \end{bmatrix} \tag{7.22}$$

By direct multiplication of $Z = XU$, we see that the first column of $Z$ is the first column of $X$ (i.e., the column of ones), the second column of $Z$ is WT2, the third column is WT9 − WT2, and the last column is WT18 − WT9.

If the linear model is $Y = X\boldsymbol{\beta} + e$, and if $U$ has an inverse, define $\boldsymbol{\alpha} = U^{-1}\boldsymbol{\beta}$. Then

$$Y = X\boldsymbol{\beta} + e$$

$$= X(UU^{-1})\boldsymbol{\beta} + e$$

$$= Z\boldsymbol{\alpha} + e \tag{7.23}$$

The least squares estimator of $\boldsymbol{\alpha}$ is $\hat{\boldsymbol{\alpha}} = (Z^T Z)^{-1} Z^T Y$, and the least squares estimator of $\boldsymbol{\beta}$ is $\hat{\boldsymbol{\beta}} = U\hat{\boldsymbol{\alpha}}$.

**Principal components.** By proper choice of $U$, we can obtain a transformed data matrix $Z$ with desirable properties. The most important example of this is to find a $U$ such that $Z^T Z = U^T(X^T X)U = D$, where $D$ is a diagonal matrix with positive diagonal elements given by $\lambda_1, \lambda_2, \ldots, \lambda_p$; we shall assume that the $\lambda$'s are ordered, with $\lambda_1 \geqslant \lambda_2 \geqslant \cdots \geqslant \lambda_p$. $\lambda_p$ will be zero only if $X^T X$ is singular. One can show the following facts:

1. The matrix $U$ is an orthogonal matrix, $U^T U = UU^T = I$, and if all the $\lambda$'s are different, $U$ is unique.

2. The columns of $U$ are the *eigenvectors* of the matrix $X^T X$; the $\lambda$'s are called *eigenvalues*. The columns of $Z = XU$ are called *principal components*.

3. The eigenvalues and eigenvectors may be different depending on the scaling of the data. Thus, computations based on $X^T X$, $\mathscr{X}^T \mathscr{X}$, and the sample correlation matrix may all lead to different eigenvalues and eigenvectors. For most other regression calculations, results do not change in meaningful ways when the data are scaled to give different versions of the cross product matrix. Generally, eigenvalues and eigenvectors of either $\mathscr{X}^T \mathscr{X}$ or more frequently of the sample correlation matrix are used.

Computational methods for finding eigenvalues and eigenvectors are discussed by Stewart (1974) and Seber (1977). Fortran subroutines are available both in the IMSL and Eispack libraries; the manual for the latter, Smith et al. (1976), provides listings of computer code. Many statistical packages, such as Minitab (Ryan, Joiner and Ryan, 1985), will compute eigenvalues, eigenvectors and principal components, and allow them to be used in other calculations.

### Example 7.4    Berkeley Guidance Study

For the boys in the Berkeley Guidance Study, using only variables WT2, WT9, and WT18, the matrix of eigenvectors $\mathbf{U}$ and the eigenvalues of $\mathscr{X}^T\mathscr{X}$ are

$$\mathbf{U} = \begin{pmatrix} 0.0354 & -0.3551 & 0.9341 \\ 0.2754 & -0.8951 & -0.3507 \\ 0.9607 & 0.2697 & 0.0662 \end{pmatrix}$$

$$(\lambda_1 \ \lambda_2 \ \lambda_3) = (3604. \ 246.2 \ 34.30)$$

The three principal components $Z_1$, $Z_2$, and $Z_3$ are defined by

$$Z_1 = 0.0354 \ \text{WT2} + 0.2754 \ \text{WT9} + 0.9607 \ \text{WT18}$$

$$Z_2 = -.3551 \ \text{WT2} - .8951 \ \text{WT9} + 0.2697 \ \text{WT18}$$

$$Z_3 = 0.9341 \ \text{WT2} - 0.3507 \ \text{WT9} + 0.0662 \ \text{WT18}$$

One can show that the principal components are defined so that, if the model

$$Y = \alpha_0 + \alpha_1 Z_1 + \alpha_2 Z_2 + \alpha_3 Z_3 + e$$

were fit with $\text{var}(e) = \sigma^2$, the estimate $\hat{\alpha}_1$ would have smaller variance than would the coefficient estimate for any other possible linear combination (of norm 1) of the $X$'s and the variance of $\hat{\alpha}_1$ will be $\sigma^2/\lambda_1$. Similarly, among all linear combinations of the $X$'s orthogonal to $Z_1$, the one specified by $Z_2$ has an estimate $\hat{\alpha}_2$ with variance $\sigma^2/\lambda_2$, smaller than for any other linear combination. This is repeated for all other principal components. In this example, $\text{var}(\hat{\alpha}_1) = \sigma^2/3604 = 2.77 \times 10^{-4}\sigma^2$, while $\text{var}(\hat{\alpha}_3) = \sigma^2/34.30 = 2.9 \times 10^{-2}\sigma^2$, and much more information is available in the data concerning $\alpha_1$ than concerning $\alpha_3$.

If rather than study the principal components based on $\mathscr{X}^T\mathscr{X}$, we compute them from the sample correlation matrix, completely different results are possible. For the example, the matrix of the transformation $\mathbf{U}^*$ and the eigenvalues $\lambda_1^*$, $\lambda_2^*$, $\lambda_3^*$ are

$$\mathbf{U}^* = \begin{pmatrix} 0.4925 & -0.7809 & -0.3843 \\ 0.6648 & 0.0525 & 0.7452 \\ 0.5618 & 0.6224 & -0.5450 \end{pmatrix}$$

$$(\lambda_1^* \; \lambda_2^* \; \lambda_3^*) = (2.028 \;\; 0.7890 \;\; 0.1829)$$

Approximately, then, by rounding of coefficients, the three principal components are

$$Z_1 = 0.4925 \; \text{WT2} + 0.6648 \; \text{WT9} + 0.5618 \; \text{WT18}$$

$$\cong \tfrac{1}{2}(\text{WT2} + \text{WT9} + \text{WT18})$$

$$= \text{measure of average weight}$$

$$Z_2 = -0.7809 \; \text{WT2} + 0.0525 \; \text{WT9} + 0.6224 \; \text{WT18}$$

$$\cong 0.7(-\text{WT2} + \text{WT18})$$

$$= \text{linear weight gain from age 2 to age 18}$$

$$Z_3 = -0.3843 \; \text{WT2} + 0.7452 \; \text{WT9} - 0.5450 \; \text{WT18}$$

$$\cong 0.4(-\text{WT2} + 2 \; \text{WT9} - \text{WT18})$$

$$= \text{quadratic weight gain}$$

Thus in the correlation form, the principal components have very simple explanations, while simple explanations are not available for the computations based on $\mathscr{X}^T\mathscr{X}$. Those computations suggest that the first principal component—the combination of $X$'s for which a coefficient is most precisely estimated—is approximately equal to $0.3[\text{WT9} + 3(\text{WT18})]$. The importance of WT18 in these results is due to the fact that the sample variance of WT18 is largest among the variances of WT2, WT9, and WT18, and larger variances give WT18 more weight in the principal component. In the correlation form, differences between variances have been removed, and all three variables are used in a seemingly useful way.

---

**Principal components of many variables.**   Principal components can be found for any set of variables, such as the predictors of SOMA in the Berkeley Guidance Study. However, the resulting variables will be linear combinations of dissimilar variables—heights, weights, and strengths—and the new variables may have no meaning. An alternative approach that would apply to the guidance study would be to find the principal components for the height variables separately from the others, the weights separately from the others, and so on, using these transformed predictors in further analysis.

## Problems

**7.1**  As an alternative to fitting the model

$$Y = \beta_0 + \beta_1 X_1 + \beta_2 X_2 + \beta_{12} X_1 X_2 + \text{error}$$

consider fitting the model

$$Y = \beta_0 + \beta_1 X_1 + \beta_2 X_2 + \beta_3 (X_1 - X_2) + \text{error}$$

Discuss the differences between these two models. Under what circumstances is the second one appropriate? When is the first one better? (*Hint:* What happens to the response in each model if $X_1$ is changed to $X_1 + \delta_1$ and/or $X_2$ is changed to $X_2 + \delta_2$?)

**7.2**  Compare the regression lines for Forbes' data (Example 1.1) and Hooker's data (problem 1.2).

**7.3**  In the Berkeley Guidance Study data, problem 2.1, consider modeling HT18 by the data collected at ages 2 and 9. Fit the same model for boys and girls and perform appropriate tests to compare the fitted regression planes.

**7.4**  For the girls in the Berkeley Guidance Study data, find the principal components for the three weight variables, paralleling the computations for boys given in the text. Comment on the meaning, if any, of the principal components.

**7.5**  For the physics data (problem 4.2), formally compare the regression lines suggested by problem 4.2.1.

**7.6**  For the apple shoot data, Example 4.2, perform a complete data analysis.

**7.7**  The data in Table 7.8 give the results of another experiment concerning tool life in cutting steel on a lathe. The experiment has three factors: speed (in feet per minute), feed rate (in thousandths of an inch), and the radius of the drill bit (also in thousandths of an inch). The response is tool life (in minutes). This experiment was also completely randomized, but it has been laid out as a complete factorial design with three levels of radius and two each of feed and speed. Each (feed, speed, radius) combination was repeated three times. Using any of the methods learned so far, find a model that adequately gives an estimate of tool life as a function of the three factors in the experiment.

Table 7.8   A second experiment with a lathe

| Speed | Feed | Radius | Tool Life (3 replicates) | | |
|-------|------|--------|------|------|------|
| 750 | 5 | 1 | 100.7 | 60.0 | 75.9 |
| 750 | 5 | 4 | 25.0 | 17.5 | 20.5 |
| 750 | 5 | 7 | 14.9 | 18.0 | 17.0 |
| 750 | 10 | 1 | 39.5 | 42.0 | 47.0 |
| 750 | 10 | 4 | 15.1 | 16.6 | 19.3 |
| 750 | 10 | 7 | 11.0 | 14.0 | 14.5 |
| 950 | 5 | 1 | 35.3 | 17.7 | 29.0 |
| 950 | 5 | 4 | 17.0 | 12.5 | 13.8 |
| 950 | 5 | 7 | 7.0 | 9.0 | 11.5 |
| 950 | 10 | 1 | 18.3 | 17.0 | 13.1 |
| 950 | 10 | 4 | 10.0 | 8.0 | 9.0 |
| 950 | 10 | 7 | 9.5 | 8.0 | 8.6 |

*Source:* M. R. Delozier, Kennametal, Inc., Latrobe, Pennsylvania.

**7.8**   Use the Box and Cox method to determine an appropriate scale for the response in the cloud seeding data (Example 7.2) and qualitatively compare the results of the analysis with those given in the text.

**7.9   Gothic and Romanesque Cathedrals.**   The data in Table 7.9 give $X =$ nave height and $Y =$ total length, both in feet, for medieval English cathedrals. In addition, the cathedrals can be classified according to their architectural style, either Romanesque or, later, Gothic. Some cathedrals have both a Gothic and a Romanesque part, each of differing height; these cathedrals are included twice.

Use these data to decide if the relationship between the response $Y =$ length and $X =$ height is the same for each of the two architectural styles. If they differ, describe briefly the differences.

**7.10   Land valuation.**   Assessors in the metropolitan area of Minneapolis–St. Paul, Minnesota are bound by law to value farmland enrolled in a "Green Acres" program only with respect to its value as productive farmland; the fact that a shopping center or industrial park has been built nearby cannot enter into the valuation. This creates difficulties because almost all sales, which are the basis for setting assessed values, are priced according to the development potential of the land, not its value as farmland. As an aid to setting assessed values, a method of "equalizing" valuation of land of comparable quality was needed.

The data in Table 7.10 represent one possible method of equalization,

Table 7.9    Height and length of English medieval cathedrals

| | Height | Length | | Height | Length |
|---|---|---|---|---|---|
| *Romanesque* | | | *Gothic* | | |
| Durham | 75 | 502 | York | 100 | 519 |
| Canterbury | 80 | 522 | Bath | 75 | 225 |
| Gloucester | 68 | 425 | Bristol | 52 | 300 |
| Hereford | 64 | 344 | Chichester | 62 | 418 |
| Norwich | 83 | 407 | Exeter | 68 | 409 |
| Peterborough | 80 | 451 | Gloucester | 86 | 425 |
| St. Albans | 70 | 551 | Lichfield | 57 | 370 |
| Winchester | 76 | 530 | Lincoln | 82 | 506 |
| Ely | 74 | 547 | Norwich | 72 | 407 |
| | | | Ripon | 88 | 295 |
| | | | Southwark | 55 | 273 |
| | | | Wells | 67 | 415 |
| | | | St. Asaph | 45 | 182 |
| | | | Winchester | 103 | 530 |
| | | | Old St. Paul | 103 | 611 |
| | | | Salisbury | 84 | 473 |

*Source:* Stephen Jay Gould.

based on a computed soil productivity score, a number between 1 and 100, with higher numbers corresponding to better land. The unit of analysis is a township, roughly 6 miles $\times$ 6 miles. For each township with tillable land, the average soil productivity score $P$ and the 1981 and 1982 average assessed value per acre were recorded. The data are from four counties (Le Sueur, Meeker, McLeod, Sibley) located south and west of Minneapolis where development pressures will have little effect on assessed value of land.

Use these data to see if the soil productivity scores provide a basis for determining assessed values. Examine county differences and year differences. Summarize your results.

**7.11    Sex discrimination.**    The data in Table 7.11 concern salary and other characteristics of all faculty in a small Midwestern college. The data have been collected for presentation in legal proceedings for which discrimination against women in salary was at issue. All persons in the data hold tenured or tenure track positions; temporary faculty are not included. The data were collected from personnel files, and consist of the following quantities:

SX = Sex, coded 1 for female and 0 for male

Table 7.10    Soil productivity scores (*P*) and average 1981 and 1982
assessed value (in dollars per acre) of farmland in four southern Minnesota
counties

| | Le Sueur Value | | | Sibley Value | | | McLeod Value | | | Meeker Value | |
|---|---|---|---|---|---|---|---|---|---|---|---|
| *P* | 1981 | 1982 | *P* | 1981 | 1982 | *P* | 1981 | 1982 | *P* | 1981 | 1982 |
| 51 | 1495 | 1719 | 75 | 1652 | 1982 | 71 | 1524 | 1752 | 31 | 1047 | 1548 |
| 54 | 1222 | 1405 | 78 | 1544 | 1865 | 73 | 1448 | 1665 | 32 | 814 | 895 |
| 55 | 1200 | 1380 | 81 | 1536 | 1843 | 76 | 1483 | 1705 | 35 | 1143 | 1257 |
| 57 | 1254 | 1442 | 83 | 1541 | 1849 | 76 | 1489 | 1712 | 35 | 1263 | 1389 |
| 63 | 1358 | 1562 | 84 | 1570 | 1884 | 77 | 1489 | 1712 | 35 | 1318 | 1450 |
| 67 | 1416 | 1628 | 85 | 1554 | 1865 | 78 | 1465 | 1685 | 37 | 995 | 1095 |
| 69 | 1428 | 1642 | 85 | 1601 | 1921 | 79 | 1427 | 1641 | 38 | 1286 | 1415 |
| 71 | 1146 | 1318 | 87 | 1587 | 1904 | 79 | 1493 | 1717 | 40 | 1000 | 1100 |
| 75 | 1347 | 1549 | 89 | 1435 | 1722 | 79 | 1497 | 1721 | 44 | 1036 | 1140 |
| 75 | 1474 | 1695 | 89 | 1676 | 2011 | 79 | 1455 | 1673 | 45 | 1251 | 1376 |
| 78 | 1382 | 1589 | 90 | 1599 | 1919 | 79 | 1496 | 1720 | 55 | 1308 | 1439 |
| 78 | 1392 | 1601 | 90 | 1647 | 1976 | 82 | 1449 | 1666 | 56 | 1059 | 1165 |
| 80 | 1441 | 1659 | 92 | 1643 | 1972 | 83 | 1481 | 1703 | 60 | 1413 | 1554 |
| | | | 93 | 1656 | 1987 | 84 | 1419 | 1632 | 68 | 1309 | 1440 |
| | | | 94 | 1592 | 1919 | | | | 73 | 1404 | 1544 |
| | | | 94 | 1619 | 1943 | | | | 75 | 1282 | 1410 |
| | | | | | | | | | 79 | 1450 | 1595 |

*Source:* Douglas Tiffany.

RK = Rank, coded 1 for Assistant Professor, 2 for Associate Professor
and 3 for Full Professor

YR = Number of years in current rank

DG = Highest degree, coded 1 if Doctorate, 0 if Masters

YD = Number of years since highest degree was earned

SL = Academic year salary in dollars

**7.11.1.** Obtain a test of the hypothesis that salary adjusted for years in
current rank, highest degree, and years since highest degree is the same for
each of the three ranks, versus the alternative that the salaries are not the
same.

**7.11.2.** Using all the variables, show by suitable diagnostics the need to
transform the response, salary, to some other scale, and suggest an appropriate transformation.

## Table 7.11 Salary data

| Row | SX | RK | YR | DG | YD | SL | Row | SX | RK | YR | DG | YD | SL |
|---|---|---|---|---|---|---|---|---|---|---|---|---|---|
| 1 | 0 | 3 | 25 | 1 | 35 | 36350 | 27 | 0 | 2 | 11 | 1 | 14 | 24800 |
| 2 | 0 | 3 | 13 | 1 | 22 | 35350 | 28 | 1 | 3 | 5 | 1 | 16 | 25500 |
| 3 | 0 | 3 | 10 | 1 | 23 | 28200 | 29 | 0 | 2 | 3 | 0 | 7 | 26182 |
| 4 | 1 | 3 | 7 | 1 | 27 | 26775 | 30 | 0 | 2 | 3 | 0 | 17 | 23725 |
| 5 | 0 | 3 | 19 | 0 | 30 | 33696 | 31 | 1 | 1 | 10 | 0 | 15 | 21600 |
| 6 | 0 | 3 | 16 | 1 | 21 | 28516 | 32 | 0 | 2 | 11 | 0 | 31 | 23300 |
| 7 | 1 | 3 | 0 | 0 | 32 | 24900 | 33 | 0 | 1 | 9 | 0 | 14 | 23713 |
| 8 | 0 | 3 | 16 | 1 | 18 | 31909 | 34 | 1 | 2 | 4 | 0 | 33 | 20690 |
| 9 | 0 | 3 | 13 | 0 | 30 | 31850 | 35 | 1 | 2 | 6 | 0 | 29 | 22450 |
| 10 | 0 | 3 | 13 | 0 | 31 | 32850 | 36 | 0 | 2 | 1 | 1 | 9 | 20850 |
| 11 | 0 | 3 | 12 | 1 | 22 | 27025 | 37 | 1 | 1 | 8 | 1 | 14 | 18304 |
| 12 | 0 | 2 | 15 | 1 | 19 | 24750 | 38 | 0 | 1 | 4 | 1 | 4 | 17095 |
| 13 | 0 | 3 | 9 | 1 | 17 | 28200 | 39 | 0 | 1 | 4 | 1 | 5 | 16700 |
| 14 | 0 | 2 | 9 | 0 | 27 | 23712 | 40 | 0 | 1 | 4 | 1 | 4 | 17600 |
| 15 | 0 | 3 | 9 | 1 | 24 | 25748 | 41 | 0 | 1 | 3 | 1 | 4 | 18075 |
| 16 | 0 | 3 | 7 | 1 | 15 | 29342 | 42 | 0 | 1 | 3 | 0 | 11 | 18000 |
| 17 | 0 | 3 | 13 | 1 | 20 | 31114 | 43 | 0 | 2 | 0 | 1 | 7 | 20999 |
| 18 | 0 | 2 | 11 | 0 | 14 | 24742 | 44 | 1 | 1 | 3 | 1 | 3 | 17250 |
| 19 | 0 | 2 | 10 | 0 | 15 | 22906 | 45 | 0 | 1 | 2 | 1 | 3 | 16500 |
| 20 | 0 | 3 | 6 | 0 | 21 | 24450 | 46 | 0 | 1 | 2 | 1 | 1 | 16094 |
| 21 | 0 | 1 | 16 | 0 | 23 | 19175 | 47 | 1 | 1 | 2 | 1 | 6 | 16150 |
| 22 | 0 | 2 | 8 | 0 | 31 | 20525 | 48 | 1 | 1 | 2 | 1 | 2 | 15350 |
| 23 | 0 | 3 | 7 | 1 | 13 | 27959 | 49 | 0 | 1 | 1 | 1 | 1 | 16244 |
| 24 | 1 | 3 | 8 | 1 | 24 | 38045 | 50 | 1 | 1 | 1 | 1 | 1 | 16686 |
| 25 | 0 | 2 | 9 | 1 | 12 | 24832 | 51 | 1 | 1 | 1 | 1 | 1 | 15000 |
| 26 | 0 | 3 | 5 | 1 | 18 | 25400 | 52 | 1 | 1 | 0 | 1 | 2 | 20300 |

**7.11.3.** After transforming the response, test for nonconstant variance (a) as a function of salary and (b) as a function of the sex indicator.

**7.11.4.** Test to see if the sex differential in transformed salary is the same in each rank.

**7.11.5.** Using all the predictors, analyze these data with regard to the question of differential salaries for men and women faculty, and summarize your results in a fashion that might be useful in court.

**7.11.6.** Finkelstein (1980), in a discussion of the use of regression in discrimination cases, wrote, " ... [a] variable may reflect a position or status bestowed by the employer, in which case if there is discrimination in the award of the position or status, the variable may be 'tainted'." Thus, for example, if discrimination is at work in promotion of faculty to higher ranks, using rank to adjust salaries before comparing the sexes may not be acceptable to the courts.

Fit exactly the same model you found in 7.11.5, except leave out the effects of rank. Summarize and compare the results of leaving out rank effects on inferences concerning differential in pay by sex.

# 8

# MODEL BUILDING II: COLLINEARITY AND VARIABLE SELECTION

When the predictors are related to each other, regression modeling can be very confusing. Estimated effects can change magnitude or even sign, depending on the other predictors in the model. It is important to understand the effects of relationships between the predictors on an analysis. This complex topic, usually called the problem of collinearity, or multicollinearity, is the first topic of this chapter. We then turn to the more general problem of selecting among many potential predictors for a model.

## 8.1   What is collinearity?

Two predictors $X_1$ and $X_2$ are exactly collinear if there is a linear equation such as

$$c_1 X_1 + c_2 X_2 = c_0 \tag{8.1}$$

for some constants $c_0$, $c_1$, and $c_2$ that is true for all cases in the data. For example, suppose that $X_1$ and $X_2$ are amounts of two chemicals and are set so that $X_1 + X_2 = 50$ ml. Since for any given value of $X_1$ in the experiment $X_2 = 50 - X_1$, knowing $X_1$ is exactly the same as knowing both $X_1$ and $X_2$. Exact collinearities most often occur by accident when, for example, weight in pounds and weight in kilograms are both included in a model, or with sets of dummy variables.

**196**

Approximate collinearity is obtained if the equation (8.1) holds approximately for the observed data. A common though not completely adequate measure of the degree of collinearity between $X_1$ and $X_2$ is the square of their sample correlation, $r_{12}^2$. Exact collinearity corresponds to $r_{12}^2 = 1$; noncollinearity corresponds to $r_{12}^2 = 0$. As $r_{12}^2$ approaches 1, approximate collinearity becomes generally stronger. Usually, the adjective *approximate* is dropped, and we would say that $X_1$ and $X_2$ are collinear if $r_{12}^2$ is large.

The definition extends naturally to $p > 2$ predictors. A set of predictors, $X_1, X_2, \ldots, X_p$ are collinear if, for constants $c_0, c_1, \ldots, c_p$,

$$c_1 X_1 + c_2 X_2 + \cdots + c_p X_p = c_0 \tag{8.2}$$

holds approximately. This suggests that at least one of the $X_k$ can be approximately determined from the others,

$$X_k \cong \left( c_0 - \sum_{j \neq k} c_j X_j \right) \bigg/ c_k \tag{8.3}$$

A simple diagnostic analogous to the squared correlation for the two-variable case is the square of the multiple correlation between $X_k$ and the other $X$'s, which we will call $R_k^2$; this number is computed from the regression of $X_k$ on the other $X$'s. If the largest $R_k^2$ is near one, we would tentatively diagnose approximate collinearity.

**Additional comments.**  When a set of predictors is exactly collinear, one or more predictors must be deleted or else unique least squares estimates of coefficients do not exist. Since the deleted predictor contains no information after the others, nothing is lost by this process. When collinearity is approximate, a usual remedy is again to delete variables from the model, with loss of information expected to be minimal.

Correlations are useful measures of collinearity primarily when the scatter of points is elliptical, corresponding roughly to approximate normality of the $X$'s. In many problems, a normality assumption for the $X$'s is unacceptable. Also, correlations are very sensitive to extreme or unusual cases. Consequently, they may be inadequate diagnostics for collinearity.

Unfortunately, suggesting improved measures is not easy. The problem is due in large part to the imprecise definition of collinearity. We have required only that a linear relationship hold to a good approximation, but have not specified how good or how it will be measured. Starting with (8.2), it is not hard to make up examples where the equation is nearly satisfied, but if one of the $X$'s is multiplied by a constant, correlations are unchanged, but no $c$'s can be found to satisfy (8.2) to the same degree of approximation. We will return to this question in Section 8.3.

## 8.2   Why is collinearity a problem?

Collinear predictors will typically lead to large variances for estimated coefficients. For example, consider a regression with two predictors,

$$Y = \beta_0 + \beta_1 X_1 + \beta_2 X_2 + e \qquad (8.4)$$

and suppose that the sample correlation between $X_1$ and $X_2$ is $r_{12}$. Define the symbol $SX_jX_j = \sum (X_j - \bar{X}_j)^2$. It is an exercise (problem 8.9) to show that

$$\text{var}(\hat{\beta}_j) = \sigma^2 \left( \frac{1}{1 - r_{12}^2} \right) \left( \frac{1}{SX_jX_j} \right) \qquad j = 1, 2 \qquad (8.5)$$

The variances of $\hat{\beta}_1$ and $\hat{\beta}_2$ are minimized if $r_{12}^2 = 0$, while as $r_{12}^2$ nears 1, these variances are greatly inflated; for example, if $r_{12}^2 = .95$ but $SX_1X_1$ stays the same, the variance of $\hat{\beta}_1$ is 20 times as large as if $r_{12}^2 = 0$. Thus the use of collinear predictors can lead to unacceptably variable estimated coefficients compared to problems with no collinearity.

The situation when $p > 2$ is analogous to the $p = 2$ case. One can then show (problem 8.10) that the variance of the $j$th coefficient is

$$\text{var}(\hat{\beta}_j) = \sigma^2 \left( \frac{1}{1 - R_j^2} \right) \left( \frac{1}{SX_jX_j} \right) \qquad j = 1, \ldots, p \qquad (8.6)$$

The quantity $1/(1 - R_j^2)$ is called the $j$th *variance inflation factor*, or $\text{VIF}_j$ (Marquardt, 1970). Assuming that the $X_j$'s could have been sampled to make $R_j^2 = 0$ while keeping $SX_jX_j$ constant, the VIF represents the increase in variance due to the correlation between the predictors and, hence, collinearity.

Collinearity will also affect the variances of predictions, although the effects are less clear. For some predictions, collinear predictors can actually have lower variance than would orthogonal predictors with the same values for $SX_jX_j$; this will not be so for all possible predictions.

### Example 8.1   The Picket Fence

---

Hocking and Pendelton (1983) have given a very useful characterization of collinearity in terms of the "picket fence" shown in Figure 8.1. This figure represents one possible configuration of points that corresponds to two collinear predictors $X_1$ and $X_2$. The length of each picket gives the value of the response $Y$ for the given values of $X_1$ and $X_2$. Fitting a regression model (8.4) is like trying to balance a fitted plane on the pickets. The plane will be unstable in the direction perpendicular to

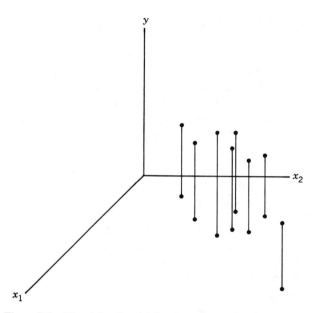

**Figure 8.1**   The picket fence (after Hocking and Pendleton, 1983).

the fence row; if the pickets were exactly on a straight line, the tilt of the plane perpendicular to the fence row would be arbitrary. Predictions off the fence would be highly variable. On the other hand, predictions for points along the fence are likely to be relatively precise, and for those predictions, collinearity is not a problem. Estimating coefficients is roughly analogous to prediction along the coordinate axes. As shown, the picket fence is at about a 45° angle with both the $X_1$ and $X_2$ axes, so we can think of estimation of coefficients for $X_1$ and $X_2$ as predictions in directions not along the fence row, leading to variable estimates for both.

## 8.3   Measuring collinearity

Measuring the degree of collinearity between a set of predictors is only easy when the $X$'s are sampled from a multivariate normal distribution; then, correlations and VIFs are sensible diagnostic statistics. An alternative approach is to measure the applicability of (8.2) directly. In matrix terms, (8.2) requires that there is a unit vector $\mathbf{c}$ such that $\mathbf{Xc}$ is nearly equal to $\mathbf{0}$, or, in terms of lengths of vectors, $(\mathbf{Xc})^T(\mathbf{Xc}) = \mathbf{c}^T(\mathbf{X}^T\mathbf{X})\mathbf{c}$ is sufficiently close to zero for some $\mathbf{c}$. One can show that for any choice of $\mathbf{c}$, $\mathbf{c}^T(\mathbf{X}^T\mathbf{X})\mathbf{c}$ is greater than or equal to the smallest eigenvalue of $(\mathbf{X}^T\mathbf{X})$; equality occurs if $\mathbf{c}$ is taken to be the eigenvector corresponding to the smallest eigenvalue. Collinearity may be diagnosed if the smallest eigenvalue is sufficiently small relative to the other eigenvalues. Judging the size of the smallest eigenvalue is difficult because it depends on the scaling of the columns of $\mathbf{X}$. As pointed out in Section 7.6, the eigenvalues of $(\mathbf{X}^T\mathbf{X})$, $(\mathscr{X}^T\mathscr{X})$, or the sample correlation matrix may all be substantially different. Usually, the eigenvalues of $(\mathscr{X}^T\mathscr{X})$ are used in preference to $(\mathbf{X}^T\mathbf{X})$, but see Belsey (1984) for a different opinion. If the units of measurement of the columns of $\mathbf{X}$ are arbitrary, then the eigenvalues of the correlation matrix are usually computed. If $(\mathscr{X}^T\mathscr{X})$ is used, columns of $\mathbf{X}$ with large sample variances will be more important in determining the eigenvalues than will columns of $\mathbf{X}$ with small sample variances, and these differences will be reflected as strongly in the eigenvalues as will collinearity. Basing computations on the correlation matrix eliminates the effect of differences in variances.

One common summary based on the eigenvalues is called the _condition number_ $\kappa$, defined by

$$\kappa = (\text{largest eigenvalue/smallest eigenvalue})^{1/2} \qquad (8.7)$$

$\kappa$ is greater than or equal to one, with large values suggesting collinearity. $\kappa$ arises as a natural summary in the analysis of numerical properties of computer algorithms, but its use as a statistical summary of collinearity is far from clear. Rules such as declaring collinearity a problem if $\kappa \geqslant 30$ have been proposed, but have little theoretical justification.

Computation of $\kappa$ generally requires finding the largest and smallest eigenvalues of a matrix, an operation that is relatively complex for linear regression problems. Berk (1977) has shown that $\kappa$ must be at least as large as the square root of the maximum $\text{VIF}_j$, so this latter statistic could be used in place of $\kappa$ without much loss of information.

Many collinearity diagnostics can be put into a common framework. Following Cook (1984), suppose that interest centers on estimating linear combinations of the parameter vector $\boldsymbol{\beta}$. Consider two regression problems: one observed, with model

$$\mathbf{Y}_1 = \mathbf{X}\boldsymbol{\beta} + \mathbf{e}_1$$

and an idealized or standard problem,

$$\mathbf{Y}_2 = \mathbf{Z}\boldsymbol{\beta} + \mathbf{e}_2$$

In each problem, the errors have the same distribution, and $\boldsymbol{\beta}$ is the same. The difference between these models is that the standard takes data at the rows of $\mathbf{Z}$ rather than at the rows of $\mathbf{X}$. The design $\mathbf{Z}$ represents an idealization of what might have been observed; often, it will correspond to an orthogonal design, since such designs exhibit no collinearity. Let $\mathbf{d}$ be a $p' \times 1$ vector and $\mathbf{d}^T\boldsymbol{\beta}$ be a linear combination of the elements of $\boldsymbol{\beta}$. The increase in variance of $\mathbf{d}^T\hat{\boldsymbol{\beta}}$ due to using the actual design $\mathbf{X}$ rather than the standard design $\mathbf{Z}$ can be measured by the ratio

$$A(\mathbf{d}|\mathbf{Z}) = \frac{\text{var}(\mathbf{d}^T\hat{\boldsymbol{\beta}}|\text{actual})}{\text{var}(\mathbf{d}^T\hat{\boldsymbol{\beta}}|\text{standard})} = \frac{\mathbf{d}^T(\mathbf{X}^T\mathbf{X})^{-1}\mathbf{d}}{\mathbf{d}^T(\mathbf{Z}^T\mathbf{Z})^{-1}\mathbf{d}} \tag{8.8}$$

Large values of (8.8) would suggest that, at least for the $\mathbf{d}$ chosen, the actual design gives much higher variance than would the idealized design. The choice of $\mathbf{Z}$ and of $\mathbf{d}$ must be made before collinearity diagnostics can be usefully interpreted and used.

Both the variance inflation factors and the condition number can be derived from (8.8) with different choices for $\mathbf{d}$ but with the same standard design $\mathbf{Z}$ chosen so that $\mathbf{Z}^T\mathbf{Z}$ and $\mathbf{X}^T\mathbf{X}$ have the same first row and column (the first row and column corresponds to the column of ones for the inter-cept) and the same main diagonal. The two designs differ in that the remaining off-diagonal elements of $\mathbf{X}^T\mathbf{X}$ may be nonzero, while $\mathbf{Z}$ is chosen so that these elements of $\mathbf{Z}^T\mathbf{Z}$ are all zero. In a sense, $\mathbf{Z}$ is the orthogonal design closest to $\mathbf{X}$.

To get the $j$th variance inflation factor, choose $\mathbf{d}$ to have a one in the $j$th place, and zeroes elsewhere. The condition number will result if we consider all possible $\mathbf{d}$s of the form $(0, \mathbf{d}_1^T)^T$, which will describe the set of all possible linear combinations of the elements of $\boldsymbol{\beta}$ excluding the intercept. One can show that the ratio of the maximum possible value of (8.8) to its minimum value as $\mathbf{d}$ varies over this class is just the square of the condition number based on the correlation matrix derived from $\mathbf{X}$.

### Example 8.2  An agricultural experiment

---

This example was suggested by Dennis Cook. An agriculturalist wishes to do a small experiment to estimate the effects of adding two chemicals, $X_1$ and $X_2$ to the soil on the yield $Y$ of a certain crop. A research assistant is sent to do the experiment, as a $2 \times 2$ factorial design with $X_1 = 0$ or 1 kg and $X_2 = 0$ or 1 kg. The planned design is

$$\mathbf{X} = \begin{pmatrix} 1 & 0 & 0 \\ 1 & 1 & 0 \\ 1 & 0 & 1 \\ 1 & 1 & 1 \end{pmatrix}$$

which is an orthogonal design that exhibits no collinearity at all. Three months later, the research assistant, having carried out the experiment, returns with an added bit of information: since the chemicals both came in 50-kg sacks, and since an additional plot of land was available, a fifth experiment was carried out with $X_1 = X_2 = 48$ kg. Thus the actual design was

$$\mathbf{X} = \begin{pmatrix} 1 & 0 & 0 \\ 1 & 1 & 0 \\ 1 & 0 & 1 \\ 1 & 1 & 1 \\ 1 & 48 & 48 \end{pmatrix}$$

The actual design is now apparently highly collinear, $r_{12}^2 = .9991$, and $VIF_1 = VIF_2 = 1/(1 - r_{12}^2) \cong 574$. Also, the condition number $\kappa = 67.9$ based on $\mathbf{X}^T \mathbf{X}$, and 60.1 based on either $\mathcal{X}^T \mathcal{X}$ or the sample correlation matrix. The collinearity diagnostics are all very large. In this problem, the standard design used by these diagnostics can be shown to be approximately

$$\mathbf{Z} = \begin{pmatrix} 1 & 10 & 10 \\ 1 & -11.25 & -11.25 \\ 1 & -11.25 & 11.25 \\ 1 & 11.25 & -11.25 \\ 1 & 11.25 & 11.25 \end{pmatrix}$$

This is an orthogonal design, centered at the same point (10 kg, 10 kg) as the actual design, with the column sum of squares the same as in the actual design. The standard collinearity diagnostics all agree that, relative to this design, the observed design is very poor, and collinearity is a problem.

This standard design is irrelevant for many reasons. First, we cannot take observations with $X_1$ or $X_2$ negative, since they are both amounts added. Second, we took no data near 10 kg, so a standard centered there has little to do with the original problem. The real difficulty with the observed five-point experiment is *not* collinearity, but our likely belief that, while an additive first-order model may work well for the original design, it may not be expected to work well for such large values of $X_1$ and $X_2$. Indeed, if the model did hold for all five points, then given the first four, the research assistant obtained the greatest amount of information from a single additional point by making $X_1$ and $X_2$

as large as possible. For the design actually carried out, $\text{var}(\hat{\beta}_1) = \sigma^2/3.25$, while if the fifth point were taken at $X_1 = X_2 = 0.5$, the resulting design would have been orthogonal, but with $\text{var}(\hat{\beta}_1) = \sigma^2$, more than three times as large.

It is an interesting exercise to devise a more sensible standard design for this problem and to measure the effects of collinearity relative to that design. This is left as an exercise for the reader.

## 8.4  Variable selection

Deletion of predictors from a model can improve a model and reduce apparent collinearity. Consider the choice between the models

$$Y = \beta_0 + \beta_1 X_1 + \beta_2 X_2 + e \tag{8.9}$$

with $p = 2$ predictors, and the subset model with $p = 1$ predictor

$$Y = \beta_0 + \beta_1 X_1 + e \tag{8.10}$$

obtained from (8.9) by deleting variable $X_2$. Suppose that estimation of $\beta_1$ is of interest. We assume that model (8.9) is the correct description of the dependence of $Y$ on the predictors, and that errors are independent with constant variance. For simplicity, we also assume that $SX_1X_1 = SX_2X_2 = 1$. For model (8.9), the usual least squares estimate of $\beta_1$ is unbiased, and, using (8.5),

$$\text{var}[\hat{\beta}_1 \,|\, \text{full model (8.9)}] = \frac{\sigma^2}{1 - r_{12}^2} \tag{8.11}$$

For model (8.10), more care is required, since $\beta_1$ now has a different meaning: it is the effect of $X_1$ on $Y$ ignoring $X_2$, while in (8.9) it is the effect of $X_1$ on $Y$ adjusting for $X_2$. Consequently, if these two are not the same, then the estimate of $\beta_1$ from (8.10) will be biased; one can show that

$$E(\hat{\beta}_1 \,|\, \text{subset model}) = \beta_1 + r_{12}\beta_2 \tag{8.12}$$

The bias in estimating $\beta_1$ from the subset model is $\beta_1 - (\beta_1 + r_{12}\beta_2) = -r_{12}\beta_2$, and the variance of $\hat{\beta}_1$ from the subset model can be shown to be

$$\text{var}(\hat{\beta}_1 \,|\, \text{subset model}) = \sigma^2 \tag{8.13}$$

which does not depend upon $r_{12}$. To compare estimation of $\beta_1$ from the full and subset models, we must compute the mean square error of $\hat{\beta}_1$ given the subset model, $\mathrm{mse}(\hat{\beta}_1|\text{subset model})$, where the mse is defined by

$$\mathrm{mse}(\hat{\beta}_1\,|\,\text{subset model}) = \mathrm{var}(\hat{\beta}_1\,|\,\text{subset model}) + (\text{bias})^2$$

Then

$$\mathrm{mse}(\hat{\beta}_1\,|\,\text{subset model}) = \sigma^2 + (r_{12}\beta_2)^2 \tag{8.14}$$

Comparing (8.11) and (8.14), we see that the subset model will estimate $\beta_1$ more precisely (i.e., $\mathrm{mse}(\hat{\beta}_1|\text{subset model}) < \mathrm{var}(\hat{\beta}_1|\text{full model})$) whenever

$$\frac{|\beta_2|}{\sigma} < \frac{1}{\sqrt{1-r_{12}^2}} \tag{8.15}$$

The result (8.15) is shown graphically in Figure 8.2. If $r_{12}^2$ is near 1, the subset model is almost always better than the full model, while for any value of $r_{12}^2$, the subset model will be preferred if $|\beta_2| < \sigma$. Thus when collinearity is observed in the data, more precise estimation of $\beta_1$ can generally be obtained from the subset model than from the full model, unless the coefficient for the deleted variable is very large. If the $\beta$'s and $\sigma^2$ were known, deletion of variables with small $|\beta_j|/\sigma$ would be desirable. Since these quantities are generally unknown, selection techniques should have the property of deleting variables with $|\beta_j|/\sigma$ probably small.

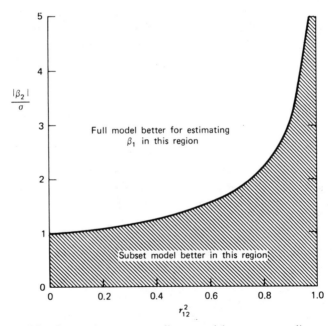

**Figure 8.2**  Comparing a one-predictor model to a two-predictor model.

Collinearity is not the only reason for variable selection. In many problems, we may seek a relatively small set of predictors that have nearly the same information as the full set; further analysis can then concentrate on this subset of predictors, and possibly simplify results. Selection is thus a part of many, but not all regression analyses.

### Example 8.3  Highway data

This data set, given in Table 8.1 and taken from an unpublished master's paper in civil engineering by Carl Hoffstedt, relates the automobile accident rate, in accidents per million vehicle miles ($Y$) to 13 potential independent variables. The data include 39 sections of large highways in the state of Minnesota in 1973. The variables in the order given in Table 8.1, are:

$Y=$ RATE $= 1973$ accident rate per million vehicle miles.

$X1 =$ LEN $=$ length of the segment in miles.

$X2 =$ ADT $=$ average daily traffic count in thousands (estimated).

$X3 =$ TRKS $=$ truck volume as a percent of the total volume.

$X4 =$ SLIM $=$ speed limit (in 1973, before the 55 mph limits).

$X5 =$ LWID $=$ lane width in feet.

$X6 =$ SHLD $=$ width in feet of outer shoulder on the roadway.

$X7 =$ ITG $=$ number of freeway-type interchanges per mile in the segment.

$X8 =$ SIGS $=$ number of signalized interchanges per mile in the segment.

$X9 =$ ACPT $=$ number of access points per mile in the segment.

$X10 =$ LANE $=$ total number of lanes of traffic in both directions.

$X11 =$ FAI $= 1$ if federal aid interstate highway, 0 otherwise.

$X12 =$ PA $= 1$ if principal arterial highway, 0 otherwise.

$X13 =$ MA $= 1$ if major arterial highway, 0 otherwise.

Two of the highway segments, numbers 38 and 39, were neither interstate highways, principal arterial, nor major arterial highways, but were classified as major collectors (MC) and are coded by FAI $=$ PA $=$ MA $= 0$. A separate variable was not used for MC because the resulting data matrix would give an exact collinearity and one of the dummy variables would have to be deleted to obtain estimates. Basic summary statistics are given in Tables 8.2 and 8.3.

## Table 8.1  Highway data

| | Y | $X_1$ | $X_2$ | $X_3$ | $X_4$ | $X_5$ | $X_6$ | $X_7$ | $X_8$ | $X_9$ | $X_{10}$ | $X_{11}$ | $X_{12}$ | $X_{13}$ |
|----|------|-------|----|----|----|----|----|------|------|-------|----|---|---|---|
| 1  | 4.58 | 4.99  | 69 | 8  | 55 | 12 | 10 | 1.20 | 0    | 4.60  | 8  | 1 | 0 | 0 |
| 2  | 2.86 | 16.11 | 73 | 8  | 60 | 12 | 10 | 1.43 | 0    | 4.40  | 4  | 1 | 0 | 0 |
| 3  | 3.02 | 9.75  | 49 | 10 | 60 | 12 | 10 | 1.54 | 0    | 4.70  | 4  | 1 | 0 | 0 |
| 4  | 2.29 | 10.65 | 61 | 13 | 65 | 12 | 10 | 0.94 | 0    | 3.80  | 6  | 1 | 0 | 0 |
| 5  | 1.61 | 20.01 | 28 | 12 | 70 | 12 | 10 | 0.65 | 0    | 2.20  | 4  | 1 | 0 | 0 |
| 6  | 6.87 | 5.97  | 30 | 6  | 55 | 12 | 10 | 0.34 | 1.84 | 24.80 | 4  | 0 | 1 | 0 |
| 7  | 3.85 | 8.57  | 46 | 8  | 55 | 12 | 8  | 0.47 | 0.70 | 11.00 | 4  | 0 | 1 | 0 |
| 8  | 6.12 | 5.24  | 25 | 9  | 55 | 12 | 10 | 0.38 | 0.38 | 18.50 | 4  | 0 | 1 | 0 |
| 9  | 3.29 | 15.79 | 43 | 12 | 50 | 12 | 4  | 0.95 | 1.39 | 7.50  | 4  | 0 | 1 | 0 |
| 10 | 5.88 | 8.26  | 23 | 7  | 50 | 12 | 5  | 0.12 | 1.21 | 8.20  | 4  | 0 | 1 | 0 |
| 11 | 4.20 | 7.03  | 23 | 6  | 60 | 12 | 10 | 0.29 | 1.85 | 5.40  | 4  | 0 | 1 | 0 |
| 12 | 4.61 | 13.28 | 20 | 9  | 50 | 12 | 2  | 0.15 | 1.21 | 11.20 | 4  | 0 | 1 | 0 |
| 13 | 4.80 | 5.40  | 18 | 14 | 50 | 12 | 8  | 0    | 0.56 | 15.20 | 2  | 0 | 1 | 0 |
| 14 | 3.85 | 2.96  | 21 | 8  | 60 | 12 | 10 | 0.34 | 0    | 5.40  | 4  | 0 | 1 | 0 |
| 15 | 2.69 | 11.75 | 27 | 7  | 55 | 12 | 10 | 0.26 | 0.60 | 7.90  | 4  | 0 | 1 | 0 |
| 16 | 1.99 | 8.86  | 22 | 9  | 60 | 12 | 10 | 0.68 | 0    | 3.20  | 4  | 0 | 1 | 0 |
| 17 | 2.01 | 9.78  | 19 | 9  | 60 | 12 | 10 | 0.20 | 0.10 | 11.00 | 4  | 0 | 1 | 0 |
| 18 | 4.22 | 5.49  | 9  | 11 | 50 | 12 | 6  | 0.18 | 0.18 | 8.90  | 2  | 0 | 1 | 0 |
| 19 | 2.76 | 8.63  | 12 | 8  | 55 | 13 | 6  | 0.14 | 0    | 12.40 | 2  | 0 | 1 | 0 |
| 20 | 2.55 | 20.31 | 12 | 7  | 60 | 12 | 10 | 0.05 | 0.99 | 7.80  | 4  | 0 | 1 | 0 |
| 21 | 1.89 | 40.09 | 15 | 13 | 55 | 12 | 8  | 0.05 | 0.12 | 9.60  | 4  | 0 | 1 | 0 |
| 22 | 2.34 | 11.81 | 8  | 8  | 60 | 12 | 10 | 0    | 0    | 4.30  | 2  | 0 | 1 | 0 |
| 23 | 2.83 | 11.39 | 5  | 9  | 50 | 12 | 8  | 0    | 0.09 | 11.10 | 2  | 0 | 1 | 0 |
| 24 | 1.81 | 22.00 | 5  | 15 | 60 | 12 | 7  | 0    | 0    | 6.80  | 2  | 0 | 1 | 0 |
| 25 | 9.23 | 3.58  | 23 | 6  | 40 | 12 | 2  | 0.56 | 2.51 | 53.00 | 4  | 0 | 0 | 1 |
| 26 | 8.60 | 3.23  | 13 | 6  | 45 | 12 | 2  | 0.31 | 0.93 | 17.30 | 2  | 0 | 0 | 1 |
| 27 | 8.21 | 7.73  | 7  | 8  | 55 | 12 | 8  | 0.13 | 0.52 | 27.30 | 2  | 0 | 0 | 1 |
| 28 | 2.93 | 14.41 | 10 | 10 | 55 | 12 | 6  | 0    | 0.07 | 18.00 | 2  | 0 | 0 | 1 |
| 29 | 7.48 | 11.54 | 12 | 7  | 45 | 12 | 3  | 0.09 | 0.09 | 30.20 | 2  | 0 | 0 | 1 |
| 30 | 2.57 | 11.10 | 9  | 8  | 60 | 12 | 7  | 0    | 0    | 10.30 | 2  | 0 | 0 | 1 |
| 31 | 5.77 | 22.09 | 4  | 8  | 45 | 11 | 3  | 0    | 0.14 | 18.20 | 2  | 0 | 0 | 1 |
| 32 | 2.90 | 9.39  | 5  | 10 | 55 | 13 | 1  | 0    | 0    | 12.30 | 2  | 0 | 0 | 1 |
| 33 | 2.97 | 19.49 | 4  | 13 | 55 | 12 | 4  | 0    | 0    | 7.10  | 2  | 0 | 0 | 1 |
| 34 | 1.84 | 21.01 | 5  | 12 | 55 | 10 | 8  | 0    | 0.10 | 14.00 | 2  | 0 | 0 | 1 |
| 35 | 3.78 | 27.16 | 2  | 10 | 55 | 12 | 3  | 0.04 | 0.04 | 11.30 | 2  | 0 | 0 | 1 |
| 36 | 2.76 | 14.03 | 3  | 8  | 50 | 12 | 4  | 0.07 | 0    | 16.30 | 2  | 0 | 0 | 1 |
| 37 | 4.27 | 20.63 | 1  | 11 | 55 | 11 | 4  | 0    | 0    | 9.60  | 2  | 0 | 0 | 1 |
| 38 | 3.05 | 20.06 | 3  | 11 | 60 | 12 | 8  | 0    | 0    | 9.00  | 2  | 0 | 0 | 0 |
| 39 | 4.12 | 12.91 | 1  | 10 | 55 | 12 | 3  | 0    | 0    | 10.40 | 2  | 0 | 0 | 0 |

Table 5.2  Summary statistics

### (a) Averages and variances

| Variable | N | Average | Variance | Standard Deviation | Minimum | Maximum |
|---|---|---|---|---|---|---|
| RATE | 39 | 3.933 | 3.944 | 1.986 | 1.610 | 9.230 |
| LEN | 39 | 12.88 | 57.91 | 7.610 | 2.960 | 40.09 |
| ADT | 39 | 19.62 | 346.4 | 18.61 | 1.000 | 73.00 |
| TRKS | 39 | 9.333 | 5.544 | 2.355 | 6.000 | 15.00 |
| SLIM | 39 | 55.00 | 34.21 | 5.849 | 40.00 | 70.00 |
| LWID | 39 | 11.95 | 0.2078 | 0.4559 | 10.00 | 13.00 |
| SHLD | 39 | 6.872 | 9.220 | 3.036 | 1.000 | 10.00 |
| ITG | 39 | 0.2964 | 0.1691 | 0.4112 | 0 | 1.540 |
| SIGS | 39 | 0.4005 | 0.4012 | 0.6334 | 0 | 2.510 |
| ACPT | 39 | 12.16 | 86.83 | 9.318 | 2.200 | 53.00 |
| LANE | 39 | 3.128 | 1.852 | 1.361 | 2.000 | 8.000 |
| FAI | 39 | 0.1282 | 0.1147 | 0.3387 | 0 | 1.000 |
| PA | 39 | 0.4872 | 0.2564 | 0.5064 | 0 | 1.000 |
| MA | 39 | 0.3333 | 0.2281 | 0.4776 | 0 | 1.000 |

### (b) Sample correlation matrix

| | RATE | LEN | ADT | TRKS | SLIM | LWID | SHLD | ITG | SIGS | ACPT | LANE | FAI | PA | MA |
|---|---|---|---|---|---|---|---|---|---|---|---|---|---|---|
| RATE | 1.00 | | | | | | | | | | | | | |
| LEN | -0.47 | 1.00 | | | | | | | | | | | | |
| ADT | -0.03 | -0.27 | 1.00 | | | | | | | | | | | |
| TRKS | -0.51 | 0.50 | -0.10 | 1.00 | | | | | | | | | | |
| SLIM | -0.68 | 0.19 | 0.24 | 0.30 | 1.00 | | | | | | | | | |
| LWID | -0.01 | -0.31 | 0.13 | -0.15 | -0.10 | 1.00 | | | | | | | | |
| SHLD | -0.39 | -0.10 | 0.46 | 0.00 | 0.69 | -0.04 | 1.00 | | | | | | | |
| ITG | -0.02 | -0.25 | 0.90 | -0.07 | 0.24 | 0.10 | 0.38 | 1.00 | | | | | | |
| SIGS | 0.56 | -0.32 | 0.15 | -0.45 | -0.41 | 0.04 | -0.13 | -0.07 | 1.00 | | | | | |
| ACPT | 0.75 | -0.24 | -0.22 | -0.36 | -0.68 | -0.04 | -0.43 | -0.20 | 0.50 | 1.00 | | | | |
| LANE | -0.03 | -0.20 | 0.82 | -0.15 | -0.26 | 0.10 | 0.48 | 0.70 | 0.25 | -0.21 | 1.00 | | | |
| FAI | -0.20 | -0.03 | 0.76 | 0.14 | 0.46 | 0.04 | 0.40 | 0.81 | -0.25 | -0.34 | 0.59 | 1.00 | | |
| PA | -0.16 | -0.15 | -0.03 | -0.05 | 0.04 | 0.23 | 0.37 | -0.13 | 0.30 | -0.23 | 0.17 | -0.37 | 1.00 | |
| MA | -0.34 | 0.13 | -0.46 | 0.10 | -0.42 | -0.28 | -0.62 | 0.36 | -0.07 | 0.51 | -0.51 | -0.27 | -0.69 | 1.00 |
| | RATE | LEN | ADT | TRKS | SLIM | LWID | SHLD | ITG | SIGS | ACPT | LANE | FAI | PA | MA |

207

Table 8.3   Regression for full model

| Variable | Estimate | Standard Error | $t$-value |
|---|---|---|---|
| Intercept | 13.7 | 6.87 | 1.99 |
| LEN | $-0.065$ | 0.033 | $-1.94$ |
| ADT | $-0.004$ | 0.034 | $-0.12$ |
| TRKS | $-0.100$ | 0.115 | $-0.87$ |
| SLIM | $-0.124$ | 0.082 | $-1.52$ |
| LWID | $-0.134$ | 0.598 | $-0.22$ |
| SHLD | 0.014 | 0.162 | 0.09 |
| ITG | $-0.475$ | 1.28 | $-0.37$ |
| SIGS | 0.713 | 0.525 | 1.36 |
| ACPT | 0.067 | 0.043 | 1.56 |
| LANE | 0.027 | 0.283 | 0.09 |
| FAI | 0.543 | 1.72 | 0.31 |
| PA | $-1.01$ | 1.11 | $-0.91$ |
| MA | $-0.548$ | 0.976 | $-0.56$ |

$\hat{\sigma}^2 = 1.44$, d.f. $= 25$, $R^2 = 0.76$, $RSS = 35.89367$.

An interesting feature of Table 8.3, the summary of the regression in the full model, is that none of the coefficients have $t$-values exceeding 2 in absolute value, in spite of the fact that $R^2 = 0.76$. Thus none of the predictors adjusted for the others is clearly important even though, taken as a group, they are useful for predicting accident rates. Usually this is evidence that several of the predictors can be removed from the model without degrading the fit.

For this data set, all of the standard collinearity diagnostics have moderate values. The condition number based on the correlation matrix is $\kappa = (4.50/0.051)^{1/2} = 9.4$, while the VIFs vary from 1.9 for TRKS to 9.1 for FAI. Collinearity can be expected to be, at worst, a moderate problem for these data.

## 8.5   Assumptions and notation

On each of $n$ cases, we observe $k$ predictors $X_1, \ldots, X_k$, and a response $Y$. In earlier chapters, the number of predictors is $p$; in this chapter we reserve $p$ for the number of predictors in a selected subset possibly including the intercept, and let $k' = k + 1$ if the model includes a constant term and $k' = k$ if regression is through the origin.

The full model using all the predictors is written in matrix terms as

$$\mathbf{Y} = \mathbf{X}\boldsymbol{\beta} + \mathbf{e} \begin{cases} \mathbf{X} : n \times k' \\ \boldsymbol{\beta} : k' \times 1 \\ \mathbf{e} : n \times 1 \quad (\text{var } (\mathbf{e}) = \sigma^2 \mathbf{I}) \end{cases} \tag{8.16}$$

If $n < k'$, we have more predictors than cases, and we will not be able to find a unique estimate of $\boldsymbol{\beta}$; however, model (8.16) still can provide a description of a relationship between $\mathbf{Y}$ and the $\mathbf{X}$'s. Now, a subset model can be specified by partitioning $\mathbf{X}$ into two matrices $\mathbf{X}_1$ and $\mathbf{X}_2$, so that $\mathbf{X}_1$ is $n \times p$ with $p \leqslant n$ (and rank$(\mathbf{X}_1) = p$) and $\mathbf{X}_2$ is $n \times (k' - p)$. $\mathbf{X}_1$ consists of the variables in the subset model, and $\mathbf{X}_2$ consists of the variables not included in the subset model.

Corresponding to the partition of $\mathbf{X}$ into $\mathbf{X}_1$ and $\mathbf{X}_2$ we partition $\boldsymbol{\beta}$ into $\boldsymbol{\beta}_1$ and $\boldsymbol{\beta}_2$ where $\boldsymbol{\beta}_1$ is $p \times 1$ and $\boldsymbol{\beta}_2$ is $(k' - p) \times 1$. The full model (8.16) can be rewritten as

$$\mathbf{Y} = \mathbf{X}_1 \boldsymbol{\beta}_1 + \mathbf{X}_2 \boldsymbol{\beta}_2 + \mathbf{e} \tag{8.17}$$

Next, a specific subset model is obtained by deleting the term $\mathbf{X}_2 \boldsymbol{\beta}_2$ to give

$$\mathbf{Y} = \mathbf{X}_1 \boldsymbol{\beta}_1 + \mathbf{e}^* \tag{8.18}$$

The two models (8.17) and (8.18) are identical only if $\boldsymbol{\beta}_2 = \mathbf{0}$. In general, these two models are quite different, as the estimate of the parameter $\boldsymbol{\beta}_1$ will usually be different in each of the models, and the interpretation of the parameter depends on which model is used. Also, by restricting ourselves to subsets with $p \leqslant n$, we can always obtain unique estimates for the parameters in (8.18) even though unique estimates do not exist for (8.17) if $k' > n$.

All selection methods proceed as if the full model includes all relevant predictors correctly transformed, and perhaps extra unimportant predictors. The goal of selection is to delete predictors that are irrelevant or, given other predictors, are not very useful.

In practice, transformations and possibly cross products or other combinations of predictors are needed, and the analyst must decide both on relevant transformations and on selecting variables. As a general approach, choice of transformation and diagnosing other problems can precede subset selection. However, repeating the diagnostics after selection will always be prudent.

In the subset model, we will estimate $\boldsymbol{\beta}_1$ by $\hat{\boldsymbol{\beta}}_1 = (\mathbf{X}_1^T \mathbf{X}_1)^{-1} \mathbf{X}_1^T \mathbf{Y}$, as if least squares were being used and no other $X$'s were observed.

## 8.6    Selecting subsets on substantive grounds

The single most important tool in selecting a subset of variables for use in a model is the analyst's knowledge of the substantive area under study and of each of the variables, including expected sign and magnitude of the coefficient.

In the highway accident data, there are $k = 13$ potential predictors, so there are $2^{13} = 8192$ possible subset models including the full model and the model containing only the intercept. However, the 13 predictors can be divided into several types. First of all, variables 11, 12, and 13, namely, FAI, PA, and MA, are simply indicator variables that taken together indicate the type of highway. It may not be reasonable to include one of these variables while omitting the other two (this is partly because the fourth type of highway, major collectors, is indicated by $FAI = PA = MA = 0$). Thus we may consider models with either all three included or all three excluded. Possibly, these may have special importance, since the type of roadway is defined by the source of financial support that the Highway Department uses to maintain the roads. We may even think of this problem as being an analysis of covariance, where we explore the possibility of a difference between highway types adjusted for the other predictors.

Also, the variable LEN should be treated differently from the others, since its inclusion in the prediction equation may be required by the way highway segments are defined. Suppose that highways consist of "safe stretches" and "bad spots," and that most accidents occur at the bad spots. If we were to lengthen a highway segment in our study by 1 mile, it is unlikely that we would add another bad spot to the section, assuming bad spots are rare. However, as a result of lengthening the roadway in the study, the computed response, accidents per million vehicle miles on the section of roadway, would have a lower value since the number of miles driven would go up, and the number of accidents would stay about the same. Thus the response and LEN should be negatively correlated, and we should consider only models that include LEN.

The number of possible subset models for the highway data is now reduced from 8192 to 512 with LEN and the type indicators, and 512 with LEN but without the type indicators, which is a more manageable problem, although still quite large.

**Redefining variables.**    Another important method of reducing the number of predictors is to define new variables that are combinations of the old variables. For example, in some studies, the two variables height and weight, which are often highly correlated, may be replaced by a combination of them to give a single height-weight index, or two IQ tests as predictors can be replaced by their average.

## 8.7 Finding subsets I: Stepwise methods

In many problems, a point is reached at which it is necessary to use the data to find subsets of the predictors. Two basic methods for this purpose have evolved. The first method, generally called *stepwise* regression, uses a convenient computational algorithm to limit possible models to a relatively small number. It is very heavily used in practice, although it does not correspond to any specific criterion for choosing a model. The second method uses a criterion statistic computed for all possible subsets of predictors. This has become practical with the development of several fast algorithms for computing all possible regressions. We discuss stepwise methods first.

The stepwise procedures provide a systematic technique for examining only a few subsets of each size. A path through the possible models is chosen, looking first at a subset of one size, and then looking only at models obtained from preceding ones by adding or deleting variables.

There are three basic algorithms for stepwise regression, generally called forward selection (FS), backward elimination (BE), and stepwise (SW). In the FS procedure, predictors are added at each step. In BE, predictors are eliminated. In SW, a step may be either an addition, an elimination, or an interchange of an "in" variable and an "out" variable.

The FS procedure works as follows. Begin with the simple regression model with that single predictor that has the biggest sample correlation in absolute value with the response $Y$. This is the first step. Next, we add to the model that predictor that meets three equivalent criteria: (1) it has the highest sample partial correlation in absolute value with the response, adjusting for the predictors in the equation already; (2) adding the variable will increase $R^2$ more than any other single variable; and (3) the variable added would have the largest $t$- or $F$-statistic of any of the variables that are not already in the model. Thus in FS, we start with a subset of size 1, and, at each step, we add another variable to the model according to the criterion. We continue adding one variable at a time, until a stopping rule is met. The possible stopping rules are:

**FS.1** Stop with a subset of a predetermined size $p^*$.

**FS.2** Stop if the $F$-test for each of the variables not yet entered would be less than some predetermined number, say $F$–IN (or, equivalently, stop if the absolute value of the $t$ statistic would be less than $(F–IN)^{1/2}$).

**FS.3** Stop when the addition of the next predictor will make the set of predictors too close to collinear. This is called a tolerance check and is usually related to the square of the multiple correlation between the next predictor to be added and the predictors already included in the equation. Berk (1977) provides details on a computational method for carrying out this check (see also problem 2.7). The tolerance check serves to identify some extreme collinearities and to protect against round-off errors in computations.

The backward elimination (BE) method is similar to FS, except that we start with the full model, and, at each step, remove one variable. The variable to be removed is the one that has the smallest $t$ or $F$ value of all the variables in the equation. This is equivalent to removing the variable that causes the smallest change in $R^2$, or has the smallest absolute partial correlation with $Y$ adjusting for all the other variables left in the model. The stopping rules for BE are also similar to those for FS:

**BE.1**   Stop with subset of predetermined size $p^*$.

**BE.2**   Stop if the $F$-test for all the variables now in the model is bigger than some predetermined number, say $F$–OUT.

The stepwise algorithm starts as in FS. At each step after the first, we consider four alternatives: add a variable, delete a variable, exchange two variables, or stop. The rules for SW can be summarized as:

**SW.1**   If there are at least two variables in the current model, and one or more has a value of $F$ less than $F$–OUT, the variable with the smallest value of $F$ is removed from the model.

**SW.2**   If there are two or more variables in the model, the one with the smallest $F$-value is removed if its removal results in a value of $R^2$ that is larger than the $R^2$ obtained for the same number of variables previously. This can happen in the SW procedure since variables are added and deleted at the various steps.

**SW.3**   If two or more variables are in the model, one of them will be exchanged with a variable not in the model if the exchange increases $R^2$.

**SW.4**   A variable is added to the model if it has the highest $F$-value as in FS, provided that $F$ is greater than $F$–IN and the tolerance criterion is satisfied.

Variants on these algorithms are obtained by applying the rules in different orders, or by changing the values of $F$–IN or $F$–OUT. If $F$–IN is very small, say 0.01, then the last step of FS will generally exclude only those predictors that fail the tolerance test. Some analysts use a small value of $F$–IN to order the predictors; predictors entered at an earlier step are presumed to be more important. No theoretical justification for this practice exists. Butler (1984), using a Bonferroni inequality, presents a method for estimating a $p$-value for entering a predictor in an equation when using stepwise procedures. This requires that $F$–IN be changed at each step. Usually, fixed values of $F$–IN and $F$–OUT between 2 and 4 are used.

**Highway accident data (continued).**  We shall now apply the various algorithms to the highway accident data given in Table 8.1. For the exposition here, we choose $F$–IN $= F$–OUT $= 2.0$. Default values for these parameters vary from program to program. First we use the FS method.

The initial task in FS is to find the single variable which is most highly correlated with the response. From Table 8.2 the correlation between $Y$ and ACPT is biggest, with a value of 0.75, so this becomes the first variable in the model. The regression is given in the first column of Table 8.4. The partial correlations of RATE adjusted for ACPT with the other potential predictors vary from the smallest (in absolute value) of 0.015 with FAI to a maximum of $-0.45$ with LEN. Thus LEN is added next to the model as shown in Table 8.4. In this manner, we add, in order, SLIM, SIGS, and then PA. The next variable considered is TRKS; however, we see that TRKS

**Table 8.4    FS method**

| | Estimate and ($t$-values) at Step Number | | | | | |
|---|---|---|---|---|---|---|
| Variable | 1 | 2 | 3 | 4 | 5 | 6 |
| Intercept | 1.98 | 3.19 | 9.325 | 8.81 | 9.94 | 10.56 |
| | (5.64) | (6.21) | (3.56) | (3.38) | (3.85) | (3.96) |
| ACPT | 0.160 | 0.145 | 0.101 | 0.089 | 0.064 | 0.0628 |
| | (6.94) | (6.71) | (3.72) | (3.17) | (2.12) | (2.07) |
| LEN | | −0.079 | −0.077 | −0.0685 | −0.074 | −0.0635 |
| | | (−2.99) | (−3.10) | (−2.72) | (−3.02) | (−2.35) |
| SLIM | | | −0.103 | −0.096 | −0.105 | −0.103 |
| | | | (−2.39) | (−2.26) | (−2.54) | (−2.49) |
| SIGS | | | | 0.485 | 0.797 | 0.701 |
| | | | | (1.42) | (2.16) | (1.83) |
| PA | | | | | −0.774 | −0.743 |
| | | | | | (−1.89) | (−1.80) |
| TRKS | | | | | | −0.089 |
| | | | | | | (−0.95) |
| d.f. | 37 | 36 | 35 | 34 | 33 | 32 |
| $\hat{\sigma}^2$ | 1.76 | 1.45 | 1.28 | 1.24 | 1.16 | 1.16 |
| $R^2$ | 0.56 | 0.65 | 0.70 | 0.72 | 0.74 | 0.75 |

Table 8.5   BE for highway data

| Step | Removed | $R^2$ | Decrease in $R^2$ | $F$ of Removed Variable |
|------|---------|-------|-------------------|-------------------------|
| 0 | | 0.7605 | | |
| 1 | LANE | 0.7604 | 0.0001 | 0.01 |
| 2 | ADT | 0.7604 | 0.0000 | 0.01 |
| 3 | SHLD | 0.7603 | 0.0001 | 0.01 |
| 4 | FAI | 0.7592 | 0.0012 | 0.14 |
| 5 | LWID | 0.7579 | 0.0012 | 0.15 |
| 6 | ITG | 0.7562 | 0.0017 | 0.22 |
| 7 | MA | 0.7521 | 0.0041 | 0.52 |
| 8 | TRKS | 0.7450 | 0.0070 | 0.90 |
| 9 | PA | 0.7521 | 0.0071 | 3.57 |

has $t = 0.95$, which is less than $(F{-}IN)^{1/2}$. Hence we stop at step 5 before adding TRKS. If we had used a different value of $F{-}IN$, we might have ended with a different model; for example, if $(F{-}IN)^{1/2}$ had been 1.5 instead of 1.414, SIGS would not have been entered, and we would have stopped with the three-variable model.

For the BE algorithm, start with the full model, and compute the regression. The variable with the smallest $|t|$ or $F$ is deleted, provided that $|t| < (F{-}OUT)^{1/2}$ (or $F < F{-}OUT$). Then the regression using the remaining variables is computed, and again the variable with smallest $|t|$ or $F$ is deleted. The process is repeated until a stopping criterion is satisfied. The results for BE (or for any stepwise algorithm) are often summarized in a table like Table 8.5 using $F{-}OUT = 2.0$. The final model is the same as was found via FS.

**Discussion of stepwise procedures.**   The stepwise methods are easy to explain, inexpensive to compute, and widely used. The comparative simplicity of the results from stepwise regression seems to appeal to many analysts. But stepwise methods must be used with caution. The model selected in a stepwise fashion need not optimize any reasonable criterion function for choosing a model. The apparent ordering of the predictors is an artifact of the method and need not reflect relationships of substantive interest. Finally, stepwise regression may seriously overstate significance of results.

Consider a simulated example. A data set of $n = 100$ cases with a response $Y$ and 50 predictors $X_1, \ldots, X_{50}$ was generated using standard normal random deviates, so all the $\beta_j$ are known to be zero, and the true multiple correlation between $Y$ and the $X$'s is also exactly zero. All numbers in the data were drawn independently. The regression of $Y$ on $X_1$ to $X_{50}$ is sum-

Table 8.6   Results of a simulated example with $n = 100$, $k = 50$

| Method | $p$ | $R^2$ | $p$-value of Overall $F$ | Number of predictors with $p$-value $\leqslant$ .25 | Number of predictors with $p$-value $\leqslant$ .05 |
|---|---|---|---|---|---|
| No selection | 50 | 0.59 | .13 | 16(32%) | 6(12%) |
| $F$–IN $= 2$ | 16 | 0.48 | $< .001$ | 16(100%) | 11(69%) |
| $F$–IN $= 4$ | 4 | 0.46 | $< .001$ | 4(100%) | 4(100%) |

marized in the first line of Table 8.6. The value of $R^2 = 0.59$ may seem surprisingly large considering that all the data are independent random numbers. The overall $F$-test, which is in a scale more easily calibrated, gives a $p$-value of .13 for the data; Rencher and Pun (1980) and Freedman (1983) report similar simulations with the overall $p$-value varying from near 0 to near 1, as it should since the null hypothesis of $\boldsymbol{\beta} = \mathbf{0}$ is true. In the simulation reported here, 16 of 50 predictors had $t$-values with corresponding $p$-values less than .25, while 6 of 50 had $p$-values less than .05. Line 2 of Table 8.6 summarizes the final model obtained from FS regression using $F$–IN $= 2$. Only 16 predictors are retained, and $R^2$ drops slightly, to 0.48. The major change is in the perceived significance of the result. The overall $F$ now has a very small $p$-value, less than .001, and the $t$-values for 11 of the 16 predictors in the equation correspond to $p$-values less than .05. The third line is similar to the second, except a more stringent $F$–IN $= 4$ was used. Using only four predictors, $R^2 = 0.46$, but all of these, as well as the overall regression have very small corresponding $p$-values.

This example demonstrates many lessons. First, stepwise selection of predictors can have important effects on the apparent significance of results. The coefficients for the predictors left in the model will generally be too large in absolute value, and have $t$- or $F$-values that are too large. Second, even if the response and the predictors are unrelated, $R^2$ can be large: when $\boldsymbol{\beta} = \mathbf{0}$, the expected value of $R^2$ is $k/(n - 1)$, and when $\boldsymbol{\beta} \neq \mathbf{0}$, the expected value of $R^2$ is even larger. With selection, $R^2$ will be biased and possibly much too large.

## 8.8   Criteria for selecting a subset

Criteria-based subset selection has two parts. First, one must choose a criterion statistic for comparing subsets. Second, a computational procedure must be available to find subsets that are best on the criterion. In this section, we consider criteria based on prediction errors: a subset model will be preferred if it gives predictions that are in some sense better. One possibility is to measure the total or average estimated mean square error of prediction for some future point or set of points of interest.

A subset model may produce biased predictions or fitted values. The bias may be tolerable if it is balanced by reduced variance. As a comprehensive measure of fit, therefore, we consider the mean square error for each fitted value, mse($\hat{y}_i$). Define $J_p$ for a given $p$ predictor model by

$$J_p = \frac{1}{\sigma^2} \sum_{i=1}^{n} \text{mse}(\hat{y}_i) \qquad (8.19)$$

Good subsets should have small values of $J_p$. The use of $J_p$ places special emphasis on the observed data. If one point is replicated, then the mse at that point will receive large weight in $J_p$. This is reasonable if the cases are a random sample of those for which prediction of future values is required, or if interest genuinely centers on the $n$ cases in the data. In other circumstances, other functions of mse($\hat{y}_i$) might be preferable.

The value of $J_p$ depends on several unobservable parameters, so it must be estimated using data. Several estimates of $J_p$ can be formulated. The simplest, due to Mallows, is called $C_p$ and is derived in Appendix 8A.1. $C_p$ may be written in any of the three following equivalent forms:

$$C_p = \frac{RSS_p}{\hat{\sigma}^2} + 2p - n \qquad (8.20)$$

$$= \frac{RSS_p - RSS_{k'}}{\hat{\sigma}^2} + p - (k' - p) \qquad (8.21)$$

$$= (k' - p)(F_p - 1) + p \qquad (8.22)$$

where $\hat{\sigma}^2$ is from the full model, and $F_p$ is the usual $F$ statistic for the hypothesis that the predictors left out of the subset model but included in the full model all have zero coefficients. $C_p$ has many useful properties.

**1.** From (8.20), $C_p$ depends only on usual regression calculations, namely $RSS_p$, $\hat{\sigma}^2$, $p$, and $n$, and can be easily computed. This is the basis for the use of $C_p$ in fast all-possible regression algorithms.

**2.** From (8.21), $C_p$ measures the difference in fitting errors between the full and subset models.

**3.** From (8.22), $C_p$ consists of two parts: a random part $F_p$ and a fixed penalty $p$. There is a trade-off between decreasing $F_p$ and adding variables. This is one example of a *penalized* method of choosing a model; for the linear regression problem, the use of $C_p$ is similar to most penalized methods.

**4.** For the full model, $C_{k'} = k'$, directly from (8.22).

**5.** For a subset model, if all the predictors that are left out have coefficients equal to zero, then, on the average, $C_p \cong p$ because $E(F_p) \cong 1$. Of course, $C_p$ is a random variable, whose nominal distribution is, by (8.22), closely related to the distribution of $F_p$.

**6.** Two subset models can be compared by comparing their values of $C_p$. Mallows (1973) has suggested that good models have $C_p \cong p$. Since $C_p$ is a random variable, two models are not easily distinguishable if their $C_p$ values are very close; certainly, any model with $C_p \leqslant k'$, corresponding to $F_p \leqslant 2$, will be a candidate for a good subset model. In some problems, many such models will be found.

The $C_p$ statistic is not the only method of computing an estimate of a measure like $J_p$. Important alternatives are based on the idea of cross validation, in which the data are split into several groups. The most straightforward approach is to split the data roughly in half, use one-half of the data to estimate a model, and then use the squared fitting errors in the other half to estimate $J_p$; see Snee (1977) for details of this approach.

Another cross-validation type method that is closely related to the work on diagnostics in Chapter 5 is to compute the predicted residuals,

$$\hat{e}_{(i)} = y_i - \mathbf{x}_i^T \hat{\boldsymbol{\beta}}_{(i)} = \frac{\hat{e}_i}{1 - h_{ii}} \tag{8.23}$$

and use the predicted residual sum of squares, or PRESS,

$$\text{PRESS} = \sum \hat{e}_{(i)}^2 \tag{8.24}$$

as a criterion function (Allen, 1974; Geisser and Eddy, 1979). Good models have small values of PRESS. This statistic has the aesthetic advantage of using an estimate that excludes case $i$ to predict at case $i$, but a computational difficulty is that it cannot be computed from sufficient statistics and requires much more work. One possibility for using PRESS is to find a few candidate regression models with a cheaper criterion such as $C_p$ and then compute PRESS only for those few models.

**Additional comments.**   Direct comparison of two subset models with differently transformed responses is not addressed by the $C_p$ criterion. A bibliography of methods for this problem is given by Pereira (1977). Another criterion function used to choose between subsets is called the adjusted $R^2$, or $\bar{R}^2$, defined by

$$\bar{R}^2 = 1 - \left(\frac{n-1}{n-p}\right)(1 - R^2) \tag{8.25}$$

Unlike $R^2$ itself, $\bar{R}^2$ need not increase if a predictor is added to a model because of the correction for degrees of freedom, $(n-1)/(n-p)$. Kennard (1971) has pointed out that $\bar{R}^2$ is closely related to $C_p$, and so we will use only $C_p$ in this discussion.

## 8.9    Subset selection II: All possible regressions

For problems with only a few predictors, say $k = 8$ or less, it is not difficult to compute one or two criteria for all possible subsets of predictors. Algorithms for this are described by Garside (1971); Schatzoff, Fienberg, and Tsao (1968); and by Morgan and Tartar (1972). One can thus find the few subsets that are best on the criterion of interest, and these few can be subjected to further study, including case analysis, questions of interpretability, and so on. If $k$ is large, the amount of computation needed grows very fast, so methods that find the best few subsets without computing all possible regressions are desirable. Algorithms for this have been given by Furnival and Wilson (1974), Hocking and Leslie (1967), Beale et al. (1967), and LaMotte and Hocking (1970). The Furnival and Wilson algorithm seems to be the one used most often. It uses information from regressions already computed to bound the possible value of the criterion function for regressions as yet not computed. This trick allows skipping the computation of most regressions. The algorithm has been implemented in the BMDP series of programs (BMDP9R, Dixon, 1983) and is also available as subroutine RLEAP in the IMSL library (IMSL, 1979). For $k$ of up to about 30, the cost of finding five subsets with lowest $C_p$ is about the same as the cost of computing a stepwise regression of the same size.

**Highway data (continued).**    As pointed out in Section 8.6, there are 1024 models to consider, 512 with LEN but excluding the three dummy variables for highway types, and 512 with LEN and the dummy variables. The procedure we follow is to compute $C_p$ for the best few of these models, with $\hat{\sigma}^2$ always computed from the full 13-predictor model. Table 8.7 lists the 20 models with smallest $C_p$, 10 with the dummy variables and 10 without them. These were found using the Furnival and Wilson algorithm, requiring less than 1 second on a CDC Cyber 72 computer. Listed in the table are $p$ = number of parameters in the model, $C_p$, $R^2$, $RSS_p$, and the predictors in that subset model; the intercept is included in all models in the table.

All of the minimum-$C_p$ models include LEN, which was forced into all models, and SLIM, while ACPT is included in most models. Adding any one of TRKS, SIGS, ITG, or LANE might have some useful effect, but the need for more than one of them is unlikely.

Examination of Table 8.7 makes clear the fact that there is no model that

Table 8.7   Twenty models with smallest $C_p$

| $p$ | $C_p$ | $R^2$ | $RSS_p$ | Predictors in the Model | | | | | | |
|---|---|---|---|---|---|---|---|---|---|---|
| 4 | .23 | .701 | 44.8465 | LEN | SLIM | ACPT | | | | |
| 5 | .48 | .718 | 42.3333 | LEN | SLIM | SIGS | ACPT | | | |
| 5 | .56 | .717 | 42.4400 | LEN | TRKS | SLIM | ACPT | | | |
| 5 | 1.33 | .710 | 43.5449 | LEN | SLIM | ACPT | LANE | | | |
| 5 | 1.59 | .707 | 43.9135 | LEN | ADT | SLIM | ACPT | | | |
| 5 | 1.63 | .707 | 43.9754 | LEN | SLIM | ITG | ACPT | | | |
| 5 | 1.97 | .703 | 44.4587 | LEN | SLIM | LWID | ACPT | | | |
| 6 | 1.51 | .727 | 40.9282 | LEN | TRKS | SLIM | SIGS | ACPT | | |
| 6 | 2.00 | .722 | 41.6406 | LEN | TRKS | SLIM | ACPT | LANE | | |
| 6 | 2.00 | .722 | 41.6406 | LEN | TRKS | SLIM | ITG | ACPT | | |
| 7 | 4.69 | .716 | 42.6231 | LEN | SLIM | ACPT | FAI | PA | MA | |
| 7 | 5.12 | .712 | 43.2450 | LEN | SLIM | SIGS | FAI | PA | MA | |
| 8 | 3.32 | .748 | 37.7958 | LEN | SLIM | SIGS | ACPT | FAI | PA | MA |
| 8 | 4.68 | .735 | 39.7448 | LEN | TRKS | SLIM | ACPT | FAI | PA | MA |
| 9 | 4.44 | .756 | 36.5374 | LEN | TRKS | SLIM | SIGS | ACPT | FAI | PA | MA |
| 9 | 5.10 | .750 | 37.4670 | LEN | SLIM | ITG | SIGS | ACPT | FAI | PA | MA |
| 9 | 5.12 | .750 | 37.5032 | LEN | SLIM | SHLD | SIGS | ACPT | FAI | PA | MA |
| 9 | 5.25 | .749 | 37.6839 | LEN | ADT | SLIM | SIGS | ACPT | FAI | PA | MA |
| 9 | 5.25 | .748 | 37.6934 | LEN | SLIM | LWID | SIGS | ACPT | FAI | PA | MA |
| 9 | 5.30 | .748 | 37.7629 | LEN | SLIM | SIGS | ACPT | LANE | FAI | PA | MA |

is obviously best, and there are many equally good models. To make a decision as to which model should be adopted, further analysis is clearly indicated. However, SLIM and ACPT in addition to LEN are certainly useful. Also, the importance of the dummy variables can be judged by an $F$-test for these variables adjusted for SLIM, ACPT, and LEN—since $F = [(44.85 - 42.62)(32)]/[(42.62)(3)] = 0.56$ with (3, 32) degrees of freedom, there is little evidence that the dummy variables are important predictors of accident rate.

A reasonable approach to finding a model is to begin with the model using only LEN, SLIM, and ACPT as predictors. A diagnostic analysis, left to the interested reader, is then in order. The regression summary for this model is given in Table 8.8. One interesting feature of this model is the sign of the estimated coefficient for SLIM—higher speed limits (before the 55 mph maximum limit) were associated with lower accident rates. It would be easy—and incorrect—to assert that higher-speed driving lowers accident rates. Rather, one response of a highway department to higher accident rates is to lower speed limits. Thus high accident rates may cause lower speed limits, not the other way around.

The added variable plots first introduced in Section 2.4 can provide a graphic aid in deciding if additional predictors should be added to the model summarized in Table 8.8. As an example of this, the effect of adding SIGS to the model including LEN, SLIM, and ACPT is shown in Figure 8.3. The solid line has slope 0.485, as would be estimated for SIGS if it were added to the model. From the graph, it appears that little systematic information will be gained if SIGS is included; the $t$-value of 1.42 for SIGS in the four-variable model supports this conclusion. Thus we choose not to include SIGS in the model.

**Table 8.8    Regression summary**

| Variable | Estimate | Standard Error | $t$-value |
|---|---|---|---|
| Intercept | 9.32 | 2.617 | 3.56 |
| LEN | − 0.0771 | 0.0249 | − 3.10 |
| SLIM | − 0.102 | 0.0429 | − 2.39 |
| ACPT | 0.101 | 0.0276 | 3.72 |

$\hat{\sigma}^2 = 1.281$, d.f. $= 35$, $R^2 = 0.70$

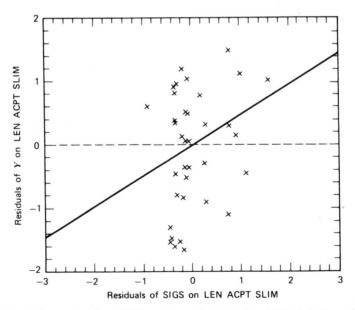

**Figure 8.3**   Effects of adding SIGS to the model including LEN, ACPT, and SLIM.

**The final model.**    There is no final model, only a group of possible models that are all judged nearly equally useful. If a model is to be used for prediction, we may be indifferent to the exact predictors included. Under these circumstances, the parsimony principle—less is better—can often be applied. In other problems, the analyst's knowledge of a problem must be used to reach conclusions.

## Problems

**8.1.** Using the following "data" apply the BE and FS algorithms. Also, find $C_p$ for all possible models, and compare results. What is the "correct model"? (Mantel, 1970)

| $Y$ | $X_1$ | $X_2$ | $X_3$ |
|-----|-------|-------|-------|
| 5  | 1    | 1004 | 6.0  |
| 6  | 200  | 806  | 7.3  |
| 8  | −50  | 1058 | 11.0 |
| 9  | 909  | 100  | 13.0 |
| 11 | 506  | 505  | 13.1 |

**8.2**    In the highway accident data, for the fit of the full model, estimate the effect on accident rate of changing a highway classification (1) from MC to MA, (2) MA to PA, and (3) PA to FAI.

**8.3**    Perform a diagnostic analysis on the highway accident data. Are transformations needed? If so, transform the data and find the models with lowest $C_p$.

**8.4**    For the cloud seeding data of Example 7.2, find the five models with minimum $C_p$.

**8.5**    For the boys in the Berkeley Guidance Study (problem 2.1) find a model for SOMA as a function of the other variables. Perform a complete analysis, including diagnostic analysis, and summarize your results.

**8.6**    Using the methodology outlined by Schatzoff et al. (1968), write a program to compute $C_p$ for all possible regressions.

**8.7**    An experiment was conducted to model oxygen uptake (O2UP), in milligrams of oxygen per minute, from five chemical measurements: biological oxygen demand (BOD), total Kjeldahl nitrogen (TKN), total solids

Table 8.9    Data from oxygen uptake experiment (Moore, 1975)

| Day | BOD | TKN | TS | TVS | COD | O2UP | Log(O2UP) |
|-----|-----|-----|-----|-----|-----|------|-----------|
| 0 | 1125. | 232. | 7160. | 85.9 | 8905. | 36.0 | 1.5563 |
| 7 | 920. | 268. | 8804. | 86.5 | 7388. | 7.9 | 0.8976 |
| 15 | 835. | 271. | 8108. | 85.2 | 5348. | 5.6 | 0.7482 |
| 22 | 1000. | 237. | 6370. | 83.8 | 8056. | 5.2 | 0.7160 |
| 29 | 1150. | 192. | 6441. | 82.1 | 6960. | 2.0 | 0.3010 |
| 37 | 990. | 202. | 5154. | 79.2 | 5690. | 2.3 | 0.3617 |
| 44 | 840. | 184. | 5896. | 81.2 | 6932. | 1.3 | 0.1139 |
| 58 | 650. | 200. | 5336. | 80.6 | 5400. | 1.3 | 0.1139 |
| 65 | 640. | 180. | 5041. | 78.4 | 3177. | 0.6 | −0.2218 |
| 72 | 583. | 165. | 5012. | 79.3 | 4461. | 0.7 | −0.1549 |
| 80 | 570. | 151. | 4825. | 78.7 | 3901. | 1.0 | 0.0000 |
| 86 | 570. | 171. | 4391. | 78.0 | 5002. | 1.0 | 0.0000 |
| 93 | 510. | 243. | 4320. | 72.3 | 4665. | 0.8 | −0.0969 |
| 100 | 555. | 147. | 3709. | 74.9 | 4642. | 0.6 | −0.2218 |
| 107 | 460. | 286. | 3969. | 74.4 | 4840. | 0.4 | −0.3979 |
| 122 | 275. | 198. | 3558. | 72.5 | 4479. | 0.7 | −0.1549 |
| 129 | 510. | 196. | 4361. | 57.7 | 4200. | 0.6 | −0.2218 |
| 151 | 165. | 210. | 3301. | 71.8 | 3410. | 0.4 | −0.3979 |
| 171 | 244. | 327. | 2964. | 72.5 | 3360. | 0.3 | −0.5229 |
| 220 | 79. | 334. | 2777. | 71.9 | 2599. | 0.9 | −0.0458 |

(TS), total volatile solids (TVS), which is a component of TS, and chemical oxygen demand (COD), each measured in milligrams per liter (Moore, 1975). The data were collected on samples of dairy wastes kept in suspension in water in a laboratory for 220 days. All observations were on the same sample over time. We desire an equation relating log(O2UP) to the other variables. The goal is to find variables that should be further studied with the eventual goal of developing a prediction equation (day cannot be used as a predictor). The data are given in Table 8.9. Table 8.10 gives, for all possible regressions, the summary statistics $C_p$, $R^2$, RSS.

A convenient summary for the $C_p$ statistics when the total number of variables considered is small (here, $k=5$) is called a $C_p$ plot, in which the values of $C_p - p$ are plotted against $p$ ($C_p$ plots are often recommended as just $C_p$ versus $p$, but the method suggested here makes interpretation somewhat easier). Good models will have $C_p - p$ generally less than zero.

**8.7.1.** Draw a $C_p$ plot. Find the model with minimum $C_p$. Identify all models for which the $F$-value for the variables not included is less than 2. Summarize the results.

**8.7.2.** Complete the analysis of these data, including a complete diagnostic analysis. What diagnostic indicates the need for transforming O2UP to a logarithmic scale?

**Table 8.10    All possible regressions with log(O2UP) as the response**

| $p$ | $C_p$ | $R^2$ | $RSS$ | Model | | | | |
|---|---|---|---|---|---|---|---|---|
| 2 | 6.29 | .697 | 1.5370 | TS | | | | |
| 2 | 6.57 | .693 | 1.5563 | COD | | | | |
| 2 | 13.50 | .598 | 2.0338 | BOD | | | | |
| 2 | 20.33 | .505 | 2.5044 | TVS | | | | |
| 2 | 56.84 | .008 | 5.0219 | TKN | | | | |
| 3 | 1.74 | .786 | 1.0850 | TS | COD | | | |
| 3 | 5.27 | .738 | 1.3287 | TVS | COD | | | |
| 3 | 6.87 | .716 | 1.4388 | TKN | COD | | | |
| 3 | 6.88 | .716 | 1.4397 | BOD | TS | | | |
| 3 | 7.16 | .712 | 1.4590 | TS | TVS | | | |
| 3 | 7.33 | .710 | 1.4707 | TKN | TS | | | |
| 3 | 7.70 | .704 | 1.4963 | BOD | COD | | | |
| 3 | 9.09 | .686 | 1.5921 | BOD | TKN | | | |
| 3 | 11.33 | .655 | 1.7462 | BOD | TVS | | | |
| 3 | 21.36 | .518 | 2.4381 | TKN | TVS | | | |
| 4 | 2.32 | .805 | 0.9871 | TKN | TS | COD | | |
| 4 | 3.42 | .790 | 1.0634 | TS | TVS | COD | | |
| 4 | 3.44 | .790 | 1.0644 | BOD | TS | COD | | |
| 4 | 5.66 | .760 | 1.2178 | TKN | TVS | COD | | |
| 4 | 6.25 | .752 | 1.2582 | BOD | TKN | TS | | |
| 4 | 6.51 | .748 | 1.2764 | BOD | TKN | COD | | |
| 4 | 7.15 | .739 | 1.3204 | BOD | TVS | COD | | |
| 4 | 8.15 | .726 | 1.3894 | BOD | TS | TVS | | |
| 4 | 8.16 | .726 | 1.3900 | TKN | TS | TVS | | |
| 4 | 8.68 | .718 | 1.4257 | BOD | TKN | TVS | | |
| 5 | 4.00 | .809 | 0.9653 | TKN | TS | TVS | COD | |
| 5 | 4.32 | .805 | 0.9871 | BOD | TKN | TS | COD | |
| 5 | 5.07 | .795 | 1.0388 | BOD | TS | TVS | COD | |
| 5 | 6.78 | .772 | 1.1565 | BOD | TKN | TVS | COD | |
| 5 | 7.70 | .759 | 1.2199 | BOD | TKN | TS | TVS | |
| 6 | 6.00 | .809 | 0.9652 | BOD | TKN | TS | TVS | COD |

**8.8**    Use (8.21) to obtain a test of the NH: $J_p \leqslant p$ versus the alternative AH: $J_p > p$, for a fixed $p$ parameter subset model.

**8.9**    Prove the result (8.5). This can be done directly in the deviations-from-averages form of the regression model using the following fact concerning the inverse of a $2 \times 2$ symmetric matrix:

$$\begin{pmatrix} a & c \\ c & b \end{pmatrix}^{-1} = \frac{1}{ab-c^2} \begin{pmatrix} b & -c \\ -c & a \end{pmatrix}$$

**8.10**    Prove result (8.6). (*Hint:* To avoid tedious algebra, use the Sweep algorithm, outlined in problem 2.7.) Without careful consideration, this problem can be very long and uninformative.

**8.11**   The Galápagos Islands off the coast of Ecuador provide an excellent laboratory for studying the factors that influence the development and survival of different life species. Johnson and Raven (1973) have presented

**Table 8.11    Galápagos Island species data**

| Island | Observed Species | | Area (km²) | Elevation (m) | Distance (km) | | Area of Adjacent Island (km²) |
|---|---|---|---|---|---|---|---|
| | Number of Species | Endemics | | | From Nearest Island | From Santa Cruz | |
| Baltra | 58 | 23 | 25.09 | – | 0.6 | 0.6 | 1.84 |
| Bartolomé | 31 | 21 | 1.24 | 109 | 0.6 | 26.3 | 572.33 |
| Caldwell | 3 | 3 | 0.21 | 114 | 2.8 | 58.7 | 0.78 |
| Champion | 25 | 9 | 0.10 | 46 | 1.9 | 47.4 | 0.18 |
| Coamano | 2 | 1 | 0.05 | – | 1.9 | 1.9 | 903.82 |
| Daphne Major | 18 | 11 | 0.34 | – | 8.0 | 8.0 | 1.84 |
| Darwin | 10 | 7 | 2.33 | 168 | 34.1 | 290.2 | 2.85 |
| Eden | 8 | 4 | 0.03 | – | 0.4 | 0.4 | 17.95 |
| Enderby | 2 | 2 | 0.18 | 112 | 2.6 | 50.2 | 0.10 |
| Espanola | 97 | 26 | 58.27 | 198 | 1.1 | 88.3 | 0.57 |
| Fernandina | 93 | 35 | 634.49 | 1494 | 4.3 | 95.3 | 4669.32 |
| Gardner[a] | 58 | 17 | 0.57 | 49 | 1.1 | 93.1 | 58.27 |
| Gardner[b] | 5 | 4 | 0.78 | 227 | 4.6 | 62.2 | 0.21 |
| Genovesa | 40 | 19 | 17.35 | 76 | 47.4 | 92.2 | 129.49 |
| Isabela | 347 | 89 | 4669.32 | 1707 | 0.7 | 28.1 | 634.49 |
| Marchena | 51 | 23 | 129.49 | 343 | 29.1 | 85.9 | 59.56 |
| Onslow | 2 | 2 | 0.01 | 25 | 3.3 | 45.9 | 0.10 |
| Pinta | 104 | 37 | 59.56 | 777 | 29.1 | 119.6 | 129.49 |
| Pinzon | 108 | 33 | 17.95 | 458 | 10.7 | 10.7 | 0.03 |
| Las Plazas | 12 | 9 | 0.23 | – | 0.5 | 0.6 | 25.09 |
| Rabida | 70 | 30 | 4.89 | 367 | 4.4 | 24.4 | 572.33 |
| San Cristóbal | 280 | 65 | 551.62 | 716 | 45.2 | 66.6 | 0.57 |
| San Salvador | 237 | 81 | 572.33 | 906 | 0.2 | 19.8 | 4.89 |
| Santa Cruz | 444 | 95 | 903.82 | 864 | 0.6 | 0.0 | 0.52 |
| Santa Fé | 62 | 28 | 24.08 | 259 | 16.5 | 16.5 | 0.52 |
| Santa Maria | 285 | 73 | 170.92 | 640 | 2.6 | 49.2 | 0.10 |
| Seymour | 44 | 16 | 1.84 | – | 0.6 | 9.6 | 25.09 |
| Tortuga | 16 | 8 | 1.24 | 186 | 6.8 | 50.9 | 17.95 |
| Wolf | 21 | 12 | 2.85 | 253 | 34.1 | 254.7 | 2.33 |

[a]Near Espanola.
[b]Near Santa Maria.
*Source:* Johnson and Raven (1973).

the data in Table 8.11 giving the number of species and related variables for 29 different islands. Counts are given for both the total number of species and the number of species that occur only on that one island (the endemics).

Use these data to find factors that seem to influence diversity, as measured by some function of the number of species and the number of endemic species, and summarize your results. One complicating factor is that elevation is not recorded for six very small islands, so some provision must be made for this. Four possibilities are: (1) find the elevations; (2) delete these six islands from the data; (3) ignore elevation as a predictor of diversity, or (4) substitute a plausible value for the missing data. Examination of large-scale maps suggests that none of these elevations exceed 200 m.

# 9

# PREDICTION

One of the most important uses of regression is to predict future values of the response variable for given values of the predictor. Regression methods seem to be ideally suited for this task, since the fitted equation is designed to give an expected value for the response given the predictors. We can then proceed almost as if the estimated prediction function represented a true relationship between the variables. For example, suppose a heavy object is dropped from a tall building. The predicted distance the object will fall in $t$ seconds is $4.9t^2$ meters. As long as $t$ is small enough or the building is high enough, the prediction will be quite accurate. The parameter 4.9 meters per second per second, half the acceleration due to gravity, is a known constant. Predictions of distance made with this equation may not agree perfectly with actual observed values because of measurement errors or neglected factors, such as friction.

If the acceleration due to gravity were unknown, predictions could be obtained by collecting data to estimate it, perhaps by dropping objects and measuring the distances they fall in various times. Letting $d$ denote distance and $t$ denote time, we would fit the model $d = \gamma t^2$ to observed data and estimate the unknown value of $\gamma$. Predictions made from equations with estimated parameters will be more variable due to uncertainty in the estimates, but as long as the functional form of the equation is correct, they will be nonetheless reliable.

The more usual regression problem is far more complex than this, since the correct functional form is rarely known. With an empirically chosen model, the usefulness of predictions is much less clear. Generally, we rely on the fact that over limited ranges for the predictors, many models will behave in nearly the same way, so even if we fit the wrong model, we may obtain useful predictions. For some values of the predictors, however, predictions may be useless; see Figure 9.1. In the falling object example, if the true functional form were unknown to us, we might decide, using the methods in this book, to fit a linear model with perhaps log($t$) as a predictor. The resulting predictions can be expected to be reasonably good as long as we predict for values of $t$ roughly in the same range as the values of $t$ in the sample used for model building, because over modest ranges for $t$, $d = \gamma t^2$ can be approximated by $d = \beta_0 + \beta_1 \log(t)$.

If a functional relationship between variables is known, the distinction between estimation of parameters and prediction is blurred, and parameters may have physical interpretations, such as the acceleration due to gravity discussed previously. When the functional relationship is not known, the difference between estimation and prediction is both meaningful and instructive. We estimate parameters with the purpose of obtaining predicted values. The parameters themselves rarely have intrinsic meaning. If we fit $\hat{d} = \hat{\beta}_0 + \hat{\beta}_1 \log(t)$ to a set of falling object data, $\beta_0$ and $\beta_1$ are artifacts of the equation fit, not of the relationship between time and distance. The estimated values $\hat{\beta}_0$ and $\hat{\beta}_1$ depend on the range of the predictor used in estimation, so the parameters $\beta_0$ and $\beta_1$ represent quantities that depend on the data and on the relationship between time and distance; they are "variable constants." However, predictions based on these variable constants may be useful if the new case is from the same population as the original data.

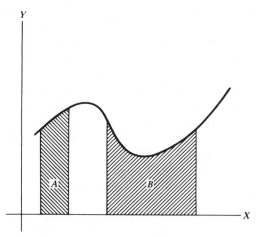

**Figure 9.1**  A linear approximation in range $A$ may be adequate and a quadratic approximation may be reasonable for $B$. Neither would be adequate outside of the range of fitting.

## 9.1   Making predictions

**Where do the data come from?**   Ideally, the data for construction of a prediction function and the future cases to be predicted are, in some sense, a sample from a well-defined population. When this is so, usual modes of statistical inference from a sample to a population apply. Often, however, one cannot guarantee these conditions. For example, in a chemical experiment, cases may be obtained by repeating a procedure with varied combinations of the predictors. If the experiment were done in one day, or by one technician, or in one laboratory, there is no general principle that the estimated prediction equation will apply to experimental results from a different day, technician, or laboratory. One must either assume that the effect due to these factors is negligible, or else collect data from several technicians or laboratories to assess the effects. In other problems, the construction sample used to estimate the prediction function is by design representative of only a small segment of a larger population. For example, a dietary study using only healthy rats may be inappropriate for making predictions for all rats including those that are not healthy.

**Interpolation versus extrapolation.**   Generally, a prediction equation can be expected to apply only for some values of the predictors. In a chemical experiment, for example, changing to an extreme temperature may make the actual outcomes completely different from those predicted from experiments at moderate temperatures. Consequently, in most prediction problems, there is an important question of determining a range of validity of predictions. Predictions in that range are called *interpolations*, while outside that range we get *extrapolations*. It is a sad fact of the predictor's art that extrapolations, though often much less reliable, can be more interesting to the analyst.

**Making predictions.**   As outlined in this book, regression modeling is as much an art as a science. No two problems are likely to be handled in exactly the same way, and, consequently, we cannot easily describe the process that leads to obtaining a prediction equation. Generally, however, we can outline a three-step process: data collection, model selection and estimation, and validation of the model. The data collection is a crucial part of this process, since without good data, the other steps are useless.

Model selection consists of choosing predictors and transformations, including using the diagnostic methods described earlier in this book. The $C_p$ criterion described in Chapter 8 makes the role of prediction explicit, since minimizing $C_p$ is like trying to choose a model to make prediction errors as small as possible. Once a functional form is chosen, the parameters of that function need to be estimated; we will continue to use least squares estimates computed as if the chosen model were decided upon a priori, although other estimation methods may be preferable (Copas, 1983). Thus we assume that $Y$ is the response to be used, and the predictors chosen are given in the $n \times p'$ matrix $\mathbf{X}$. The linear model is

$$\mathbf{Y} = \mathbf{X}\boldsymbol{\beta} + \mathbf{e} \tag{9.1}$$

and the estimate of $\boldsymbol{\beta}$ is taken to be $\hat{\boldsymbol{\beta}} = (\mathbf{X}^T\mathbf{X})^{-1}\mathbf{X}^T\mathbf{Y}$. For a new case with predictors $\mathbf{x}_*$, we wish to predict the as yet unobserved value of the response, $y_*$. The predicted value is $\tilde{y}_* = \mathbf{x}_*^T\hat{\boldsymbol{\beta}}$. If the linear model is correct, the prediction is an unbiased point estimate of $E(y_*)$. The variance of the prediction, *again assuming that the model is correct*, will be $\mathrm{varpred}(\tilde{y}_* \mid \mathbf{x}_*) = \sigma^2(1 + \mathbf{x}_*^T(\mathbf{X}^T\mathbf{X})^{-1}\mathbf{x}_*) = \sigma^2(1 + h_*)$, say. Since $\sigma^2$ is generally estimated by the residual mean square $\hat{\sigma}^2$ given the model (9.1), the estimated standard error of prediction is given by

$$\mathrm{sepred}(\tilde{y}_* \mid \mathbf{x}_*) = \sigma\sqrt{1 + h_*} \tag{9.2}$$

When the model (9.1) is chosen through data analysis, the formula (9.2) for the standard error of a prediction is likely to underestimate the prediction errors. First, the model is often selected to make the estimate of $\sigma^2$ relatively small, so the residual variance may be underestimated. Also, when the functional form is unknown, predictions from (9.1) are likely to be biased, requiring the addition of a squared bias component to get a mean square error to replace (9.2).

Unlike parameters, future values have the useful characteristic of being *observable*. We can use a fitted model to obtain predictions, and then see how well they do, giving a direct estimate of prediction errors. This brings us to the third step in making predictions: validation of the model. This can be done in several ways, but in each there is at least an implicit requirement that the fitted model must be used to obtain predictions for data not used for estimation. We have already encountered two methods for this in Chapter 8, using cross validation and the PRESS or sample reuse statistic to measure average prediction errors.

In *cross validation*, the data are split into two or more subsets. One of the subsets is called the construction set, and is used for estimation. Predictions for the cases in the other subset, called the validation set, can be obtained from the model fit to the construction set. These can be compared to the observed values of the response. One useful criterion function is the square root of the average squared error of prediction. This is an attractive method whenever large amounts of data are available. The data can be divided into subsets randomly, or in some optimal way (Snee, 1977).

The extreme case for cross validation would divide the data into $n$ overlapping subsets, each subset consisting of $n-1$ cases. Estimates from the $n-1$

cases can then be used to predict a value for the deleted case. This leads to the PRESS statistic described in Section 8.8. A sensible estimate of average prediction error, assuming that the data at hand are a sample from the population of future values, is then $(\text{PRESS}/n)^{1/2}$.

*Example 9.1    Predicting the interval to eruption of Old Faithful Geyser*

---

A *geyser* is a hot spring that occasionally becomes unstable and erupts hot water and steam into the air. Different geysers will erupt for various lengths of time and intervals. Probably the most famous geyser is Old Faithful, in Yellowstone National Park, Wyoming. The intervals between eruptions of Old Faithful range from about 30 to 90 minutes. Water shoots to heights generally over 35 meters, with eruptions lasting from 1 to $5\frac{1}{2}$ minutes (Marler, 1969). Because of its regularity and beauty, Old Faithful Geyser is a major tourist attraction, and it is not uncommon for several thousand people to watch an eruption on a summer afternoon.

Prediction of the time of the next eruption of this geyser is of interest both to the Park Service and to visitors. In fact, prediction of the time of the next eruption is posted in prominent locations near the geyser. Given in Table 9.1 are $y =$ interval to the next eruption and $x =$ duration of an eruption, both in minutes, for all eruptions of Old Faithful Geyser between 6 a.m. and midnight for August 1 to August 8, 1978.‡

Data on Old Faithful has been collected for many years by ranger/ naturalists in the park, using a stopwatch. The duration measurements have been rounded to the nearest 0.1 minutes or 6 seconds, while intervals are reported to the nearest minute. The National Park Service uses values of $x$ to predict future values for $y$, and we shall follow their example. It is unlikely that a causal relationship exists between $x$ and $y$, rather both are probably results of some other unobserved cause or set of causes. Modeling the observed association between $x$ and $y$ may lead to useful predictions, but the fitted relationship may have no geological meaning.

---

‡The data given in the first edition of this book were, unfortunately, garbled, and have been corrected for this edition. These data were provided by Roderick A. Hutchinson, Yellowstone Park Geologist.

Table 9.1 Eruptions of Old Faithful Geyser, August 1 to August 8
1978, $x$ = duration, $y$ = interval (both in minutes)

| Date | $y$ | $x$ | Date | $y$ | $x$ | Date | $y$ | $x$ | Date | $y$ | $x$ |
|---|---|---|---|---|---|---|---|---|---|---|---|
| 1 | 78 | 4.4 | 2 | 80 | 4.3 | 3 | 76 | 4.5 | 4 | 75 | 4.0 |
| 1 | 74 | 3.9 | 2 | 56 | 1.7 | 3 | 82 | 3.9 | 4 | 73 | 3.7 |
| 1 | 68 | 4.0 | 2 | 80 | 3.9 | 3 | 84 | 4.3 | 4 | 67 | 3.7 |
| 1 | 76 | 4.0 | 2 | 69 | 3.7 | 3 | 53 | 2.3 | 4 | 68 | 4.3 |
| 1 | 80 | 3.5 | 2 | 57 | 3.1 | 3 | 86 | 3.8 | 4 | 86 | 3.6 |
| 1 | 84 | 4.1 | 2 | 90 | 4.0 | 3 | 51 | 1.9 | 4 | 72 | 3.8 |
| 1 | 50 | 2.3 | 2 | 42 | 1.8 | 3 | 85 | 4.6 | 4 | 75 | 3.8 |
| 1 | 93 | 4.7 | 2 | 91 | 4.1 | 3 | 45 | 1.8 | 4 | 75 | 3.8 |
| 1 | 55 | 1.7 | 2 | 51 | 1.8 | 3 | 88 | 4.7 | 4 | 66 | 2.5 |
| 1 | 76 | 4.9 | 2 | 79 | 3.2 | 3 | 51 | 1.8 | 4 | 84 | 4.5 |
| 1 | 58 | 1.7 | 2 | 53 | 1.9 | 3 | 80 | 4.6 | 4 | 70 | 4.1 |
| 1 | 74 | 4.6 | 2 | 82 | 4.6 | 3 | 49 | 1.9 | 4 | 79 | 3.7 |
| 1 | 75 | 3.4 | 2 | 51 | 2.0 | 3 | 82 | 3.5 | 4 | 60 | 3.8 |
|   |    |     |   |    |     |   |    |     | 4 | 86 | 3.4 |
| 5 | 71 | 4.0 | 6 | 55 | 1.8 | 7 | 81 | 3.5 | 8 | 77 | 4.2 |
| 5 | 67 | 2.3 | 6 | 75 | 4.6 | 7 | 53 | 2.0 | 8 | 73 | 4.4 |
| 5 | 81 | 4.4 | 6 | 73 | 3.5 | 7 | 89 | 4.3 | 8 | 70 | 4.1 |
| 5 | 76 | 4.1 | 6 | 70 | 4.0 | 7 | 44 | 1.8 | 8 | 88 | 4.1 |
| 5 | 83 | 4.3 | 6 | 83 | 3.7 | 7 | 78 | 4.1 | 8 | 75 | 4.0 |
| 5 | 76 | 3.3 | 6 | 50 | 1.7 | 7 | 61 | 1.8 | 8 | 83 | 4.1 |
| 5 | 55 | 2.0 | 6 | 95 | 4.6 | 7 | 73 | 4.7 | 8 | 61 | 2.7 |
| 5 | 73 | 4.3 | 6 | 51 | 1.7 | 7 | 75 | 4.2 | 8 | 78 | 4.6 |
| 5 | 56 | 2.9 | 6 | 82 | 4.0 | 7 | 73 | 3.9 | 8 | 61 | 1.9 |
| 5 | 83 | 4.6 | 6 | 54 | 1.8 | 7 | 76 | 4.3 | 8 | 81 | 4.5 |
| 5 | 57 | 1.9 | 6 | 83 | 4.4 | 7 | 55 | 1.8 | 8 | 51 | 2.0 |
| 5 | 71 | 3.6 | 6 | 51 | 1.9 | 7 | 86 | 4.5 | 8 | 80 | 4.8 |
| 5 | 72 | 3.7 | 6 | 80 | 4.6 | 7 | 48 | 2.0 | 8 | 79 | 4.1 |
| 5 | 77 | 3.7 | 6 | 78 | 2.9 |   |    |     |   |    |     |

The scatter plot of $y$ versus $x$ for the 107 cases in Table 9.1 is given as
Figure 9.2. The plot generally shows that shorter eruptions are followed
by shorter intervals, and longer eruptions by longer intervals. The points
seem to fall into two clusters in the upper right and lower left of the plot.
The general linearity of the graph suggests use of a simple linear regression model,

$$y = \beta_0 + \beta_1 x + \text{error} \qquad (9.3)$$

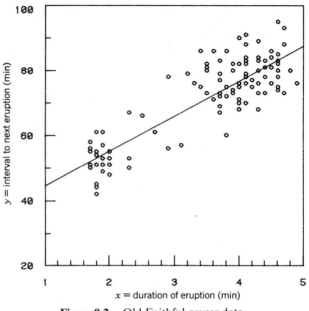

**Figure 9.2**   Old Faithful geyser data.

The fitted model is

$$y = 33.8 + 10.7x \qquad (9.4)$$

with other summary statistics shown in Table 9.2. The standard error of prediction, from (9.2) is

$$\mathrm{sepred}(\tilde{y}_* | x_*) = 6.68 \left[ 1 + \frac{1}{107} + \frac{(x_* - 3.461)^2}{113.83} \right]^{1/2} \qquad (9.5)$$

**Table 9.2   Summary statistics for simple regression, Old Faithful Geyser data**

| Variable | $n$ | Average | Standard Deviation | Minimum | Maximum |
|---|---|---|---|---|---|
| $x$ | 107 | 3.4607 | 1.0363 | 1.7 | 4.9 |
| $y$ | 107 | 71.000 | 12.967 | 42 | 95 |

| Regression of $y$ on $x$ | | | |
|---|---|---|---|
| Variable | Estimate | Standard Error | $t$-Value |
| Intercept | 33.83 | 2.26 | 14.96 |
| $x$ | 10.74 | 0.63 | 17.15 |

$\hat{\sigma} = 6.68$, d.f. $= 105$, $R^2 = .74$

For example, if $x_* = 4.5$ minutes, $\tilde{y}_* = 33.8 + 10.7(4.5) = 82.2$ minutes, with standard error of prediction 6.8 minutes. A 95% prediction interval is obtained using a multiplier from the $t(105)$ distribution, as

$$82.2 - t(0.05; 105)6.8 \leqslant y_* \leqslant 82.2 + t(0.05; 105)6.8$$

$$68.7 \leqslant y_* \leqslant 95.7$$

⚹ Validation of the simple regression model for these data will be done in two ways. First, we can use $(\text{PRESS}/n)^{1/2}$ to estimate the average prediction error; for these data and this model $(\text{PRESS}/n)^{1/2} = (4848/107)^{1/2} = 6.7$ minutes. To use the cross-validation approach, we need more data. Provided in Table 9.3 are the results of 115 eruptions of Old Faithful for the period August 16 to August 23, 1979, a year after the original data. For cross validation to make sense, we must assume that the underlying process that drives $x$ and $y$ has not changed between the two time periods, so we can pretend that the data in Tables 9.1 and 9.3 are drawn from the same population. When model (9.4) is applied to each value of $x$ in the validation data, the square root of the average squared fitting error is $(4029.1/115)^{1/2} = 5.9$ minutes. Given the close agreement between these two methods of validation, we can reasonably expect prediction standard errors of about 6 minutes using (9.4) to predict eruption times of Old Faithful Geyser.

Although the simple model (9.4) will provide predictions that are accurate to within 10 minutes about 90% of the time, some improvements to the model may be desirable. If the data points are examined in time sequence, it will be seen that a short interval, less than 60 minutes, is always followed by a long interval. Long intervals however, often follow long intervals. This suggests that the usual assumption of independence of cases may be violated. One approximate way to incorporate this additional information into the model is to use *past* values of $y$ and $x$ to predict the current values of $y$. For example, we can model $y$, the interval to the next eruption, as

$$y = \beta_0 + \beta_1 x + \beta_2(\text{duration of last eruption})$$
$$+ \beta_3(\text{interval between last two eruptions}) + \text{error} \qquad (9.6)$$

These two additional variables are called *lagged variables*.

To fit model (9.6), the data in Table 9.1 are reduced to 99 cases, since for the first recorded eruption on each day, the lagged interval and

Table 9.3    Eruptions of Old Faithful Geyser, August 1979

| Date | y | x | Date | y | x | Date | y | x | Date | y | x |
|---|---|---|---|---|---|---|---|---|---|---|---|
| 16 | 82 | 4.1 | 17 | 91 | 4.8 | 18 | 58 | 2.2 | 19 | 84 | 4.5 |
| 16 | 80 | 4.2 | 17 | 66 | 4.1 | 18 | 82 | 4.8 | 19 | 72 | 3.8 |
| 16 | 76 | 4.5 | 17 | 71 | 4.0 | 18 | 77 | 4.3 | 19 | 89 | 4.3 |
| 16 | 56 | 1.9 | 17 | 75 | 4.0 | 18 | 75 | 3.8 | 19 | 75 | 4.4 |
| 16 | 82 | 4.7 | 17 | 81 | 4.4 | 18 | 77 | 4.0 | 19 | 57 | 2.2 |
| 16 | 47 | 2.0 | 17 | 77 | 4.1 | 18 | 77 | 4.1 | 19 | 81 | 4.8 |
| 16 | 76 | 4.7 | 17 | 74 | 4.3 | 18 | 53 | 1.8 | 19 | 49 | 1.9 |
| 16 | 61 | 2.5 | 17 | 70 | 4.0 | 18 | 75 | 4.4 | 19 | 87 | 4.7 |
| 16 | 75 | 4.3 | 17 | 83 | 3.9 | 18 | 78 | 4.0 | 19 | 43 | 1.8 |
| 16 | 72 | 4.4 | 17 | 53 | 3.2 | 18 | 51 | 2.2 | 19 | 94 | 4.8 |
| 16 | 74 | 4.4 | 17 | 82 | 4.5 | 18 | 81 | 5.1 | 19 | 45 | 2.0 |
| 16 | 69 | 4.3 | 17 | 62 | 2.2 | 18 | 52 | 1.9 | 19 | 81 | 4.4 |
| 16 | 78 | 4.6 | 17 | 73 | 4.7 | 18 | 76 | 5.0 | 19 | 59 | 2.5 |
| 16 | 52 | 2.1 | 17 | 84 | 4.6 | 18 | 73 | 4.4 | 19 | 82 | 4.3 |
| 20 | 80 | 4.4 | 21 | 84 | 3.7 | 22 | 72 | 3.0 | 23 | 51 | 1.7 |
| 20 | 54 | 1.9 | 21 | 58 | 1.8 | 22 | 54 | 2.1 | 23 | 83 | 4.4 |
| 20 | 75 | 4.7 | 21 | 90 | 4.7 | 22 | 75 | 4.6 | 23 | 76 | 4.2 |
| 20 | 73 | 4.3 | 21 | 82 | 4.5 | 22 | 74 | 4.0 | 23 | 51 | 2.2 |
| 20 | 57 | 2.2 | 21 | 71 | 4.5 | 22 | 51 | 2.2 | 23 | 90 | 4.7 |
| 20 | 80 | 4.7 | 21 | 80 | 4.8 | 22 | 91 | 5.1 | 23 | 71 | 4.0 |
| 20 | 51 | 2.3 | 21 | 51 | 2.0 | 22 | 60 | 2.9 | 23 | 49 | 1.8 |
| 20 | 77 | 4.6 | 21 | 80 | 4.8 | 22 | 80 | 4.3 | 23 | 88 | 4.7 |
| 20 | 66 | 3.3 | 21 | 62 | 1.9 | 22 | 54 | 2.1 | 23 | 52 | 1.8 |
| 20 | 77 | 4.2 | 21 | 84 | 4.7 | 22 | 80 | 4.7 | 23 | 79 | 4.5 |
| 20 | 60 | 2.9 | 21 | 51 | 2.0 | 22 | 70 | 4.5 | 23 | 61 | 2.1 |
| 20 | 86 | 4.6 | 21 | 81 | 5.1 | 22 | 60 | 1.7 | 23 | 81 | 4.2 |
| 20 | 62 | 3.3 | 21 | 83 | 4.3 | 22 | 86 | 4.2 | 23 | 48 | 2.1 |
| 20 | 75 | 4.2 | 21 | 84 | 4.8 | 22 | 78 | 4.3 | 23 | 84 | 5.2 |
| 20 | 67 | 2.6 | | | | | | | 23 | 63 | 2.0 |
| 20 | 69 | 4.6 | | | | | | | | | |

Table 9.4    Regression with lagged variables

| Variable | Estimate | Standard Error | t-Value |
|---|---|---|---|
| Intercept | 64.15 | 6.96 | 9.21 |
| x | 88.85 | 0.80 | 11.12 |
| y (lagged) | −0.54 | 0.10 | −5.12 |
| x (lagged) | 4.08 | 1.17 | 3.48 |

$\hat{\sigma} = 6.07$, d.f. $= 95$, $R^2 = .80$

duration are unknown. The data for August 1, for example, are

| y | x | y(lagged) | x(lagged) |
|---|---|---|---|
| 74 | 3.9 | 78 | 4.4 |
| 68 | 4.0 | 74 | 3.9 |
| 76 | 4.0 | 68 | 4.0 |
| ⋮ | ⋮ | ⋮ | ⋮ |
| 74 | 4.6 | 58 | 1.7 |
| 75 | 3.4 | 74 | 4.6 |

The regression of $y$ on $x$, $x$(lagged), and $y$(lagged) is summarized in Table 9.4. All the predictors have large $t$-statistics. The model appears to be a modest improvement over the model with just $x$, since the residual variance is somewhat smaller and the clustering of cases in Figure 9.1 or in a residual plot versus predicted values would disappear in the equivalent plot for model (9.6); these plots are omitted. To validate this model, we can again use both PRESS and cross-validation. Both these approaches give an average prediction error of about 6.2 minutes. For predictions, there is no apparent reason to use the more complicated (9.6) in place of the simple regression model (9.4).

## 9.2 Interpolation versus extrapolation

Interpolation means prediction for new cases with predictor variables not too different from the values of the predictor variables in the construction sample. For a simple linear regression model, interpolation generally occurs when the predictor is in the range observed in the construction sample; outside of this range, we would call a prediction an extrapolation. In the Old Faithful Geyser data, we should be reluctant to make predictions for $y$ if $x$ is very short (say a few seconds) or very long (say 10 minutes) since, from our data, we have no information about the relationship between $y$ and $x$ under those conditions.

The usefulness of the distinction between interpolation and extrapolation is illustrated in Figure 9.3. If the form of the prediction function is not known a priori, then we have no information on the relationship outside the observed range for the predictor. Any of the dotted paths shown in the figure may be appropriate and, without further information, we cannot know which should

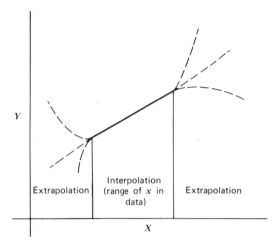

**Figure 9.3**   Extrapolation.

be used. For an interpolation to be useful, only a few assumptions must be satisfied, principally that the new case behaves like those in the construction sample. For extrapolation, the assumption that the estimated prediction function is relevant to cases with predictors outside of the interpolatory range is also needed. For example, rainfall generally increases crop yields, and data collected over a reasonable range of rainfalls might give a good prediction function for yield from rainfall. This function, however, might be useless for extrapolations. According to the equation, zero rainfall might imply negative crop yields and very large rainfalls might result in predictions of very large yields—even though after a point more rainfall will decrease yield.

In multiple regression, it is difficult to define a range of validity for prediction. Clearly, it must depend on the observed data in the construction sample. Suppose, as an example, we consider the Old Faithful Geyser data, except use only $x$ and $x$(lagged) as predictors so a two-dimensional picture is possible. Figure 9.4 gives a scatter plot of $x$ versus $x$(lagged). A prediction would be called an interpolation if it had $(x, x$(lagged)$)$ like those used to construct the prediction equation. This suggests a possible definition of the range of validity to be the smallest closed figure that includes all the points. We shall consider two approximations to this region, one easy to find, one not so easy.

Contours of constant $h = \mathbf{x}^T(\mathbf{X}^T\mathbf{X})^{-1}\mathbf{x}$ are ellipsoids. If $h_{max}$ is the largest $h_{ii}$ in the construction sample, the set of all points $\mathbf{x}_*$ such that $\mathbf{x}_*^T(\mathbf{X}^T\mathbf{X})^{-1}\mathbf{x}_*$ $\leqslant h_{max}$ is an ellipsoid that includes all the construction sample. This ellipsoid is drawn on Figure 9.4 as a solid curve. It is centered at point 1 on the graph.

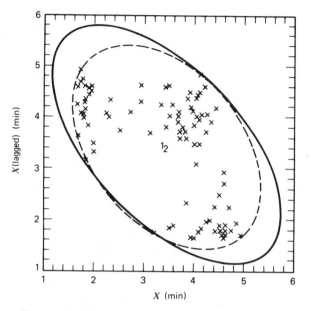

**Figure 9.4** Approximations to interpolatory regions.

For a prediction at point $x_*$, if $h_* = x_*^T(X^TX)^{-1}x_* > h_{max}$, then $x_*$ is not in the ellipsoid, and the prediction can be regarded as an extrapolation.

In Figure 9.4, it is clear that the set defined by the solid curve is much larger than the smallest closed figure containing the observed data. To provide a better approximation, the smallest-volume ellipsoid containing these points may be found. This is called the minimum covering ellipsoid or MCE and is given by the dashed curve centered at point 2. One approach to extrapolation would call any point outside the MCE an extrapolation. Titterington (1978) gives an algorithm for finding the MCE.

## 9.3 Additional comments

**Predictions in a different scale.** A prediction equation may yield predicted values in a transformed scale, and occasionally it may be desirable to change back to the original units as in Example 6.1, where the relationship between brain and body weight of mammal species was discussed, and the equation

$$\log(\text{brain weight}) = 0.9271 + 0.7517 \log(\text{body weight})$$

was found to represent the data well. If a new species has average body weight of 10 kilograms, the predicted log(brain weight) is $0.9271 + 0.7517 \log(10) =$

1.6788. Assuming errors in the log scale to be symmetric, as they would be under a normality assumption, 1.6788 is an estimate of the *mean* of the *predictive distribution* for log(brain weight) given body weight of 10 kg. A naive estimate of brain weight for such a species would be the antilogarithm, $10^{1.6788} = 47.73$ grams. Rather than estimating the mean of the predictive distribution for brain weight, however, this estimates the median. This is a sensible one-number summary since, if errors of log(brain weight) are symmetric, then errors for brain weight are asymmetric, and the median can provide a better summary of the center of an asymmetric distribution.

Confidence and prediction intervals are easily converted from one scale to another by simply reexpressing the limits of a computed confidence statement in the new scale. In the brain weight example, a 95% prediction interval for log(brain weight) of a 10-kg species is given by the set $1.0704 \leqslant$ log(brain weight) $\leqslant 2.2872$. The corresponding 95% confidence interval for brain weight is $10^{1.0704} \leqslant$ brain weight $\leqslant 10^{2.2872}$, or between 11.8 and 193.7 grams, which is a very wide interval. Also, the interval is not symmetric about the naive predictor 47.73 grams, corresponding to the asymmetry of the predictive distribution.

This method of getting confidence statements is always correct in that, given the assumptions, the actual confidence level is equal to the stated confidence level and should therefore be adequate for most purposes. However, the interval may not be optimal in the sense of having the shortest length; see Land (1974).

Before any reexpression of scale is done, the careful analyst should consider whether the new scale is any better than the old. In the brain weight/ body weight example, expression of the weights using logarithms may be more natural if in this scale weight is linearly related to other interesting quantities.

**Prediction and model selection.** There is a close connection between the problems of prediction and of model selection as presented in Chapter 8. The approach taken in this chapter is to use standard output from regression analysis to get the prediction function, but given prediction as a goal, alternative fitting procedures have been derived. Copas (1983) gives an interesting non-Bayesian approach to these alternative methods for prediction and modeling. Picard and Cook (1984) discuss the problems in validating a model chosen using the methods in this book. A Bayesian or predictivist view of these questions, in a more general setting, is given by Aitchison and Dunsmore (1975), Geisser (1980) and by Stone (1974). The Bayesian approach to prediction is potentially very rich, as one can focus attention on the whole predictive distribution of unobserved future values, given the predictors and prior data and information. This can be more informative than just looking at point predictors. For example, if a predictive model is to be used to estimate the severity of an illness, the predictive distribution provides a tool for estimating the probability that a patient's severity exceeds some value that would require action by a physician. Action could then be taken if the probability is estimated to be sufficiently high.

**Computations.**    Both PRESS and the cross-validation sum of squares are fairly easy to compute using most statistical packages. In Minitab, for example, equation (8.23) can be used to compute the $\hat{e}_{(i)}$ from the $\hat{e}_i$ and $h_{ii}$, and equation (8.24) can be used to compute PRESS. Other programs, such as BMDP2R or BMDP9R, automatically compute PRESS whenever the residuals are computed.

Prediction errors for a new set of data may be found in many ways. One useful method is to append the validation set to the construction set, and define a column of weights for weighted least squares with value one for each case in the construction set, and value zero for each case in the validation set. Most programs will ignore zero weight cases in computing estimates, but will use them to compute residuals and related statistics. The cross-validation sum of squares is computed from the residuals for the zero weight cases.

## Problems

**9.1**    In the Old Faithful Geyser data, Table 9.1, any possible effects of day-to day variation have been neglected. How could possible day-to-day variation be included in the model? Under what circumstances would day-to-day variation make predictions impossible? Under what circumstances would predictions be possible but more variable? Fit a model that takes day-to-day effects into account and summarize results.

**9.2**    Consider adding lag-two variables (i.e., the values of $x$ and $y$ for the eruption before last) to model (9.6). Are these variables useful?

**9.3**    (John Rice). In heart catheterization a Teflon tube (or catheter) of 3-millimeter diameter is passed into a major vein or artery at the femoral region and moved into the heart. The catheter can be maneuvered into specific regions to provide information concerning the physiology and function of the heart. This procedure is sometimes done on children with congenital heart defects where the physician must guess the proper length of the catheter.

In a small study of 12 patients the proper length $Y$ of the catheter was determined by checking with a fluoroscope (X-ray) that the catheter tip had reached the aortic valve. Each patient's $X_1 =$ height (inches) and $X_2 =$ weight (in pounds) were also recorded to see if these could be helpful in predicting catheter length (in centimeters). The data are given below.

| $Y$ | $X_1$ | $X_2$ |
|-----|-------|-------|
| 37 | 42.8 | 40.0 |
| 50 | 63.5 | 93.5 |
| 34 | 37.5 | 35.5 |
| 36 | 39.5 | 30.0 |
| 43 | 45.5 | 52.0 |
| 28 | 38.5 | 17.0 |
| 37 | 43.0 | 38.5 |
| 20 | 22.5 | 8.5 |
| 34 | 37.0 | 33.0 |
| 30 | 23.5 | 9.5 |
| 38 | 33.0 | 21.0 |
| 47 | 58.0 | 79.0 |

Construct prediction equations using $X_1$ alone, $X_2$ alone, and both $X_1$ and $X_2$. If an error in prediction of $\pm 2$ centimeters is tolerable, are any of the prediction equations adequate? You must decide what the "$\pm 2$-centimeters" means as a probability statement. Summarize your results.

**9.4** Beginning in the mid-1970s, law and other professional schools have had many more applicants for admission than places for students. As a result, equitable methods of choosing between candidates for admission are very important. Regression ideas are applied in the following way. First data are collected on students currently enrolled in the school. Typical measurements taken are undergraduate grade point average ($X_1$), score on a standardized test such as the Law School Aptitude Test ($X_2$), and a measure of performance, such as first-year grade point average ($Y$). Then, a prediction function of the form $\hat{Y} = \hat{\beta}_0 + \hat{\beta}_1 X_1 + \hat{\beta}_2 X_2$ is estimated (although the estimates may not be obtained by least squares; see Rubin, 1980). In the pool of applicants both $X_1$ and $X_2$ are available, so a value of $\hat{Y}$ can be computed for each applicant, and applicants with large values of $\hat{Y}$ are admitted (although not all schools decide admission solely by $\hat{Y}$).

Discuss the problems in using this methodology. In particular, compare the population for construction to the target population. Suppose the correlation between $X_1$ and $X_2$ for current students turned out to be negative (as it occasionally does). Explain how this could happen. What, if anything, does such a correlation mean? (See also Aitken, 1934, and Lawley, 1943.)

**9.5** Property taxes on a house are supposedly dependent on the current market value of the house. Since houses actually sell only rarely, the sale price of each house must be estimated every year when property taxes are set. Regression methods are sometimes used to make up a prediction function (Renshaw, 1958).

The data in Table 9.5 are for $n = 27$ houses in Erie, Pennsylvania, that were actually sold (Narula and Wellington, 1977). The variables are:

$X_1 = $ current taxes (local, school, and county) $\div 100$ (dollars).

$X_2 = $ number of bathrooms.

Table 9.5 House data[a]

| $X_1$ | $X_2$ | $X_3$ | $X_4$ | $X_5$ | $X_6$ | $X_7$ | $X_8$ | $X_9$ | $Y$ |
|---|---|---|---|---|---|---|---|---|---|
| 4.9176 | 1.0 | 3.4720 | .9980 | 1.0 | 7 | 4 | 42 | 0 | 25.9 |
| 5.0208 | 1.0 | 3.5310 | 1.5000 | 2.0 | 7 | 4 | 62 | 0 | 29.5 |
| 4.5429 | 1.0 | 2.2750 | 1.1750 | 1.0 | 6 | 3 | 40 | 0 | 27.9 |
| 4.5573 | 1.0 | 4.0500 | 1.2320 | 1.0 | 6 | 3 | 54 | 0 | 25.9 |
| 5.0597 | 1.0 | 4.4550 | 1.1210 | 1.0 | 6 | 3 | 42 | 0 | 29.9 |
| 3.8910 | 1.0 | 4.4550 | .9880 | 1.0 | 6 | 3 | 56 | 0 | 29.9 |
| 5.8980 | 1.0 | 5.8500 | 1.2400 | 1.0 | 7 | 3 | 51 | 1 | 30.9 |
| 5.6039 | 1.0 | 9.5200 | 1.5010 | 0 | 6 | 3 | 32 | 0 | 28.9 |
| 15.4202 | 2.5 | 9.8000 | 3.4200 | 2.0 | 10 | 5 | 42 | 1 | 84.9 |
| 14.4598 | 2.5 | 12.8000 | 3.0000 | 2.0 | 9 | 5 | 14 | 1 | 82.9 |
| 5.8282 | 1.0 | 6.4350 | 1.2250 | 2.0 | 6 | 3 | 32 | 0 | 35.9 |
| 5.3003 | 1.0 | 4.9883 | 1.5520 | 1.0 | 6 | 3 | 30 | 0 | 31.5 |
| 6.2712 | 1.0 | 5.5200 | .9750 | 1.0 | 5 | 2 | 30 | 0 | 31.0 |
| 5.9592 | ·1.0 | 6.6660 | 1.1210 | 2.0 | 6 | 3 | 32 | 0 | 30.9 |
| 5.0500 | 1.0 | 5.0000 | 1.0200 | 0 | 5 | 2 | 46 | 1 | 30.0 |
| 8.2464 | 1.5 | 5.1500 | 1.6640 | 2.0 | 8 | 4 | 50 | 0 | 36.9 |
| 6.6969 | 1.5 | 6.9020 | 1.4880 | 1.5 | 7 | 3 | 22 | 1 | 41.9 |
| 7.7841 | 1.5 | 7.1020 | 1.3760 | 1.0 | 6 | 3 | 17 | 0 | 40.5 |
| 9.0384 | 1.0 | 7.8000 | 1.5000 | 1.5 | 7 | 3 | 23 | 0 | 43.9 |
| 5.9894 | 1.0 | 5.5200 | 1.2560 | 2.0 | 6 | 3 | 40 | 1 | 37.5 |
| 7.5422 | 1.5 | 4.0000 | 1.6900 | 1.0 | 6 | 3 | 22 | 0 | 37.9 |
| 8.7951 | 1.5 | 9.8900 | 1.8200 | 2.0 | 8 | 4 | 50 | 1 | 44.5 |
| 6.0931 | 1.5 | 6.7265 | 1.6520 | 1.0 | 6 | 3 | 44 | 0 | 37.9 |
| 8.3607 | 1.5 | 9.1500 | 1.7770 | 2.0 | 8 | 4 | 48 | 1 | 38.9 |
| 8.1400 | 1.0 | 8.0000 | 1.5040 | 2.0 | 7 | 3 | 3 | 0 | 36.9 |
| 9.1416 | 1.5 | 7.3262 | 1.8310 | 1.5 | 8 | 4 | 31 | 0 | 45.8 |
| 12.0000 | 1.5 | 5.0000 | 1.2000 | 2.0 | 6 | 3 | 30 | 1 | 41.0 |

*Source:* Narula and Wellington (1977).

[a]In the original data, one case was accidentally given twice; this case has been deleted here.

$X_3 =$ lot size $\div 1000$ (square feet).

$X_4 =$ living space $\div 1000$ (square feet).

$X_5 =$ number of garage spaces.

$X_6 =$ number of rooms.

$X_7 =$ number of bedrooms.

$X_8 =$ age of house (years).

$X_9 =$ number of fireplaces.

$Y =$ actual sale price $\div 1000$ (dollars).

Use the data to estimate a function to predict $Y$ from the $X$'s and functions of the $X$'s. (In practice, the data set used to estimate sale prices would be much larger than the data set available here, and other variables for neighborhood indicators such as quality of schools, etc. would be included.)

# 10

# INCOMPLETE DATA

---

In many data sets, some variables will be unrecorded for some cases. Indeed, in large studies, complete data are more the exception than the rule. Since the standard methods of analysis can be applied directly only to complete data sets, additional techniques are needed. The most common techniques require that the data set be modified, either by deleting partially observed cases or variables, or by filling in guesses for unobserved values. A usual analysis is then performed, adjusting where necessary to account for the modifications made in the data. Alternatively, methods for analysis of incomplete data without filling-in or deleting have been developed. Use of these generally relies on strong assumptions. As we shall see, neither of these general approaches is wholly satisfactory.

The statistical literature on incomplete data problems is very large, and the treatment of the subject given here is not comprehensive. A bibliography on the subject is given by Afifi and Elashoff (1966) and in the more recent papers cited in this chapter.

## 10.1  Missing at random

Most of the methodology for analyzing incomplete data uses the assumption that the cause of values being unobserved is unrelated to the relationships under study. For example, data that are unobserved because of a dropped test tube or a lost coding sheet would ordinarily satisfy this assumption. Under circumstances like these, analysis that ignores the causes of failure to observe values is justified. On the other hand, if the reason for not observing values depends on the values that would have been observed, then the analysis of

**243**

data must include modeling the cause of the failure to observe values.

Rubin (1976) has made a precise distinction between the two types of incomplete data. For this discussion, an incomplete data set will have values that are *missing at random* (MAR) if the failure to observe a value does not depend on the value that would have been observed. Determining whether an assumption of MAR is appropriate for a particular data set is an important step in the analysis of incomplete data. The following examples illustrate the application of the definition.

**Dart-throwing.**  Imagine that each of the numbers in a data matrix were attached to a dart board, and a random number of darts were thrown at the board. Each number hit by a dart would be missing. In this example, each observation is missing with probability that does not depend on the values observed. This model applies to keypunching errors, among others. The data are clearly MAR.

**Missing pretest.**  In a study of educational achievement a teacher forgets to give one of the pretests. The missing pretest scores are missing at random.

**Randomized experiment.**  In an experiment to compare a treatment and a control, each unit in the study is assigned at random to one group or the other. All units assigned to the treatment group are missing the score they would have obtained had they been given the control rather than the treatment. Their control score is then missing at random. In this sense, all randomized experiments have data that are missing at random, and analysis of the data ignoring the processes that caused missing data, randomization, is permissible.

**Turkey tenderness.**  In an experiment to study a method of grading turkeys for tenderness, a sample of $n = 17$ turkeys had $X =$ empirical estimate of tenderness measured, where $X$ is a score from 1 (very tender) to 5 (very tough). The turkeys were then dressed, frozen, and, after a fixed time thawed and cooked. A laboratory measure of $Y =$ actual tenderness was obtained. However, during storage, three of the turkeys, all rated $X = 1$, were stolen, so their value of $Y$ could not be recorded. The response $Y$ is MAR because the process that caused the missingness may be a function of $X$, but it is not a function of $Y$, even though $X$ and $Y$ are presumably related.

The following are examples where MAR fails.

**Censoring.**  Suppose that one variable in a study is a time to failure, but some units do not fail during the experimental period, so their failure times are not recorded. Missing at random is clearly inappropriate here.

**Small reactions.**   A chemical concentration may not be measurable on some units if the actual value is less than the minimum amount measurable on the equipment being used. Again, failure to measure depends on the values that would be observed, so MAR does not hold.

The following is an example for which the MAR assumption is in doubt.

**Subjects dropping out of an experiment.**   At the beginning of an experiment, subjects are divided into several groups at random. Within each group, a different treatment is applied. At the end of the experiment, perhaps several weeks later, post-treatment scores must be obtained, but some of the subjects are no longer available. For those subjects, the response is missing.

If the reason the subjects dropped out of the experiment was unrelated to the missing variable, MAR is reasonable. Human subjects who move away will generally result in data that are MAR. On the other hand, subjects who drop out because they are not doing well in the experiment will usually violate the MAR assumption.

The importance of the distinction between MAR and non-MAR is this: *If observed data are not MAR, then any inferences that ignore the causes of the missing data may be seriously in error.* For non-MAR data, a reliable analysis will require building a model to account for the unobserved data. The major example of a systematic approach to a non-MAR problem is in the study of censored and survival data, for which book-length treatments are available (Kalbfleisch and Prentice, 1980, and Cox and Oakes, 1984). When the data are MAR, the process causing missing data can be ignored, and the techniques and methods described here are useful.

Careful consideration of the process that generated the data is the best diagnostic tool for studying the assumption of MAR, as illustrated in each of the preceding examples.

## 10.2   Handling incomplete data by filling-in or deleting

The simplest method for analyzing data with a few missing observations is to delete cases, variables, or a combination of the two, and obtain a data set that is complete. As long as the assumption of MAR is tenable, usual methods, with some adjustments, can then be applied to the complete data because the data set obtained by deletion will be representative of the whole. Of course, if the MAR assumption is questionable, then inferences from the data set obtained by deletion are suspect.

If only a few cases have unobserved values, case deletion is the most attractive method available, since the fewest assumptions are required. Cases missing either the response or all the predictors should be deleted.

Cases with several unobserved predictors are strong candidates for deletion since they often contain little useful information.

Variable deletion is more complicated. Were the linear model under consideration known to be exactly true, then deletion of a partially observed variable might be difficult to justify, since an incorrect model would result. Since most linear models are used only as approximations, however, deletion of partially observed variables will often be useful. If a combination of fully observed predictors is highly correlated with a partially observed predictor, the fully observed predictors can be used in place of the partially observed predictor. Here, collinearity works to the advantage of the investigator.

**Fill-in methods.**   Filling in for the missing data achieves more or less the same outcome as deleting: a complete data set results and, with some modification, usual methods of estimation can be applied (testing and confidence statements are not as clear). The problem is deciding on values to fill in for the missing data. The prudent approach to this problem is to use all available information to obtain plausible values for the missing data, fill in various combinations of them, and attempt to monitor the effect of the missing data on the estimation of parameters and on model building.

If additional information external to the data about the missing values is available, it should be used to help choose values to fill in. For data collected in time sequence, for example, missing values can be reliably estimated from observed values for that variable immediately prior to, and after, the missing value. In the Berkeley Guidance Study (problem 2.1), a longitudinal study of growth of children, measurements such as height and weight were obtained for each child at intervals of $\frac{1}{2}$ year up to age 18 years. If the weight of a child at age 8 years were missing, it could be estimated by averaging the observed weights of that child at ages $7\frac{1}{2}$ years and $8\frac{1}{2}$ years. In other problems, similar considerations in the experiment might severely restrict the set of plausible values to some relatively small set.

Lacking additional information bearing directly on the value to be estimated, data that are fully observed can be used to obtain an estimation equation for data that are only partially observed. As a simple example, consider two predictors $X_1$ and $X_2$, both of which are occasionally missing. For the fully observed data, suppose that the scatter plot of $X_1$ versus $X_2$ is as given in Figure 10.1. Actually, this figure is optimistic, since it was generated by making the $(X_1, X_2)$ pairs bivariate normal. The cloud of points is more or less elliptical, and the apparent regression of $X_2$ on $X_1$ is linear. One problem of interest is to obtain values to fill in for $X_2$ for cases in which $X_1$ is observed, but $X_2$ is not. One could compute the regression of $X_2$ on $X_1$ for the complete data and use predictions from this fitted equation to estimate $X_2$ for cases with observed values of $X_1$ but unobserved values for $X_2$. This would correspond to filling in cases so that they lie along the solid line drawn on Figure 10.1, the line of the regression of $X_2$ on $X_1$. Similarly, to fill in for $X_1$, values filled in would fall along the dashed line, the regression of $X_1$ on $X_2$. This process relies on the assumption that the cases to be filled in are similar to those that are complete, so a model fit to the latter gives information about the former.

Since this method tends to make the filled-in cases fall as near the center of the data swarm as possible, the influence of these cases on the resulting data analysis is made small. If several of the variables in a single case are filled in via regression, the influence of this case may be very small indeed. For example, in a four-variable problem, $X_1$, $X_2$, and $X_3$ could each be regressed on $X_4$ for the fully observed cases, and then fill-in values could be estimated for them for cases with $X_4$ observed, but not $X_1$, $X_2$, or $X_3$. If it turned out that the true values of the $X_j$'s for this case were far from the middle of the data, and this case should be influential, then filling in via regression will lose this information.

Filling in via regression has many variations. A model for one predictor as a function of the others could be built using the methodology derived in this book. This could include subset selection, transformation of scales, and so on. Simpler methods, such as using only one or two variable models to fill in, can be used.

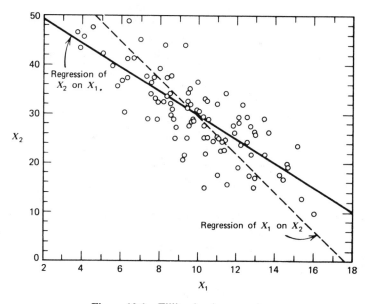

**Figure 10.1** Filling in via regression.

**Adjusting the analysis with filled-in values.** If a data set is filled in using any ad hoc method, a usual analysis can be carried out, except that the number of degrees of freedom for error should be reduced by one for each value filled in. Naturally, this limits the number of values filled in to $n - p'$, the number of degrees of freedom for error, as is quite reasonable, since $n - p'$ values may be found to make $RSS = 0$. Depending on the nature of the data filled in, tests and confidence statements may be seriously in error. Any claims of optimality for estimates, such as unbiasedness or minimum variance, generally cannot be made, in part because the filled-in values are "too good" and do not reflect the usual variability apparent in real data. In addition, very little is known about the behavior of diagnostic procedures applied to filled-in data.

## 10.3   Maximum likelihood estimates assuming normality

If we can assume that all the data, both observed and unobserved, are sampled from a multivariate distribution, usually the normal, then it is possible to compute parameter estimates that make use of all the data. Although these methods are computationally more complex than linear least squares, the maximum likelihood estimates produced are attractive whenever the assumption of multivariate normality makes sense. Two different computational methods have been suggested. Hartley and Hocking (1971) provide an algorithm using a standard iterative procedure called the method of scoring. Dempster, Laird, and Rubin (1977) provide a method called the *EM algorithm* designed specifically for incomplete data situations. Their method follows earlier work by Buck (1960), Orchard and Woodbury (1972), and Beale and Little (1975). Rubin (1974) described conditions under which noniterative computations are possible.

Although methods for finding maximum likelihood estimates with incomplete data have been available for several years, their use has been limited for several reasons. First, the assumption of multivariate normality often is not tenable, although Little (1979) shows how the assumption can be modified if the fully observed predictors are not distributed normally. Second, the more popular EM algorithm does not produce estimates of the standard errors of coefficients (see Louis, 1982 for a method of estimating standard errors). Third, little is known about the behavior of tests and inferential statements in small to moderate samples. Finally, diagnostic procedures have not been carefully studied for incomplete data problems. However, putting aside these difficulties, the maximum likelihood approach using incomplete data is potentially useful.

## 10.4   Missing observation correlation

All of the calculations of aggregate analysis depend on only a few summary statistics, namely, sample averages, variances, and covariances. A commonly applied method with missing data is to compute each of these sample statistics using all the observed data, and then proceed with the least squares analysis as if these were all computed from complete data. This method is often called the missing observation correlation method.

For example, suppose a data set consists of $n = 8$ cases on four variables as shown below.

| Case Number | $Y$ | $X_1$ | $X_2$ | $X_3$ |
|:---:|:---:|:---:|:---:|:---:|
| 1 | X | X | X |  |
| 2 | X | X | X | X |
| 3 | X | X |  |  |
| 4 | X | X |  | X |
| 5 | X | X |  |  |
| 6 | X | X |  | X |
| 7 | X |  | X | X |
| 8 | X |  | X |  |

X = observed      blank = missing

The correlation between $Y$ and $X_1$ would be computed using the first six cases, the correlation between $Y$ and $X_3$ would be based on cases 2, 4, 6 and 7, and so on. For these data, as many as six cases or as few as two cases (for the correlation between $X_2$ and $X_3$) will be used to estimate a correlation.

The success of this method depends on the assumption that the data are a sample from a population, so each correlation in the missing observation correlation matrix is an estimate of a population correlation. As long as all the correlations are small, and sample sizes are not too small, the results of this method can be expected to be reasonable. However, in the presence of large correlations, severe problems may result, including computation of negative sums of squares, or $R^2$ values larger than one. This happens because the computed correlation matrix may not be positive definite when each correlation is based on different cases. Because of this, routine use of the missing observation correlation matrix should be avoided. With it, interpretations of tests, estimates, and predictions are nearly impossible, and no meaningful case analysis has been suggested.

## 10.5    General recommendations

Since missing data cannot always be avoided, some general guidelines for handling a data set that is incomplete are needed. First, study the pattern of the missing data. This can often be done using a special program, such as BMDPAM in the BMDP series (Dixon, 1983). Much can be gained by determining which variables are partially observed, which cases have many missing variables, and the overall pattern of missing data.

If the missing at random assumption is reasonable, then decisions to delete cases or variables can be made. To the data remaining after deletion, fill-in methods may be applied. Or, better yet, several methods can be applied and compared.

We conclude with a warning: *Do not let a computer program decide what is to be done about missing data.* Many large-scale packages can handle missing data, often using one or more of the methods described here, or modifications of them. Keep in mind that the writer of the program did not know the details of your problem. The default analysis used in the program is almost certainly not adequate for your problem.

# 11

# NONLEAST SQUARES ESTIMATION

The least squares estimates used throughout most of this book are not the only estimates that can be used in linear regression problems. They are used because they are computationally simple, geometrically elegant, and, given very strict assumptions, they are, in several important senses, optimal. One main qualification of least squares estimation is that it has been used successfully for over 150 years.

Recently, especially with the availability of cheap, high-speed computing, many competing estimates have been proposed, mostly in response to some perceived deficiency with the least squares estimates. For example, we have seen in Chapter 5 that a single outlier in a data set can have profound influence in tests and on least squares estimates of coefficients. Therefore, one might wish to use an alternative estimator that is less sensitive or more *robust* to the presence of outliers. Robust regression methodology, discussed briefly in Section 11.1, has grown out of this concern for occasional bad data.

Another perceived deficiency of least squares is its behavior in the presence of collinearity. *Ridge regression* and its relatives, discussed in Section 11.2, are designed to produce better estimates for correlated predictors. These particular estimates seem to be falling out of fashion at the moment because they may not provide much advantage over least squares.

251

## 11.1   Robust regression

Good statistical procedures should work well even if the underlying assumptions are somewhat in error. This concern for *robustness*—a term first used in this context by Box (1953)—has been an important component of recent statistical research. The diagnostic methods discussed in Chapters 5 and 6 represent one approach to the question of robustness. Diagnostic procedures are designed to assess the effects on the resulting conclusions of perturbing some aspect of the assumed model, data, or assumptions. For example, Cook's distance measures the effects of perturbing the data by deleting a case. The score test for nonconstant variance is a diagnostic that gives information on the assumption of constant variance.

Robust estimation represents a different use of the ideas of robustness in statistics. Rather than using diagnostics to monitor the effects of perturbations, we seek alternative estimates that will perform reasonably well even given the perturbation. The classic example of a robust estimate is in the problem of estimating the center of a symmetric distribution based on a random sample $y_1, \ldots, y_n$. The usual estimate, the sample average $\bar{y}$, is notoriously nonrobust, since one very large $y_i$ can almost determine the estimate. On the other hand, the sample median of the $y_i$ is insensitive to a small number of large observations, and is robust. Andrews et al. (1972) presented a large study of the properties of a variety of estimates of the center of a symmetric distribution including more than just the average and median.

One important class of robust estimates, due to P. Huber, is called *M-estimates*, a shorthand for maximum likelihood type estimates. These are obtained by choosing the estimates to minimize a function of the residuals other than their sum of squares. For the single sample, let $\mu$ be the center of the symmetric distribution, and let $\tilde{\mu}$ be its estimate. Then, choose $\tilde{\mu}$ to be the value of $\mu$ that minimizes the function

$$\sum_{i=1}^{n} \rho(y_i - \mu)/\sigma \qquad (11.1)$$

where $\rho$ is a function to be specified and $\sigma$ is a scale factor. If $\rho(z) = z^2$, then (11.1) is the usual least squares criterion. Another popular choice is $\rho(z) = |z|^f$, where $f$ is often taken to be a positive constant smaller than 2. The choice of $f = 1$ gives the *least absolute deviation estimator*, which gives the median of a single sample. Smaller values of $f$ will tend to give increasingly smaller weight to large residuals. Another important choice is

$$\rho(z) = \begin{cases} z^2/2 & \text{if } |z| \leqslant c, \ c \text{ a fixed constant} \\ c|z| - c^2/2 & \text{if } |z| > c \end{cases} \qquad (11.2)$$

This choice for $\rho$ results in downweighting cases with large residuals; in the single sample case, this is the same as paying less attention to extreme $y_i$. Huber (1981) provides a complete discussion of these estimates, including the choice of $c$, methods of estimating $\sigma$, computational aspects, and properties.

For the linear regression problem, $M$-estimates are computed using (11.1), except that in (11.1) $\mathbf{x}_i^T \boldsymbol{\beta}$ is substituted for $\mu$. Otherwise all methods are the same.

The $M$-estimates are designed to be robust against outliers, unusual responses, in a data set. They are not designed to be insensitive to any of the other assumptions that are made in any regression analysis. For example, $M$-estimates are no better than least squares if the correct scales of predictors (e.g., logs or square roots) are in doubt. In addition, the $M$-estimates can be as sensitive to high potential, or large $h_{ii}$ cases, as least squares. Compared to many of the other problems in regression analysis, the need for insensitivity to outliers in the response seems unimportant. Therefore, the role of $M$-estimates in regression seems to be limited.

The most accepted use of robust regression seems to be as a check on least squares estimates. Many workers advocate performing both a least squares and a robust analysis. If they agree, more or less, then we can believe in the least squares analysis. If they disagree, try to find out why. Here, the robust estimates are advocated as diagnostics. However, if diagnostics for a specific purpose are available, these should be used.

The literature on robust methodology has become very large in the last few years. Some important references not yet cited include Andrews (1974), Mosteller and Tukey (1977), Devlin, Gnanadesikan, and Kettenring (1975), and Krasker and Welsch (1983).

## 11.2 Biased regression

A second approach to finding alternatives to least squares results in relaxing the necessity of finding unbiased estimates. Consider the linear model given in the deviations from sample averages form,

$$\mathbf{Y} = \mathbf{1}\beta_0 + \mathscr{X}\boldsymbol{\beta} + \mathbf{e} \tag{11.3}$$

where $\mathscr{X}$ is $n \times p$, $\boldsymbol{\beta}$ is $p \times 1$ excluding a term for the intercept, and $\operatorname{cov}(\mathbf{e}) = \sigma^2 \mathbf{I}$. The least squares estimator is $\hat{\boldsymbol{\beta}} = (\mathscr{X}^T \mathscr{X})^{-1} \mathscr{X}^T \mathbf{Y}$. As mentioned in Chapter 2, $\hat{\boldsymbol{\beta}}$ is the minimum variance unbiased estimator, $\operatorname{var}(\hat{\boldsymbol{\beta}}) = \sigma^2 (\mathscr{X}^T \mathscr{X})^{-1}$.

Suppose we enlarge the class of estimators considered to include biased estimators, and consider as a criterion function the sum of the mean square errors of estimating the $\beta_j$'s, SMSE, where

$$\text{SMSE} = \sum_{j=1}^{p} E(\hat{\beta}_j - \beta_j)^2$$

$$= \sum_{j=1}^{p} \{\text{var}(\hat{\beta}_j) + [\text{bias}(\hat{\beta}_j)]^2\}$$

$$= E(\hat{\boldsymbol{\beta}} - \boldsymbol{\beta})^T(\hat{\boldsymbol{\beta}} - \boldsymbol{\beta}) \tag{11.4}$$

This criterion function is different from the one that leads to the least squares estimator, so it should be no surprise that other estimators may be better on this criterion. Before describing such estimators, it is useful to study (11.4) carefully. The important characteristics are: (1) all the $\beta_j$ enter in equally, so, implicitly, interest is equal in each of them; (2) SMSE is not scale invariant, so scaling of the $X$'s is critical; (3) interest centers on estimation of parameters rather than on any other aspect of regression analysis; and (4) covariance between estimators is ignored.

To deal with the lack of invariance, each column of $\mathscr{X}$ is usually standardized so that $\mathscr{X}^T\mathscr{X}$ is the sample correlation matrix (Marquardt and Snee, 1975; Obenchain, 1975). As further notation let $\lambda_1 \geqslant \lambda_2 \geqslant \cdots \geqslant \lambda_p$ be the ordered eigenvalues of $\mathscr{X}^T\mathscr{X}$. Then, Hoerl and Kennard (1970a) have shown that, for $\hat{\boldsymbol{\beta}} =$ least squares estimate,

$$\text{SMSE} = E(\hat{\boldsymbol{\beta}} - \boldsymbol{\beta})^T(\hat{\boldsymbol{\beta}} - \boldsymbol{\beta}) = \sigma^2 \text{ trace}(\mathscr{X}^T\mathscr{X})^{-1} = \sigma^2 \sum_{j=1}^{p} \lambda_j^{-1} \tag{11.5}$$

But $E(\hat{\boldsymbol{\beta}} - \boldsymbol{\beta})^T(\hat{\boldsymbol{\beta}} - \boldsymbol{\beta}) = E(\hat{\boldsymbol{\beta}}^T\hat{\boldsymbol{\beta}}) - \boldsymbol{\beta}^T\boldsymbol{\beta}$. Substituting this into (11.5),

$$E(\hat{\boldsymbol{\beta}}^T\hat{\boldsymbol{\beta}}) = \boldsymbol{\beta}^T\boldsymbol{\beta} + \sigma^2 \sum \lambda_j^{-1}$$

$$\geqslant \boldsymbol{\beta}^T\boldsymbol{\beta} + \sigma^2\lambda_p^{-1} \tag{11.6}$$

Thus even though $\hat{\boldsymbol{\beta}}$ is unbiased for $\boldsymbol{\beta}$, $\hat{\boldsymbol{\beta}}^T\hat{\boldsymbol{\beta}}$ is not unbiased for $\boldsymbol{\beta}^T\boldsymbol{\beta}$ and if the smallest eigenvalue $\lambda_p$ is near zero, then on the average $\hat{\boldsymbol{\beta}}^T\hat{\boldsymbol{\beta}}$ will be much too big; recall that $\lambda_p$ near zero is symptomatic of collinearity. When $\lambda_p$ is small, and (11.4) is of interest, substantial gain over least squares is possible.

Most alternative estimators to be considered here have the common characteristic that they will give an estimate $\tilde{\boldsymbol{\beta}}$ of $\boldsymbol{\beta}$ that is shorter than least squares ($\tilde{\boldsymbol{\beta}}^T\tilde{\boldsymbol{\beta}} < \hat{\boldsymbol{\beta}}^T\hat{\boldsymbol{\beta}}$), so these techniques will shrink the least squares estimators, generally toward the origin $\mathbf{0}$. We have already encountered one of these shrinkers: subset selection, where enough coefficients are set to zero to make the smallest eigenvalue used in (11.6) relatively larger.

Before turning to other specific estimators, it is useful to reexpress the original problem (11.3) in canonical form, in which the columns of $\mathscr{X}$ are replaced by the $p$ orthogonal variates (the principal components). From Section 7.6, there is a $p \times p$ orthogonal matrix whose columns are the eigenvectors of $\mathscr{X}^T\mathscr{X}$, say $\mathbf{U}$ ($\mathbf{U}\mathbf{U}^T = \mathbf{U}^T\mathbf{U} = \mathbf{I}$), and a $p \times p$ diagonal matrix $\mathbf{D}$

with diagonal elements $\lambda_1 \geqslant \lambda_2 \geqslant \cdots \geqslant \lambda_p \geqslant 0$, such that

$$\mathscr{X}^T\mathscr{X} = \mathbf{U}\mathbf{D}\mathbf{U}^T \qquad (11.7)$$

Letting $\mathbf{Z} = \mathscr{X}\mathbf{U}$ (so the columns of $\mathbf{Z}$ are the principal components of $\mathscr{X}$) and $\boldsymbol{\alpha} = \mathbf{U}^T\boldsymbol{\beta}$, then

$$\begin{aligned}
\mathbf{Y} &= \mathbf{1}\beta_0 + \mathscr{X}\boldsymbol{\beta} + \mathbf{e} \\
&= \mathbf{1}\beta_0 + \mathscr{X}(\mathbf{U}\mathbf{U}^T)\boldsymbol{\beta} + \mathbf{e} \\
&= \mathbf{1}\beta_0 + \mathbf{Z}\boldsymbol{\alpha} + \mathbf{e} \qquad (11.8)
\end{aligned}$$

and model (11.8) is equivalent to (11.3). Estimates of $\boldsymbol{\beta}$ and of $\boldsymbol{\alpha}$ are related by the equation

$$(\text{estimate of } \boldsymbol{\alpha}) = \mathbf{U}^T(\text{estimate of } \boldsymbol{\beta}) \qquad (11.9)$$

so either $\hat{\boldsymbol{\alpha}}$ or $\hat{\boldsymbol{\beta}}$ can be computed. However, since $\mathbf{Z}^T\mathbf{Z} = \mathbf{D}$, $\text{var}(\hat{\boldsymbol{\alpha}}) = \sigma^2(\mathbf{Z}^T\mathbf{Z})^{-1} = \sigma^2\mathbf{D}^{-1}$, the $\hat{\alpha}_j$ are independent of each other. Also, $\text{var}(\hat{\alpha}_p) = \sigma^2\lambda_p^{-1}$, and thus $\hat{\alpha}_p$ has larger variance than any other possible estimate—the information in the data concerning the $p$th column of $\mathbf{Z}$ is less than the information for any other possible variable that is a linear combination of the original $X$'s.

**Ridge regression.** The ridge regression estimator $\tilde{\boldsymbol{\beta}}(\text{RR})$, is defined by Hoerl and Kennard (1970a; 1970b), for some $k \geqslant 0$, by

$$\tilde{\boldsymbol{\beta}}(\text{RR}) = (\mathscr{X}^T\mathscr{X} + k\mathbf{I})^{-1}\mathscr{X}^T\mathbf{Y} \qquad (11.10)$$

If $k = 0$, $\tilde{\boldsymbol{\beta}}(\text{RR}) = \hat{\boldsymbol{\beta}}$, the least squares estimator, while larger $k$ will move $\tilde{\boldsymbol{\beta}}(\text{RR})$ away from least squares, and increase the bias in the estimate. The ridge parameter $k$ indexes an infinity of possible ridge estimators.

Figure 11.1 illustrates the effect of increasing $k$ on the ridge estimator. The data used to generate this figure were also used in Figure 4.3; the 95% confidence ellipse for $(\hat{\beta}_1, \hat{\beta}_2)$ based on least squares is shown. The ridge estimates were computed from (11.10), and the curved line gives a plot of $\tilde{\boldsymbol{\beta}}(\text{RR})$ as $k$ increases from 0. At $k = 0$, $\tilde{\boldsymbol{\beta}}(\text{RR}) = \hat{\boldsymbol{\beta}}$. When $k \cong 0.6$, $\tilde{\boldsymbol{\beta}}(\text{RR})$ falls on the boundary of the 95% confidence ellipse. For any $k < 0.6$, $\tilde{\boldsymbol{\beta}}(\text{RR})$ is inside the ellipse, while larger values put $\tilde{\boldsymbol{\beta}}(\text{RR})$ outside the ellipse.

Generally, $k$ is an unknown tuning constant that can be set by the analyst. Large values of $k$ correspond to increased bias but lower variance, so a value of $k$ must be chosen to balance bias against variance. The basic benefit of ridge regression is summarized by the following result of Hoerl and Kennard (1970a): For every fixed $\mathscr{X}$ and $\boldsymbol{\beta}$ in the model (11.3), there is a $k_0$ such that for all $0 < k < k_0$, the SMSE of $\tilde{\boldsymbol{\beta}}(\text{RR})$ is less than the SMSE of $\hat{\boldsymbol{\beta}}$. However, Thisted (1978b) has shown that for any fixed $k > 0$ and any $\mathscr{X}$, there is a regression problem (i.e., an actual value $\boldsymbol{\beta}$) such that the SMSE of $\hat{\boldsymbol{\beta}}$ is less

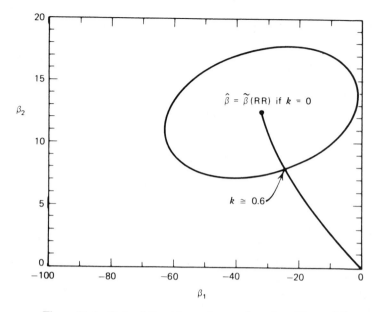

**Figure 11.1**   Path of the ridge estimator (based on Figure 4.3).

than the SMSE of $\tilde{\boldsymbol{\beta}}(RR)$. Thus if the object of using ridge regression is to minimize SMSE, $k$ must be estimated from the data. Methods of estimating $k$ have been suggested by many writers, including Hoerl and Kennard (1970a); a survey of suggested methods is given by Draper and Van Nostrand (1978).

**Canonical form.**   In the canonical form, the ridge estimate $\tilde{\alpha}_j(RR)$ of $\alpha_j$ can be shown to be equal to

$$\tilde{\alpha}_j(RR) = \frac{\lambda_j}{\lambda_j + k}\hat{\alpha}_j \qquad j = 1, 2, \ldots, p \qquad (11.11)$$

where $\hat{\alpha}_j$ is the least squares estimator. The effect of ridge regression is to leave least squares estimates nearly unchanged if $\lambda_j$ is much larger than $k$, but if $\lambda_j$ is small compared to $k$, then the corresponding $\tilde{\alpha}_j(RR)$ will be much smaller than $\hat{\alpha}_j$.

**Relationship to Bayes rules.**   Prior information about parameters is formally incorporated into a problem through the use of Bayesian methods. If we assume that the vector $\boldsymbol{\alpha}$ from (11.8) is drawn from a normal distribution

$$\boldsymbol{\alpha} \sim N(\mathbf{0}, k^{-1}\mathbf{I}) \qquad (11.12)$$

and if we also assume that $\mathbf{e} \sim N(\mathbf{0}, \sigma^2\mathbf{I})$, then the Bayes estimate of $\boldsymbol{\beta}$ in (11.3) is given by (11.10). The value $k^{-1}$, the variance of each $\alpha_j$, represents the prior variability in $\boldsymbol{\alpha}$, and, if the assumptions made are reasonable, the estimator (11.10) is very attractive. However, an assumption of **0** prior mean is rarely reasonable (of course, methods can be modified for other prior means); if it were tenable, collection and analysis of data would probably never have been done.

Examples of the use of ridge regression are given by Marquardt and Snee (1975); also, see Smith and Campbell (1980).

**Generalized ridge regression.**   One generalization of the ridge regression rule is to replace the ridge parameter $k$ by a vector of parameters $(k_1, k_2, \ldots, k_p)$. In the canonical formulation, the generalized ridge estimate $\tilde{\alpha}_j(\text{GR})$, is

$$\tilde{\alpha}_j(\text{GR}) = \frac{\lambda_j}{\lambda_j + k_j} \hat{\alpha}_j \qquad j = 1, 2, \ldots, p \qquad (11.13)$$

Generalized ridge regression rules therefore apply different ridge parameters to each $\lambda_j$, allowing for many different patterns of shrinking the estimates. In terms of the original coordinates, define **G** to be the $p \times p$ matrix such that, if **K** is the diagonal matrix with $k_1, k_2, \ldots, k_p$ on the diagonal, $\mathbf{G} = \mathbf{U}\,\mathbf{K}\mathbf{U}^T$, with **U** as defined near (11.8). Then, the generalized ridge estimator is (Hoerl and Kennard, 1970a; Bingham and Larntz, 1977)

$$\tilde{\alpha}_j(\text{GR}) = (\mathscr{X}^T \mathscr{X} + \mathbf{G})^{-1} \mathscr{X}^T \mathbf{Y} \qquad (11.14)$$

In particular if $k_1 = k_2 = \cdots = k_p = k$, then $\mathbf{K} = k\mathbf{I}$, and $\mathbf{G} = \mathbf{U}(k\mathbf{I})\mathbf{U}^T = k\mathbf{I}$, and ridge regression is a special case of generalized ridge regression.

Various methods of estimating or otherwise determining $k_1, \ldots, k_p$ have been suggested; when a criterion like (11.4) is of interest, methods suggested by Strawderman (1978), Berger (1975), and Thisted (1978a) are relevant. Other criteria will lead to alternative choices for the $k_j$'s, one of which is considered next.

**Regression on principal components.**   One special case of generalized ridge regression results in performing the regression on some of the principal component vectors. If each $k_j$ is set either to 0 or allowed to approach $+\infty$, then the regression on principal component estimators $\tilde{\alpha}_j(\text{PC}), j = 1, 2, \ldots, p$ are given by

$$\tilde{\alpha}_j(\text{PC}) = \begin{cases} \hat{\alpha}_j & \text{if} \quad k_j = 0 \\ 0 & \text{if} \quad k_j \to +\infty \end{cases} \qquad (11.15)$$

The corresponding $\hat{\boldsymbol{\beta}}(\text{PC})$ is found by substituting $\tilde{\boldsymbol{\alpha}}(\text{PC})$ into (11.9) (Marquardt, 1970; Mansfield et al., 1977). If the $k_j$'s for small $\lambda_j$ are allowed to approach $+\infty$, the regression on principal components can have much smaller SMSE than least squares (Dempster, Schatzoff and Wermuth, 1977). A computer program for regression on principal components is included in the BMDP series (BMDP4R, Dixon, 1983).

**James–Stein estimators.** These estimators are based on the result of Stein (1956) and James and Stein (1961), that in the problem of estimating the means of three or more normal distributions, the sample average vector is inadmissible—that is, there exist estimators that are always better, in some sense, than the sample averages. These improved estimators have the vector of sample averages shrunken toward zero or some other point; see Efron and Morris (1973; 1975) for more details.

The usual James–Stein estimators can be obtained in a Bayesian framework by assuming that the $\alpha_j$'s are independent and normally distributed, with mean zero and variance proportional to $\sigma^2/\lambda_j$. To obtain ridge estimators, one assumes variance proportional to $\sigma^2$ (Dempster, 1973; Goldstein and Smith, 1974; Sclove, 1968; Rolph, 1976). Then, the James–Stein estimator $\tilde{\beta}(\mathrm{JS})$, is given by

$$\tilde{\beta}(\mathrm{JS}) = (1 - \hat{B})\hat{\beta} \tag{11.16}$$

A common choice of $\hat{B}$ is

$$\hat{B} = \text{minimum} \left\{ 1, \ \frac{(p-2)(n-p')\hat{\sigma}^2}{\hat{\beta}^T\hat{\beta}(n-p'+2)} \right\}$$

This form of the James–Stein rule (there are others) has the undesirable property of proportionally shrinking each estimator. This rule can also be written

$$\tilde{\beta}(\mathrm{JS}) = [\mathscr{X}^T\mathscr{X} + \hat{B}(1-\hat{B})^{-1}\mathscr{X}^T\mathscr{X}]^{-1}\mathscr{X}^T\mathbf{Y} \tag{11.17}$$

for comparison to other estimators in this chapter. The class of proportionally shrinking James–Stein rules is generated by modifying the definition of $\hat{B}$. Copas (1983) provides an interesting non-Bayesian interpretation of estimates of this type.

**Summary of biased estimators.** All of the biased estimators can result in improved SMSE in some circumstances. Draper and Van Nostrand (1978) report that the improvement over least squares will be very small whenever the parameter $\beta$ is well estimated—that is, if collinearity is not a problem, and $\beta$ is not too close to zero. On the other hand, if $\beta$ is poorly estimated, either because of collinearity or $\beta$ being near zero, the biased estimates may provide a bigger improvement over least squares. But the importance of this improvement is far from clear. If $\beta$ is near zero, the predictors are only slightly related to the response, and more precise estimation may not be of any value when the form of the model is in doubt. If the data are collinear, then for some combinations of parameters, the data contain little reliable information to be used for estimation. While least squares result in poor estimates for these combinations, the biased methods produce somewhat more precise estimates of poorly determined quantities.

# 12

# GENERALIZATIONS OF LINEAR REGRESSION

The linear regression paradigm provides a sufficiently large and complex range of models to suit the needs of many analysts. Yet linear regression cannot be expected to be appropriate for all problems. Sometimes, the response and the predictors are related through a known nonlinear function. When errors can be considered additive and normal, this leads to *nonlinear regression*, described in Section 12.1, which differs from linear regression only in that the response varies as a nonlinear function of the parameters. Section 12.2 contains a brief introduction to *logistic regression*, which is often appropriate when the response is either a success or a failure, or the count of the number of successes in a fixed number of trials. Logistic regression is an example of a *generalized linear model*, a class of models first proposed by Nelder and Wedderburn in a landmark paper in 1972. These models are outlined briefly in Section 12.3.

## 12.1   Nonlinear regression

Consider the problem of modeling $y$ = weight gain as a function of $x$ = amount of dietary supplement supplied to young animals. How are we to model the relationship between $y$ and $x$?

A straight-line simple linear regression model is likely to be inappropriate for this problem for several reasons. The response is likely to be bounded below by the amount of growth that would occur without the supplement. The response may be bounded above by some biological maximum incremental growth that could be due to diet supplementation. Straight lines cannot mimic this behavior since they increase by a constant amount for each unit increase in the predictor and will eventually cross any lower or upper boundary. A three parameter S-shaped curve, however, may adequately describe this behavior. The three parameters could correspond to the minimum value or intercept, the maximum value for any amount of supplementation, called the *asymptote*, and a third parameter controlling the rate of increase from the minimum to the maximum. One model that has these properties is

$$y_i = \theta_1 + \theta_2[1 - \exp(\theta_3 x_i)] + e_i \qquad (12.1)$$

As long as $\theta_3 < 0$, as $x_i$ grows large, the response $y_i$ approaches $\theta_1 + \theta_2$ as an asymptote. If $x_i = 0$, $y_i$ is expected to be $\theta_1$. The third parameter, $\theta_3$, is the rate parameter. Equation (12.1) is but one of many possible models for asymptotic regression, and asymptotic regression is but one of many nonlinear models. It does, however, display the important features of the nonlinear model paradigm: (1) The function relating the response to the predictors is a nonlinear function of the parameters. Model (12.1) is nonlinear in $\theta_3$. (2) Unlike the linear model, there need not be a direct correspondence between predictors and parameters. In (12.1), we have only one predictor, but three parameters. (3) Parameterization is not unique, so many nonlinear regression models are equivalent. For example, the model

$$y_i = \eta_1 + \eta_2(\eta_3)^{x_i} + e_i \qquad (12.2)$$

is equivalent to (12.1), since it can be obtained from (12.1) by defining $\eta_1 = \theta_1 + \theta_2$, $\eta_2 = -\theta_2$, and $\eta_3 = \exp(\theta_3)$. The lack of uniqueness of parameterization for nonlinear models makes fitting and interpreting these models much more complex. (4) Like the linear regression model, the errors $e_i$ are assumed independent, and they enter the model by adding an amount to the response. An assumption of constant variance is usually made, but this can be relaxed, as in linear regression (Sec. 4.1), using weighted least squares. The normality assumption for the $e_i$ plays the same role that it does in the linear model: it is used to make inference statements.

**Estimation.** The standard method of estimating parameters in nonlinear regression is least squares. For the general nonlinear model,

$$y_i = f(\mathbf{x}_i; \boldsymbol{\theta}) + e_i \qquad (12.3)$$

where $\mathbf{x}_i$ is a $p \times 1$ vector of predictors, $\boldsymbol{\theta}$ is a $q \times 1$ vector of parameters, $f$ is a nonlinear function of $\boldsymbol{\theta}$, and $\text{var}(e_i) = \sigma^2/w_i$, with $w_i > 0$ known, the estimator $\hat{\boldsymbol{\theta}}$ is chosen to be the value of $\boldsymbol{\theta}$ that minimizes the weighted residual sum of squares function,

$$RSS(\theta) = \sum_{i=1}^{n} w_i [y_i - f(\mathbf{x}_i; \theta)]^2 \qquad (12.4)$$

If the $e_i$ are independent, $N(0, \sigma^2/w_i)$, then $\hat{\theta}$ is the maximum likelihood estimate of $\theta$. The maximum likelihood estimate of $\sigma^2$ is $\hat{\sigma}^2 = RSS(\hat{\theta})/n$, although the divisor $n - q$ is often used in place of $n$.

**Computations.** Finding least squares estimates generally requires an iterative function minimization routine. It is a well-known fact that no one algorithm will work for all possible nonlinear regression problems, and several programs for minimizing (12.4) may be needed. Convergence of an algorithm to the least squares estimate may be sensitive to choice of starting values and to the parameterization of the model. Many algorithms require computation of the first and possibly second derivatives of the function $f(\mathbf{x}_i, \theta)$ in (12.4) with respect to each of the parameters. Programs that do not require formulas for derivatives will often approximate them numerically.

**Inferential statements.** Inferential statements for nonlinear regression lean heavily on normality and are accurate only for very large samples. In smaller samples, the accuracy of the large-sample results will vary greatly from problem to problem and can depend on the choice of parameterization. Standard errors produced using usual large-sample calculations that are given by most computer packages may be seriously in error in some problems and can either understate or overstate the precision of an estimate. As a first approximation, however, the standard errors can be used as they are used in linear regression. For example, the ratio of an estimate to its standard error can provide a test statistic for the hypothesis that a parameter is equal to zero; approximate $p$-values are obtained from the standard normal distribution, not from a $t$ distribution. Tests for comparing competing models are also possible, as illustrated in Example 12.1.

**Diagnostic methods.** Procedures for studying assumptions like normality, constant variance, and so on, are as yet not well developed for the nonlinear model. However, graphical methods, such as a normal probability plot of residuals, can be suggested as aids in studying assumptions.

*Example 12.1    The effects of a dietary supplement*

---

The data used here come from an experiment using newborn turkeys to compare the effects of two sources of methionine on growth. The turkeys were divided into pens, with 15 turkeys per pen. In each pen, methionine from one of the two sources was added to the standard diet at one of five doses ranging from 0.04 to 0.44% of the total diet. The response variable was average body weight of the animals in the pen (in grams) at four weeks of age. Each dose/source combination was repeated in five pens. Reported in Table 12.1 and graphed in Figure 12.1 are the average responses for the five replicates. These data are taken from a larger study (Noll et al., 1984), in which the mean square between pens treated alike was 343.3 with 72 d.f. This provides a pure error estimate of $\sigma^2$, assuming that residual variance is the same for all treatment groups.

In analyzing this experiment, we have the joint goals of modeling growth as a function of dose and also of comparing the response curves for the two sources. We use a slight generalization of model (12.1). Let $x_{i1} =$ the dose of methionine from source A given to the $i$th group of

Table 12.1 Average body weight of four-week-old male turkeys given supplemental methionine from one of two sources

| Average Body Weight[a] $y_i$ | Dose (%) | |
|---|---|---|
| | Source A $x_{i1}$ | Source B $x_{i2}$ |
| 672 | 0.04 | 0 |
| 709 | 0.10 | 0 |
| 729 | 0.16 | 0 |
| 778 | 0.28 | 0 |
| 797 | 0.44 | 0 |
| 680 | 0 | 0.04 |
| 721 | 0 | 0.10 |
| 750 | 0 | 0.16 |
| 790 | 0 | 0.28 |
| 799 | 0 | 0.44 |

Source: Noll et al. (1984)
[a]Averages are taken over 15 birds in each pen and over 5 pens in each dose/source combination. The residual mean square between pens given the same treatment was 343.3 with 72 d.f.

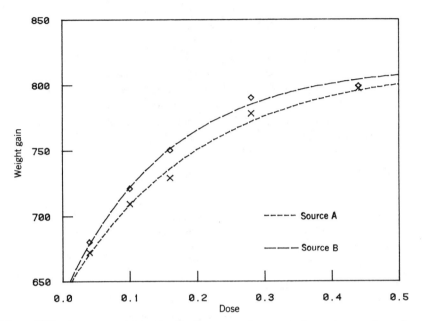

**Figure 12.1**    Average weight gains for five pens of young turkeys given various doses of methionine from two sources. $\times$ = source $A$; $\Diamond$ = source $B$.

pens, and let $x_{i2}$ = the dose of methionine from source B given to the $i$th group of pens. The $x_{i1}$ and $x_{i2}$ are given in columns 2 and 3 of Table 12.1. We then consider the model

$$y_i = \theta_1 + \theta_2[1 - \exp(\theta_3 x_{i1} + \theta_4 x_{i2})] + e_i \qquad (12.5)$$

This model forces the response curves for sources A and B to have common intercept $\theta_1$ and common asymptote $\theta_1 + \theta_2$, but they can have different rate coefficients $\theta_3$ and $\theta_4$. One could also consider a model with differing asymptotes for the two sources, but that approach is not pursued here.

Model (12.5) has been fit with the aid of the program BMDPAR (Dixon, 1983), a general function minimizer that uses numerical differentiation. To use this program, one must specify the form of the model using Fortran-like statements. In addition, one must provide initial guesses for the $\theta_j$'s. The intercept $\theta_1$ will probably be less than the values of the response $y_i$ when the dose is its minimum value of 0.04%; an initial guess of 600 is reasonable. Similarly, $\theta_1 + \theta_2$, the asymptote, should be about 800, suggesting that an initial guess for $\theta_2$ is $800 - 600 = 200$. Initial guesses for the rate parameters are more difficult without more experience; however, they must be negative. The values $-8$ and $-6$ for $\theta_3$ and $\theta_4$, respectively, were used. Although we have assumed that all groups have the same residual variance, we will use weighted least squares with all the $w_i = 5$, the number of pens per average. This will translate the scale of the residual sum of squares to use the pen as the unit of analysis rather than the average of five pens. It will then be in the same units as the pure error estimate of variance.

Using these values and all defaults given in the program, 17 iterations were required for the program to converge to least squares estimates. These are summarized in Table 12.2a, and the corresponding response curves are drawn on Figure 12.1. The standard errors reported use the 6-d.f. residual sum of squares from the regression, and not the 72-d.f. pure error, as an estimate of $\sigma^2$. The curves seem to match the data points very closely. An $F$-test for lack of fit can be obtained as in Section 4.3 as the ratio

$$F = \frac{RSS(\hat{\theta})/\text{d.f.}}{\hat{\sigma}^2(\text{p.e.})} = \frac{732.805/6}{343.3} = 0.35$$

$F$ can be compared approximately to the $F(6, 72)$ distribution in this case; since the observed value is so small, there is no evidence of lack of fit of this model.

Of interest now is to compare the two sources; this can be done in several ways. First, we can fit the model

$$y_i = \theta_1 + \theta_2\{1 - \exp[\theta_5(x_{i1} + x_{i2})]\} + e_i \tag{12.6}$$

Table 12.2 Nonlinear regression summaries

|  | (a) Model (12.5) | | (b) Model (12.6) | | (c) Model (12.7) | |
|---|---|---|---|---|---|---|
|  | Value | Standard error | Value | Standard error | Value | Standard error |
| $\theta_1$ | 638.839 | 6.588 | 640.240 | 11.462 | 638.837 | 6.587 |
| $\theta_2$ | 175.904 | 6.212 | 175.307 | 11.153 | 175.905 | 6.210 |
| $\theta_3$ | $-5.053$ | 0.620 |  |  |  |  |
| $\theta_4$ | $-6.387$ | 0.802 |  |  | $-6.387$ | 0.802 |
| $\theta_5$ |  |  | $-5.554$ | 1.204 |  |  |
| $\theta_6$ |  |  |  |  | 0.791 | 0.049 |
| RSS | 732.805 |  | 2509.595 |  | 732.805 |  |
| d.f. | 6 |  | 7 |  | 6 |  |

which differs from (12.5) only in that a common rate coefficient $\theta_5$ is fit for both sources. We again fit this model using BMDPAR, with results summarized in Table 12.2(*b*). A likelihood ratio test of model (12.5) to model (12.6), essentially comparing the rate coefficients for the two groups, is given by

$$\text{LRT} = -n\left[\ln\left(\frac{RSS(\hat{\theta}|\text{model (12.5)})}{RSS(\hat{\theta}|\text{model (12.6)})}\right)\right] = 12.31$$

The likelihood ratio statistic can be compared to the chi-squared distribution with degrees of freedom equal to the number of parameters in the larger model minus the number of parameters in the smaller model, here equal to 1. For this problem, the *p*-value is very small and model (12.5) provides a much better description of the data. We conclude that the rate parameters are different for the two sources.

Alternatively, we consider using a reparametrized model. In place of (12.5), we can fit

$$y_i = \theta_1 + \theta_2\{1 - \exp[\theta_4(\theta_6 x_{i1} + x_{i2})]\} + e_i \qquad (12.7)$$

In this model, $\theta_6$ is the *relative potency* of source A to source B; $\theta_6 = 1$ corresponds to equal potency. We can fit (12.7), and compare the two sources by a confidence statement for $\theta_6$. The results of this model are summarized in Table 12.2(*c*). The fits of models (12.5) and (12.7) are identical, except for slight differences in the answers produced by the computational procedure; for example, $\hat{\theta}_6$ is related to the $\hat{\theta}$'s from (12.5) by $\hat{\theta}_6 = \hat{\theta}_3/\hat{\theta}_4$. The advantage to the reparameterization is that it gives directly an interpretable relative potency parameter and its standard error. Based on normal theory, a 95% confidence interval for the potency of source A relative to source B is

$$(.79 - 1.96(.0488), .79 + 1.96(.0488)) = (.69, .89).$$

---

**Additional comments.** Ratkowsky (1983) provides a book-length treatment of nonlinear regression and gives many specific examples of its use in models other than asymptotic regression. The more important computing algorithms for nonlinear least squares are described by Kennedy and Gentle (1980, Chapter 10); many packaged programs will have nonlinear least squares routines, but of uneven quality.

Statisticians have recently been studying the geometrical properties of the nonlinear least squares problem. This study is based on the fact that most computational algorithms and inferential procedures can be viewed as approximating the nonlinear regression problem by a linear regression problem that, one hopes, is close to the nonlinear problem for values of $\theta$ near the true value. In some problems, this approximation works very well, but in others it can do very poorly. Beale (1960) and Box (1971) first described these problems of inference due to the poor linear approximation for non-linear models, but the more recent paper by Bates and Watts (1980) has fostered widespread interest in this area. It is likely that this continuing work will have a profound influence on the practice of fitting nonlinear models.

## 12.2    Logistic regression

In some regression problems, the response variable is categorical, often either success or failure. For such problems, the normal linear model is sure to be inappropriate, because normal errors do not correspond to a zero/one response. One important method that can be used in this situation is called _logistic regression._

Table 12.3 summarizes an experiment carried out by R. Norell. Of interest is the effect of small electrical currents on farm animals, with the eventual goal of understanding the effects of high-voltage powerlines on livestock. This experiment was carried out with 7 cows, and 6 shock intensities, 0, 1, 2, 3, 4, and 5 milliamps (shocks on the order of 15 milliamps are painful for many humans; Dalziel et al., 1941). Each cow was given 30 shocks, five at each intensity, in random order. The entire experiment was then repeated, so each cow received a total of 60 shocks. For each shock, the response, mouth movement, was either present or absent. The data in Table 12.3 give the total number of responses, out of 70 trials, at each shock level. For the

**Table 12.3   Response of seven cows to six different intensities of very small shocks**

| Current (milliamperes) | Number of Trials | Number of Responses | Proportion of Responses |
|---|---|---|---|
| 0 | 70 | 0 | .000 |
| 1 | 70 | 9 | .129 |
| 2 | 70 | 21 | .300 |
| 3 | 70 | 47 | .671 |
| 4 | 70 | 60 | .857 |
| 5 | 70 | 63 | .900 |

_Source:_ Rick Norell.

moment, we ignore cow differences and differences between blocks (experiments).

Let $y_i$ be the observed number of successes out of $n_i$ trials; the $n_i$ and $y_i$ are given in columns 2 and 3 of Table 12.3. Suppose that $y_i$ is a binomial random variable, with $n_i$ trials, and with probability of success on any trial equal to $\theta_i$, with $0 \leqslant \theta_i \leqslant 1$ unknown. In logistic regression, we model $\theta_i$ as a function of the predictors, here intensity, $x_i$. The simple linear regression model, $\theta_i = \beta_0 + \beta_1 x_i$ will not be adequate for this purpose if for no other reason than $\theta_i$ is bounded between 0 and 1, but $\beta_0 + \beta_1 x_i$ is not bounded. Furthermore, the graph in Figure 12.2 of the observed proportions of success, $y_i/n_i$, versus $x_i$ appears more S-shaped than linear. This sort of S-shaped behavior is very common in modeling a binomial response as a function of predictors. We need to use a different functional form relating $\theta_i$ to $x_i$.

The objectives outlined in the last paragraph can be met by using the logit transform of $\theta_i$, defined to be

$$\text{logit}(\theta_i) = \ln\left[\frac{\theta_i}{(1-\theta_i)}\right] \tag{12.8}$$

The logit is the logarithm of the *odds* of success, the ratio of the probability of success to the probability of failure. It has several nice properties. First, as

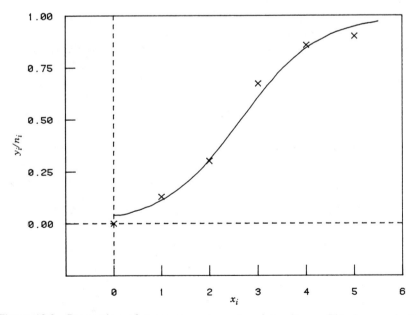

**Figure 12.2**  Proportion of successes $= y_i/n_i$ versus intensity $x_i$. Fitted curve is the logistic regression curve.

$\theta_i$ increases, so does logit($\theta_i$). Second, logit($\theta_i$) varies over the whole real line, whereas $\theta_i$ is bounded between 0 and 1. If $\theta_i < .5$, logit($\theta_i$) is negative, while if $\theta_i > .5$, logit($\theta_i$) is positive.

The logistic regression model can then be expressed in two equivalent ways. First, we can fit a linear model in the logit scale,

$$\text{logit}(\theta_i) = \beta_0 + \beta_1 x_i \tag{12.9}$$

This is almost the same as modeling the logit of the proportion of successes $y_i/n_i$ as a linear function of the predictors. Solving (12.9) for $\theta_i$, using (12.8), we get the form

$$E\left(\frac{y_i}{n_i}\right) = \theta_i = \frac{\exp(\beta_0 + \beta_1 x_i)}{1 + \exp(\beta_0 + \beta_1 x_i)} \tag{12.10}$$

Equation (12.10) expresses the model as an S-shaped curve in the original probability scale. Equations (12.9) and (12.10) are equivalent.

The structure of the logistic regression model consists of three parts. First, the response variable consists of independent binomial counts of the number of successes in a known number of trials. Second, the probability of success depends on the predictors only through a linear form. Third, there is a function, here the logit transform, that links the linear form in the predictors to the expected value of the binomial count. This three part structure is the basis for the class of generalized linear models discussed in Section 12.3.

**Estimation.** The available data consist of ($y_i$, $x_i$, $n_i$), for $i = 1, 2, \ldots,$ $N$; we use $N$ here to distinguish the number of binomials from the number $n_i$ of trials per binomial. Since $\text{var}(y_i/n_i) = \theta_i(1 - \theta_i)/n_i$, the variance for each binomial will be different. The weighted least squares regression of logit($y_i/n_i$) on the predictors with weights $w_i = n_i/[\theta_i(1 - \theta_i)]$ would seem to be appropriate; unfortunately, the $\theta_i$, and hence the $w_i$, are unknown. An iterative procedure can be used, however, first estimating the $\theta$'s, then computing the $w_i$ given the $\theta$'s. A slight elaboration on this idea will lead to maximum likelihood estimates (McCullagh and Nelder, 1983, Sec. 2.5 and Chap. 4). The algorithm works as follows:

1. Obtain initial estimates of the $\beta$'s and hence, through (12.10), of the $\theta_i$. Taking the initial estimates of the $\beta$'s to be zero is often adequate for this algorithm; this is the same as taking the initial estimates of all the $\theta_i$ to be 0.5. Call the current estimates $\tilde{\beta}_j$ and $\tilde{\theta}_i$.

2. Given the current estimates, compute an adjusted response $z_i$, where

$$z_i = \text{logit}(\tilde{\theta}_i) + \frac{(y_i - n_i\tilde{\theta}_i)}{n_i\tilde{\theta}_i(1 - \tilde{\theta}_i)}$$

Using the adjusted response in the iteration leads to the maximum likelihood estimates.

3. Set $w_i = n_i/[\tilde{\theta}_i(1 - \tilde{\theta}_i)]$. Compute the linear regression of $z_i$ on the predictors using the $w_i$ as weights. Use the resulting estimates of the $\beta_j$'s to update the estimates of the $\theta_i$.

4. Continue steps 2 and 3 until a stopping criterion is met, perhaps until the maximum change in a $\hat{\beta}_j$ is sufficiently small. Call the final estimates $\hat{\beta}_j$ and $\hat{\theta}_i$. The usual weighted standard errors are the correct large-sample standard errors for the $\hat{\beta}_j$.

**Deviance.**   The analogue to the residual sum of squares in linear regression is the *deviance*, defined for logistic regression by

$$\text{deviance} = 2 \sum_{i=1}^{N} \left[ y_i \ln \left( \frac{y_i}{n_i \hat{\theta}_i} \right) + (n_i - y_i) \ln \left( \frac{n_i - y_i}{n_i - n_i \hat{\theta}_i} \right) \right]$$

Like the usual residual sum of squares, the deviance has $N - p'$ degrees of freedom, where $p'$ is the number of $\beta$'s in the linear form. The deviance is useful in some goodness of fit tests, and changes in deviance between various models are used in significance testing.

For the data of Table 12.3, the program GLIM (Baker and Nelder, 1979) has been used to estimate coefficients. The results are shown in Table 12.4, and the resulting logistic function is graphed in Figure 12.2. A rough test of the dependence of the response on the predictor is obtained as the ratio of the estimate for $\beta_1$ to its standard error, $1.246/.1119 = 11.13$. Comparing this to the standard normal distribution would give an extremely small $p$-value, confirming that response rate increases with increasing current. A more reliable test is obtained by fitting the model

$$E(y_i) = \frac{\exp(\beta_0)}{1 + \exp(\beta_0)} \tag{12.11}$$

as summarized in the last two columns of Table 12.4. To compare two nested

**Table 12.4   Logistic regression summaries for two models**

|  | Model (12.9) | | Model (12.11) | |
|---|---|---|---|---|
|  | Estimate | Standard Error | Estimate | Standard Error |
| $\beta_0$ | $-3.301$ | .3238 | $-.0953$ | .0977 |
| $\beta_1$ | 1.246 | .1119 | | |
| Deviance | 9.35 (4 d.f.) | | 250.5 (5 d.f.) | |

models like these, we can compute the change in deviance, $250.5 - 9.4 = 241.1$, which has $5 - 4 = 1$ degree of freedom. This can then be compared to the chi-squared distribution with 1 d.f. to get $p$-values; again we have clear evidence that response rate increases with intensity. For linear models, the two approaches to testing two competing linear models that lead to $t$- and $F$-tests were equivalent; for other models such as logistic regression, they are not equivalent and could give conflicting results. The change in deviance method is to be preferred.

**Additional comments.**   The logistic regression model arises naturally in the log-linear approach to categorical data problems; see Fienberg (1981). The model is also discussed in greater detail than presented here by McCullagh and Nelder (1983) and Cox (1970). It naturally extends to more than one predictor and even to factorial designs, just as linear models extend simple regression. The change in deviance can be used to compare various sub-models; the program GLIM is admirable for this purpose. However, the use of deviance statistics in general is not completely understood.

In our example, all the $n_i$ were equal and greater than one, but there is no reason why some or even all of the $n_i$ cannot equal one. For example, one may wish to model the odds of success of an operation depending on several characteristics of a patient, such as age, sex, severity of disease, other illnesses, and the like. In this type of study it is likely that each subject will have a unique set of predictors.

In this brief introduction, diagnostics for logistic regression have been omitted, not because they are unnecessary, but rather because there is not yet an accepted body of methods. Pregibon (1980, 1981), and Landwehr, Pregibon, and Shoemaker (1984) have presented some approximate diagnostics roughly equivalent to many of the methods in Chapters 5 and 6; see also Cook and Weisberg (1982, Sec. 5.4).

Finally, the computational algorithm for logistic regression discussed in this section is not the only one for finding maximum likelihood estimates. As with nonlinear regression, algorithms that use second derivatives are available, and these may be preferable in some problems. Kennedy and Gentle (1980, Chapter 10) is again a useful reference for computational methods.

## 12.3   Generalized linear models

Logistic regression is one special case of the class of *generalized linear models* (GLMs) first proposed by Nelder and Wedderburn (1972). These models keep the idea of a linear predictor by requiring that the response depend on the predictors only through a linear form. They generalize linear models in two ways: through the specification of a *link function* that ties the expectation of response to the linear predictor and through an *error function* for the distribution of the errors. These generalizations allow many methods for linear models to be carried over to more general problems. As in linear regression, the goal is to model a response $y_i$ as a function of $p$ predictors, $x_{i1}, \ldots, x_{pi}, i = 1, \ldots, n$.

**The linear predictor.**  The expectation of the $i$th response, $E(y_i)$, may depend on $\mathbf{x}_i$ only through the linear predictor $\boldsymbol{\beta}^T \mathbf{x}_i$, where as usual $\boldsymbol{\beta}$ is a $p' \times 1$ vector of unknown parameters, possibly including an intercept.

**The link function.**  The link function specifies the relationship between the linear predictor and $E(y_i)$, providing the first generalization of the linear model. We have so far encountered two link functions. The first is the *identity link*, which states that

$$E(y_i) = \mathbf{x}_i^T \boldsymbol{\beta}$$

This is the link function used in usual linear models. The link function for logistic regression was given at (12.9),

$$\mathbf{x}_i^T \boldsymbol{\beta} = \ln \left[ \frac{E(y_i/n_i)}{1 - E(y_i/n_i)} \right] \tag{12.12}$$

When (12.12) is solved for $E(y_i/n_i)$, we get the regression model,

$$E\left(\frac{y_i}{n_i}\right) = \frac{\exp(\mathbf{x}_i^T \boldsymbol{\beta})}{1 - \exp(\mathbf{x}_i^T \boldsymbol{\beta})} \tag{12.13}$$

Several other common link functions are listed in Table 12.5. For example, the log link asserts that

$$\mathbf{x}_i^T \boldsymbol{\beta} = \ln(E(y_i))]$$

corresponding to the regression model

$$E(y_i) = \exp(\mathbf{x}_i^T \boldsymbol{\beta})$$

This may be a very sensible model if the $y_i$ are strictly positive. The second column of Table 12.5 gives the models that correspond to the link functions shown.

**The error function.**  The final part of a generalized linear model is the random component. We retain the assumption that cases are independent, but drop the assumptions of additive and normal errors. We can choose the error function from any of the *exponential family* of distributions, but the most common choices are shown in the last column of Table 12.5. Thus, for

**Table 12.5   Common link and error functions**[a]

| Link Function | Inverse Link Function (Regression Model) | Canonical Error Function |
|---|---|---|
| Identity $\mathbf{x}^T\boldsymbol{\beta} = E(y)$ | $E(y) = \mathbf{x}^T\boldsymbol{\beta}$ | Normal |
| Log $\mathbf{x}^T\boldsymbol{\beta} = ln[E(y)]$ | $E(y) = \exp(\mathbf{x}^T\boldsymbol{\beta})$ | Poisson |
| Logit $\mathbf{x}^T\boldsymbol{\beta} = \text{logit } [E(y)]$ | $E(y) = \dfrac{\exp(\mathbf{x}^T\boldsymbol{\beta})}{1 + \exp(\mathbf{x}^T\boldsymbol{\beta})}$ | Binomial |
| Inverse $\mathbf{x}^T\boldsymbol{\beta} = \dfrac{1}{E(y)}$ | $E(y) = \dfrac{1}{\mathbf{x}^T\boldsymbol{\beta}}$ | Gamma |

[a]A subscript indicating case number has been suppressed for simplicity.

example, to get the normal linear model, we would assume $y_i$ to be normally distributed with mean $\mathbf{x}_i^T\boldsymbol{\beta}$ and unknown variance $\sigma^2$. If we assume $y_i$ to be a Poisson random variable with mean $\exp(\mathbf{x}_i^T\boldsymbol{\beta})$, we get a Poisson regression model.

All three components of a GLM are present in the case of logistic regression. The probability of response is assumed to depend on predictors only through a linear function. The link function is the logit. The error function is the binomial. Fitting GLMs is usually accomplished via maximum likelihood. The algorithm outlined in the last section generalizes to fitting all GLMs and is the basis of the program GLIM (Baker and Nelder, 1979). A deviance statistic, analogous to the residual sum of squares, is a routine part of the fitting of a GLM.

The linear predictor, the link function and the error function completely specify a generalized linear model. As pointed out in the logistic regression case, the cost of generalization is increased difficulty in computation and in interpretation of results. The benefits are great. GLMs will be appropriate for a much wider class of problems. For example, GLMs provide an easy method for incorporating the analysis of discrete data into the same framework used for analyzing continuous data. Much of the intuition developed for linear regression models, such as factorial structure for analysis of variance problems, analysis of covariance, and methods for comparing regressions in different groups can be carried over to a wider class of problems. On balance, generalized linear models seem to provide an important unifying methodology for regression modeling.

**Additional comments.**   McCullagh and Nelder (1983) provide a comprehensive introduction to both the theory and practice of generalized linear models. Table 12.5 implies that links and error functions in one row must be fit with one another. While use of these pairs is often justified, other combinations may be used, as McCullagh and Nelder illustrate.

## Problems

**12.1**   The data in Table 12.6 come from an experiment similar to the one described in Section 12.1, except in this experiment, a different number of pens was used at each level of methionine, allocated according to an optimal design to minimize variance (see Noll et al., 1984). Also, a control ($x_1 = x_2 = 0$) was included. Repeat the analysis given in the text, but for these data. Based on these data, is Source A as potent as Source B?

**12.2**   The data in Table 12.7 give the detailed results of the shock intensity experiment described in Section 12.2. The data are given as the number of positive responses in five trials for each of seven cows at each of six levels of current. All of the observations in the second block were taken after all of the observations in the first block; results may differ between blocks due to

Table 12.6   Average body weight of four-week-old male turkeys given supplemental methionine from one of two sources (Noll et al., 1984)

| Average Body Weight[a] $y_i$ | Number of Pens $m_i = w_i$ | Dose (%) | |
|---|---|---|---|
| | | Source A $x_{i1}$ | Source B $x_{i2}$ |
| 674 | 10 | 0 | 0 |
| 764 | 5 | 0.12 | 0 |
| 795 | 2 | 0.22 | 0 |
| 796 | 2 | 0.32 | 0 |
| 826 | 5 | 0.44 | 0 |
| 782 | 5 | 0 | 0.12 |
| 834 | 2 | 0 | 0.22 |
| 836 | 2 | 0 | 0.32 |
| 830 | 5 | 0 | 0.44 |

[a]Averages are taken over 15 birds in each pen, and over $m_i$ pens for the $i$th dose/source combination. The pure error estimate of variance is 449.2 with 66 d.f.

Table 12.7  The complete shock data[a]

| Cow No. | Block No. | Current Level (milliamperes) | | | | | |
|---|---|---|---|---|---|---|---|
| | | 0 | 1 | 2 | 3 | 4 | 5 |
| 1 | 1 | 0 | 1 | 1 | 5 | 5 | 5 |
| 1 | 2 | 0 | 0 | 4 | 5 | 5 | 5 |
| 2 | 1 | 0 | 0 | 2 | 5 | 5 | 5 |
| 2 | 2 | 0 | 0 | 0 | 4 | 5 | 5 |
| 3 | 1 | 0 | 0 | 1 | 3 | 5 | 5 |
| 3 | 2 | 0 | 0 | 0 | 3 | 5 | 5 |
| 4 | 1 | 0 | 0 | 3 | 3 | 4 | 4 |
| 4 | 2 | 0 | 2 | 2 | 3 | 5 | 5 |
| 5 | 1 | 0 | 1 | 1 | 3 | 4 | 5 |
| 5 | 2 | 0 | 1 | 0 | 1 | 4 | 3 |
| 6 | 1 | 0 | 2 | 2 | 2 | 5 | 5 |
| 6 | 2 | 0 | 0 | 0 | 2 | 0 | 1 |
| 7 | 1 | 0 | 2 | 3 | 5 | 5 | 5 |
| 7 | 2 | 0 | 0 | 2 | 3 | 3 | 5 |

[a]Values in the table are the number of positive responses in five trials.

fatigue of the animals or due to learning. Analyze these data by fitting a logistic regression model with block and cow effects as well as current effects. Summarize results.

# APPENDIX

## 1A.1   A formal development of the simple regression model

Suppose we have two quantities, a predictor $X$ and a response $Y$, and the actual relationship between $X$ and $Y$ is determined by some unknown function $f$, so that

$$Y = f(X) \tag{1A.1}$$

By collecting data, we want to study $f$, and thus study the relationship between $X$ and $Y$. To this end, for each of $n$ units or cases, we observe values $x_i$ of $X$ and $y_i$ of $Y$, $i = 1, 2, \ldots, n$, where

$$y_i = f(x_i) + \varepsilon_i \tag{1A.2}$$

and $\varepsilon_i$ is a random error representing variability in the observational process due to measurement error, neglected factors, and the like.

Now suppose that the shape of the unknown $f$ can be approximated by a straight line. For this to be valid, the scales of $X$ and/or $Y$ may need to be changed or else the range of values of $X$ considered may be limited. In any case, view the straight line as a first try at approximating $f$, and, if later analysis suggests this model to be inadequate, other analyses should be substituted. Thus $f(x)$ is approximated by $\beta_0 + \beta_1 x_i$ for some $\beta_0$, $\beta_1$ and

$$f(x_i) = \beta_0 + \beta_1 x_i + \delta_i \tag{1A.3}$$

where $\delta_i$ is the fixed or lack of fit error reflecting the inadequacy of the straight line in matching $f$, $\delta_i = f(x_i) - \beta_0 - \beta_1 x_i$. For the simple regression model to be useful, the $\delta_i$ should be small (negligible) when compared to the $\varepsilon_i$.

**276**

Combining (1A.2) and (1A.3), and defining $e_i = \varepsilon_i + \delta_i$, we get the simple regression model

$$y_i = \beta_0 + \beta_1 x_i + e_i \qquad i = 1, 2, \ldots, n \tag{1A.4}$$

where the $e_i$'s, as in the text, consist of a fixed component and a random component.

In this development, we have taken the $x_i$'s to be measured without error. Including errors in the $X$'s complicates some of the analysis, and, whenever possible, it is useful to assume that errors in $X$ are relatively small. Methodology for checking this assumption is a topic in Chapter 3.

## 1A.2    Means and variances of random variables

Suppose we let $u_1, u_2, \ldots, u_n$ be random variables, and also let $a_0, a_1, \ldots, a_n$ be $n + 1$ known constants.

$E$ **notation.**    The symbol $E(u_i)$ is read as the expected value of the random variable $u_i$. The term "expected value" is the same as the term "mean value," or less formally, the arithmetic average for a very large sample size. The statement $E(u_i) = 0$ means that the average value we would get for $u_i$, if we sampled its distribution repeatedly, is 0; however, any specific realization of $u_i$ that we observe is likely to be nonzero.

The expectation of a sum of random variables may be symbolically expressed by the following two equations:

$$E(a_0 + a_1 u_1) = a_0 + a_1 E(u_1) \tag{1A.5}$$

$$E\left(a_0 + \sum a_i u_i\right) = a_0 + \sum a_i E(u_i) \tag{1A.6}$$

For example, suppose that $u_1, u_2, \ldots, u_n$ make up a random sample, and $E(u_i) = \mu$, a constant value for all $i = 1, 2, \ldots, n$. Then, the expected value of the sample average of the $u_i$'s, $\bar{u} = \sum u_i/n = (1/n)u_1 + (1/n)u_2 + \cdots + (1/n)u_n$, can be found from equation (1A.6) with $a_i = 1/n$, $i = 1, 2, \ldots, n$, and $a_0 = 0$. Thus

$$E(\bar{u}) = \sum \left(\frac{1}{n}\right) E(u_i) = \left(\frac{1}{n}\right) \sum \mu = \left(\frac{1}{n}\right) n\mu = \mu \tag{1A.7}$$

so that the sample average is an unbiased estimate of the mean $\mu$.

**var Notation.**    The symbol $\mathrm{var}(u_i)$ is read as the variance of $u_i$. The variance is defined by the equation $\mathrm{var}(u_i) = E[u_i - E(u_i)]^2 =$ the expected square of the difference between an observed value for $u_i$ and its average value. The larger var$(u_i)$ is, the more variable observed values for $u_i$ are likely to be. The symbol $\sigma^2$ is often used for a variance, or $\sigma_u^2$ might be used for the variance of the $u$'s when several variances are being discussed.

The general rule for the variance of a sum of random variables (if the variables are uncorrelated) is

$$\text{var}\left(a_0 + \sum a_i u_i\right) = \sum a_i^2 \, \text{var}(u_i) \tag{1A.8}$$

The $a_0$ term vanishes: the variance of a constant is zero. We can now apply this equation to find the variance of the sample average, assuming that the $u_i$'s are uncorrelated with common variance var$(u_i) = \sigma_u^2$:

$$\text{var}\left(\sum \left(\frac{1}{n}\right) u_i\right) = \sum \left(\frac{1}{n}\right)^2 \text{var}(u_i) = n \left(\frac{1}{n}\right)^2 \sigma_u^2 = \frac{\sigma_u^2}{n}$$

**cov Notation.**   The symbol cov$(u_i, u_j)$ is read to be the covariance between the random variables $u_i$ and $u_j$ and is defined by the equation cov$(u_i, u_j) = E[(u_i - E(u_i))(u_j - E(u_j))]$. The covariance describes the way two random variables jointly vary. If the two variables are independent, then they are uncorrelated, but not necessarily conversely. If $i = j$ in the definition, then we see that cov$(u_i, u_i) = \text{var}(u_i)$ by the definition of the latter symbol. The rule for covariance is

$$\text{cov}(a_0 + a_1 u_1, a_3 + a_2 u_2) = a_1 a_2 \, \text{cov}(u_1, u_2) \tag{1A.9}$$

Often, a scale-free version of the covariance is used in its place. This is called the *correlation coefficient*, abbreviated as corr$(u_i, u_j)$, and is defined by the equation

$$\text{corr}(u_i, u_j) = \frac{\text{cov}(u_i, u_j)}{\sqrt{\text{var}(u_i)\text{var}(u_j)}} \tag{1A.10}$$

The correlation does not depend on the units of measurement for the random variables and has a value between $+1$ and $-1$. If the correlation is zero, then the variables $u_i$ and $u_j$ are uncorrelated; this will happen only if cov$(u_i, u_j) = 0$.

The general form of the variance of linear combination of random variables depends both on the variances of the variables, and on their covariances, according to the following rule:

$$\text{var}(a_0 + \sum a_i u_i) = \sum_{i=1}^{n} a_i^2 \, \text{var}(u_i) + 2 \sum_{i=1}^{n-1} \sum_{j=i+1}^{n} a_i a_j \, \text{cov}(u_i, u_j) \tag{1A.11}$$

## 1A.3 Least squares

The least squares estimates of $\beta_0$ and $\beta_1$ in simple regression are those values $\hat{\beta}_0$ and $\hat{\beta}_1$ that minimize the residual sum of squares function,

$$RSS(\beta_0, \beta_1) = \sum_{i=1}^{n} (y_i - \beta_0 - \beta_1 x_i)^2 \qquad (1A.12)$$

One method of minimizing (1A.12) is to differentiate with respect to $\beta_0$ and $\beta_1$, set the derivatives equal to zero, and solve the resulting equations. Carrying out this plan,

$$\frac{\partial RSS(\beta_0, \beta_1)}{\partial \beta_0} = -2 \sum_{i=1}^{n} (y_i - \beta_0 - \beta_1 x_i) = 0$$

$$\frac{\partial RSS(\beta_0, \beta_1)}{\partial \beta_1} = -2 \sum_{i=1}^{n} x_i(y_i - \beta_0 - \beta_1 x_i) = 0 \qquad (1A.13)$$

Upon rearranging terms, (1A.13) becomes

$$\beta_0 n + \beta_1 \sum x_i = \sum y_i$$

$$\beta_0 \sum x_i + \beta_1 \sum x_i^2 = \sum x_i y_i \qquad (1A.14)$$

Equations (1A.14) are called the *normal equations* for model (1.2). The data are used only through the aggregates or sufficient statistics $\sum x_i$, $\sum y_i$, $\sum x_i^2$, and $\sum x_i y_i$, or, equivalently, through the numerically more stable $\bar{x}$, $\bar{y}$, $SXX$, and $SXY$; these are the aggregate statistics that would be found if the deviations from sample average form of the model were used. Any two sets of data with these quantities the same will have the same estimators $\hat{\beta}_0$ and $\hat{\beta}_1$. Solving the two linear equations (1A.14) gives

$$\hat{\beta}_0 = \bar{y} - \hat{\beta}_1 \bar{x}$$

$$\hat{\beta}_1 = \frac{SXY}{SXX} \qquad (1A.15)$$

## 1A.4 Means and variances of least squares estimates

The least squares estimates are linear functions of the $y_i$, $i = 1, \ldots, n$, and, since the $y_i$'s are linear functions of the $e_i$'s, we will be able to apply the results of section 1A.2 to the estimates found in 1A.3 to get the means, variances, and covariances of the estimates. In particular, assume the simple regression model

$$y_i = \beta_0 + \beta_1 x_i + e_i \qquad i = 1, 2, \ldots, n$$

is correct. By results (1A.6) and (1A.8), $E(y_i) = \beta_0 + \beta_1 x_i$, and $\mathrm{var}(y_i) = \mathrm{var}(e_i) = \sigma^2$. Now, consider the estimate $\hat{\beta}_1$, given in (1A.15). Suppose we define the constants $c_1, c_2, \ldots, c_n$ by the equation (for each $i$)

$$c_i = \frac{(x_i - \bar{x})}{SXX} \qquad i = 1, 2, \ldots, n$$

Since the $x_i$ are considered to be fixed numbers, so are the $c_i$. The estimator $\hat{\beta}_1$ is equal to $\sum c_i y_i$, a linear combination of the $y_i$'s. The mean of $\hat{\beta}_1$ is then found as

$$E(\hat{\beta}_1) = E\left(\sum c_i y_i\right) = \sum c_i E(y_i)$$

$$= \sum c_i (\beta_0 + \beta_1 x_i) = \beta_0 \sum c_i + \beta_1 \sum c_i x_i$$

But one can show by direct summation $\sum c_i = 0, \sum c_i x_i = 1$, giving

$$E(\hat{\beta}_1) = \beta_1 \tag{1A.16}$$

which shows that, as long as $E(y_i) = \beta_0 + \beta_1 x_i$, $\hat{\beta}_1$ is an unbiased estimate of $\beta_1$. Also, one can easily show that $E(\hat{\beta}_0) = \beta_0$.

The variance of $\hat{\beta}_1$ is found by

$$\mathrm{var}(\hat{\beta}_1) = \mathrm{var} \sum (c_i y_i)$$

$$= \sum c_i^2 \, \mathrm{var}(y_i) + 2 \sum_{i=1}^{n-1} \sum_{j=i+1}^{n} c_i c_j \, \mathrm{cov}(y_i, y_j)$$

but $\mathrm{cov}(y_i, y_j) = \mathrm{cov}(\beta_0 + \beta_1 x_j + e_j, \beta_0 + \beta_1 x_i + e_i) = \mathrm{cov}(e_j, e_i) = 0$ by assumption. Also $\mathrm{var}(y_i) = \mathrm{var}(e_i) = \sigma^2$ by assumption. Hence

$$\mathrm{var}(\hat{\beta}_1) = \sigma^2 \sum c_i^2$$

but $\sum c_i^2 = 1/SXX$, so that

$$\mathrm{var}(\hat{\beta}_1) = \sigma^2 \frac{1}{SXX} \tag{1A.17}$$

To find the variance of $\hat{\beta}_0$, write

$$\mathrm{var}(\hat{\beta}_0) = \mathrm{var}(\bar{y} - \hat{\beta}_1 \bar{x})$$

$$= \mathrm{var}(\bar{y}) + \bar{x}^2 \, \mathrm{var}(\hat{\beta}_1) - 2\bar{x} \, \mathrm{cov}(\bar{y}, \hat{\beta}_1)$$

Now, $\text{var}(\bar{y}) = \sigma^2/n$, $\text{var}(\hat{\beta}_1)$ is given by (1A.17), and $\text{cov}(\bar{y}, \hat{\beta}_1) = 0$. This last result can be shown by application of the rules in the last section, but it is intuitively clear because the average value $\bar{y}$ should not depend in any way on the fitted slope $\hat{\beta}_1$. Thus we get

$$\text{var}(\hat{\beta}_0) = \sigma^2 \left( \frac{1}{n} + \frac{\bar{x}^2}{SXX} \right) \tag{1A.18}$$

Finally,

$$
\begin{aligned}
\text{cov}(\hat{\beta}_0, \hat{\beta}_1) &= \text{cov}(\bar{y} - \hat{\beta}_1 \bar{x}, \hat{\beta}_1) \\
&= \text{cov}(\bar{y}, \hat{\beta}_1) - \bar{x}\, \text{cov}(\hat{\beta}_1, \hat{\beta}_1) \\
&= -\frac{\sigma^2 \bar{x}}{SXX}
\end{aligned}
\tag{1A.19}
$$

A further application of these results gives the variance of a fitted value $\hat{y}_i = \hat{\beta}_0 + \hat{\beta}_1 x_i$:

$$
\begin{aligned}
\text{var}(\hat{y}_i) &= \text{var}(\hat{\beta}_0 + \hat{\beta}_1 x_i) \\
&= \text{var}(\hat{\beta}_0) + x_i^2\, \text{var}(\hat{\beta}_1) + 2x_i\, \text{cov}(\hat{\beta}_0, \hat{\beta}_1) \\
&= \sigma^2 \left( \frac{1}{n} + \frac{\bar{x}^2}{SXX} \right) + \sigma^2 x_i^2 \frac{1}{SXX} - 2\sigma^2 x_i \frac{\bar{x}}{SXX} \\
&= \sigma^2 \left[ \frac{1}{n} + \frac{(x_i - \bar{x})^2}{SXX} \right]
\end{aligned}
\tag{1A.20}
$$

The variance of a prediction $\tilde{y}_*$ of the future value $y_*$ at $x_*$ is a bit more complicated. Both $\tilde{y}_*$ and $y_*$ have the same expectation, so we can write

$$
\begin{aligned}
\text{E}(\tilde{y}_* - y_*)^2 &= \text{E}\{[\tilde{y}_* - \text{E}(y_*)] - [y_* - \text{E}(y_*)]\}^2 \\
&= \text{E}[\tilde{y}_* - \text{E}(y_*)]^2 + \text{E}[y_* - \text{E}(y_*)]^2 \\
&\quad - 2\text{E}[\tilde{y}_* - \text{E}(y_*)][y_* - \text{E}(y_*)]
\end{aligned}
$$

Since $\tilde{y}_*$ is computed from the past data, and $y_*$ is an observation in the future, the prediction and the future value are uncorrelated, and the covariance term in the last expression is zero. The first term in square brackets is the same as (1A.20), with $x_i$ replaced by $x_*$. The second term in square brackets is just the variance of the future value, equal to $\sigma^2$. Thus the last expression is

$$\text{var}(\tilde{y}_*) = \sigma^2 \left[ \frac{1}{n} + \frac{(x_* - \bar{x})^2}{SXX} \right] + \sigma^2 \tag{1A.21}$$

which, upon rearranging terms, gives (1.36).

## 1A.5 Rounding, round-off error, and accuracy of regression calculations

Most regression calculations are done on a computer, often using standard statistical packages. The results reported by these programs will appear to be accurate to many digits; reported eight- or even ten-place accuracy is not uncommon. However, the uncritical use of such apparently accurate numbers should not be encouraged. First, solutions to statistical problems will rarely have more digits of accuracy than the accuracy of the measurements themselves. If, for example, temperatures are measured to three or four accurate digits, then estimates and predictions should give about the same accuracy. Second, some programs may use inaccurate computational procedures that will occasionally make all the reported digits incorrect. Finally, there is a real question of how many digits are required to make reported results comprehensible to others.

We consider first the problem of computational errors in regression by computing a sum of squares. Suppose $x_1 = 12541$, $x_2 = 12537$, and $x_3 = 12548$. Using the uncorrected formula for $SXX$, but rounding the result of each multiplication to seven digits *before* adding, as is done on some computers, will give

$$\sum x_i^2 = (12541)^2 + (12537)^2 + (12548)^2$$

$$= 157276700 + 157176400 + 157452300$$

$$= 471905400$$

Now

$$\bar{x} = (12541 + 12537 + 12548)/3 = 12542$$

so

$$n\bar{x}^2 = 3(12542)^2 = 3(157301800) = 471905400$$

which gives

$$SXX = \sum x_i^2 - n\bar{x}^2 = 0.$$

The correct computation gives

$$SXX = (12541 - 12542)^2 + (12537 - 12542)^2 + (12548 - 12542)^2 = 62$$

The uncorrected formula with seven digits of accuracy gives no accurate digits in the computed solution.

Chan, Golub, and LeVeque (1983) discuss several methods of computing sums of squares that are likely to be much more accurate than the "usual" method just used, but faster than the direct computation of $SXX$ by the two-pass method of first computing $\bar{x}$, and then on the second pass computing each of the $x_i - \bar{x}$, squaring and adding them up. The alternative method is called *updating*, and after reading $x_1, \ldots, x_m$, we will have computed

$$\bar{x}_m = \frac{1}{m} \sum_{i=1}^{m} x_i \quad \text{and} \quad SXX_m = \sum_{i=1}^{m} (x_i - \bar{x}_m)^2$$

When the next observation $x_{m+1}$ becomes available, we can update these estimates using the equations

$$SXX_{m+1} = SXX_m + \frac{m}{m+1} (x_{m+1} - \bar{x}_m)^2$$

$$\bar{x}_{m+1} = \bar{x}_m + \frac{1}{m+1} (x_{m+1} - \bar{x}_m)$$

If we define $\bar{x}_0 = SXX_0 = 0$, the description of the algorithm is complete. Programming this updating method on a computer is no harder than programming either the usual method or the two-pass method, but it is generally more accurate than the former and faster than the latter. The formula for updating a cross product $SXY$, given new observations $(x_{m+1}, y_{m+1})$ and current estimates $\bar{x}_m$, $\bar{y}_m$, and $SXY_m$, is

$$SXY_{m+1} = SXY_m + \frac{m}{m+1} (x_{m+1} - \bar{x}_m)(y_{m+1} - \bar{y}_m)$$

Even when computations are accurate, it is often desirable to round final results, but not intermediate results, to just a few significant digits. Most people cannot distinguish between a correlation of 0.752 and 0.773, and both may be conveniently rounded to 0.8 without any loss of essential information. Similarly, the standard error of regression in Forbes' data could be rounded to 0.38, since the response had been rounded to two decimal digits before computations.

Ehrenberg (1981) discusses the common problem that many people have with assimilating numerical information. He ascribes this problem more to the inability of analysts to present data sensibly than to a lack in the human character. He suggests several good rules for the presentation of numerical information, including rounding to two digits, and effective use of the layout of the numbers to guide the eye and to facilitate comparisons. While his rules are often violated in this book, since the computations are presented with sufficient accuracy for the interested reader to duplicate, they are sensible rules for the user of regression to follow.

## 2A.1   A brief introduction to matrices and vectors

A complete presentation of matrices and vectors is not attempted in this book. Two useful references on linear algebra with applications in statistics are Graybill (1969) and Searle (1982), although the material necessary for this book should be contained in any good linear algebra book.

A *matrix* (plural: matrices) is a rectangular array of numbers. We would say that $\mathbf{X}$ is an $r \times c$ matrix if it is an array of numbers with $r$ rows and $c$ columns. A specific $4 \times 3$ matrix $\mathbf{X}$ is

$$\mathbf{X} = \begin{bmatrix} 1 & 2 & 1 \\ 1 & 1 & 5 \\ 1 & 3 & 4 \\ 1 & 4 & 6 \end{bmatrix} = \begin{bmatrix} x_{11} & x_{12} & x_{13} \\ x_{21} & x_{22} & x_{23} \\ x_{31} & x_{32} & x_{33} \\ x_{41} & x_{42} & x_{43} \end{bmatrix} = (x_{ij})$$

An element of a matrix $\mathbf{X}$ is given by $x_{ij}$, meaning the number in the $i$th row and the $j$th column of $\mathbf{X}$. For example, in the preceding matrix, $x_{23} = 5$. The usual convention in this book is to name matrices with boldface letters and elements of a matrix by lowercase subscripted letters.

A *vector* is a matrix with 1 column. A specific $4 \times 1$ matrix $\mathbf{Y}$ (a vector of length 4) is given by

$$\mathbf{Y} = \begin{bmatrix} y_1 \\ y_2 \\ y_3 \\ y_4 \end{bmatrix} = \begin{bmatrix} 2 \\ 3 \\ -2 \\ 0 \end{bmatrix}$$

Vectors are also denoted in boldface type, and the elements of a vector are singly subscripted. Thus $y_3 = -2$ is the third element of $\mathbf{Y}$. We complicate the notation somewhat in the case of a special vector, namely, the parameter vector $\boldsymbol{\beta}$ defined by

$$\boldsymbol{\beta} = \begin{bmatrix} \beta_0 \\ \beta_1 \\ \vdots \\ \beta_p \end{bmatrix}$$

where $\boldsymbol{\beta}$ usually starts with $\beta_0$ rather than with $\beta_1$ and $\boldsymbol{\beta}$ is $(p+1) \times 1$ if the intercept is in the model

A *row vector* is a matrix with one row. In this book, all vectors are column vectors. If a vector is needed to represent a row, a transpose of a column vector will be used (see below).

A *square matrix* has the number of rows $r$ equal to the number of columns $c$. A square matrix $\mathbf{X}$ is *symmetric* if $x_{ij} = x_{ji}$ for all $i$ and $j$. A square matrix is *diagonal* if all elements off the main diagonal are zero, $x_{ij} = 0$ unless $i = j$. The matrices **C** and **D** below are symmetric and diagonal, respectively,

$$\mathbf{C} = \begin{pmatrix} 7 & 3 & 2 & 1 \\ 3 & 4 & 1 & -1 \\ 2 & 1 & 6 & 3 \\ 1 & -1 & 3 & 8 \end{pmatrix} \qquad \mathbf{D} = \begin{pmatrix} 7 & 0 & 0 & 0 \\ 0 & 4 & 0 & 0 \\ 0 & 0 & 6 & 0 \\ 0 & 0 & 0 & 8 \end{pmatrix}$$

The diagonal matrix with all elements on the diagonal equal to one is called the *identity matrix*, for which the symbol **I** is used. Sometimes the identity will be written as $\mathbf{I}_n$, indicating that the identity matrix is $n \times n$:

$$\mathbf{I}_4 = \begin{pmatrix} 1 & 0 & 0 & 0 \\ 0 & 1 & 0 & 0 \\ 0 & 0 & 1 & 0 \\ 0 & 0 & 0 & 1 \end{pmatrix}$$

A *scalar* is a $1 \times 1$ matrix, an ordinary number. Scalars are usually not subscripted.

**Operating with matrices: addition and subtraction.**    Two matrices can be added or subtracted only if they have the same number of rows and columns. If **A** and **B** are both $n \times p$ matrices, then their sum $\mathbf{C} = \mathbf{A} + \mathbf{B}$ is also $n \times p$. Addition is done elementwise:

$$\mathbf{C} = \mathbf{A} + \mathbf{B} = \begin{pmatrix} a_{11} & a_{12} \\ a_{21} & a_{22} \\ a_{31} & a_{32} \end{pmatrix} + \begin{pmatrix} b_{11} & b_{12} \\ b_{21} & b_{22} \\ b_{31} & b_{32} \end{pmatrix} = \begin{pmatrix} a_{11}+b_{11} & a_{12}+b_{12} \\ a_{21}+b_{21} & a_{22}+b_{22} \\ a_{31}+b_{31} & a_{32}+b_{32} \end{pmatrix}$$

Subtraction works the same way with $(+)$ signs changed to $(-)$ signs. The usual rules for addition of numbers apply to addition of matrices, namely, commutativity $(\mathbf{A} + \mathbf{B} = \mathbf{B} + \mathbf{A})$ and associativity $(\mathbf{A} + \mathbf{B}) + \mathbf{C} = \mathbf{A} + (\mathbf{B} + \mathbf{C}) = (\mathbf{A} + \mathbf{C}) + \mathbf{B}$.

**Multiplication by a scalar.**    Suppose that $k$ is a real number or scalar. If **A** is an $r \times c$ matrix with elements $(a_{ij})$, then $k\mathbf{A}$ is an $r \times c$ matrix with elements equal to $ka_{ij}$. This notation is used to specify the variance-covariance matrix of a random vector. The expression $\text{var}(\mathbf{e}) = \sigma^2 \mathbf{I}_n$ means that the variance-covariance matrix of **e** is obtained by multiplying the identity elementwise by $\sigma^2$, so each element of **e** has variance $\sigma^2(1) = \sigma^2$, and the covariance between two elements of **e** is $\sigma^2(0) = 0$. More generally, the variance-covariance matrix of a $p \times 1$ vector **z** is often written as $\sigma_z^2 \mathbf{\Sigma}$, and, if $s_{ij}$ is the $(i, j)$ element of $\mathbf{\Sigma}$, then the covariance between the $i$th and $j$th elements of **z** is $\sigma_z^2 s_{ij}$.

**Multiplication of two matrices.**    Multiplication of matrices follows rules that are more complicated than are the rules for addition and subtraction. For two matrices to be multiplied together in the order **AB**, the number of columns (the second dimension) of **A** must equal the number of rows (the first dimension) in **B**. For example, if **A** is $n \times p$, and **B** is $p \times q$, then we can form the product **C** = **AB**, which will be an $n \times q$ matrix. If the elements of **A** are $a_{ij}$ and those of **B** are $b_{ij}$, then the elements of **C** are given by $c_{ij}$, where the formula for $c_{ij}$ is

$$c_{ij} = \sum_{k=1}^{p} a_{ik}b_{kj}$$

In words, this formula says that $c_{ij}$ is formed by taking the $i$th row of **A** and the $j$th column of **B**, multiplying the first element of the specified row in **A** by the first element in the specified column in **B**, multiplying second elements, and so on, and then adding the products together.

Possibly the simplest case of multiplying two matrices **A** and **B** together occurs when **A** is $1 \times p$ and **B** is $p \times 1$. The resulting matrix will be $1 \times 1$, a scalar or ordinary number. For example, if **A** and **B** are

$$\mathbf{A} = (1 \quad 3 \quad 2 \quad -1) \qquad \mathbf{B} = \begin{pmatrix} 2 \\ 1 \\ -2 \\ 4 \end{pmatrix}$$

then the product **AB** is

$$\mathbf{AB} = 1(2) + 3(1) + 2(-2) + -1(4) = -3.$$

It is not true that **AB** is the same as **BA**. In fact, for the preceding matrices, the product **BA** will be a $4 \times 4$ matrix:

$$\mathbf{BA} = \begin{pmatrix} 2 & 6 & 4 & -2 \\ 1 & 3 & 2 & -1 \\ -2 & -6 & -4 & 2 \\ 4 & 12 & 8 & -4 \end{pmatrix}$$

Consider a small example. Symbolically, a $3 \times 2$ matrix **A** times a $2 \times 2$ matrix **B** is given as

$$\begin{pmatrix} a_{11} & a_{12} \\ a_{21} & a_{22} \\ a_{31} & a_{32} \end{pmatrix} \begin{pmatrix} b_{11} & b_{12} \\ b_{21} & b_{22} \end{pmatrix} = \begin{pmatrix} a_{11}b_{11}+a_{12}b_{21} & a_{11}b_{12}+a_{12}b_{22} \\ a_{21}b_{11}+a_{22}b_{21} & a_{21}b_{12}+a_{22}b_{22} \\ a_{31}b_{11}+a_{32}b_{21} & a_{31}b_{12}+a_{32}b_{22} \end{pmatrix}$$

Using numbers, an example of a multiplication of two matrices is

$$
\begin{pmatrix} 3 & 1 \\ -1 & 0 \\ 2 & 2 \end{pmatrix} \begin{pmatrix} 5 & 1 \\ 0 & 4 \end{pmatrix} = \begin{pmatrix} 15+0 & 3+4 \\ -5+0 & -1+0 \\ 10+0 & 2+8 \end{pmatrix} = \begin{pmatrix} 15 & 7 \\ -5 & -1 \\ 10 & 10 \end{pmatrix}
$$

In this example, not only is $\mathbf{AB} \neq \mathbf{BA}$, but, for the matrices given, $\mathbf{BA}$ is not defined. However, the associative law holds: $\mathbf{A(BC)} = \mathbf{(AB)C}$.

**Transpose of a matrix.**  The transpose of an $r \times c$ matrix $\mathbf{X}$ is that $c \times r$ matrix called $\mathbf{X}^T$ such that if $x_{ij}$ are the elements of $\mathbf{X}$ and $x'_{ij}$ are the elements of $\mathbf{X}^T$, then $x_{ij} = x'_{ji}$. For $\mathbf{X}$ previously given,

$$
\mathbf{X}^T = \begin{pmatrix} 1 & 1 & 1 & 1 \\ 2 & 1 & 3 & 4 \\ 1 & 5 & 4 & 6 \end{pmatrix}
$$

The transpose of a column vector is a row vector. The transpose of a product $(\mathbf{AB})^T$ is the product of the transposes, in *opposite order*, $(\mathbf{AB})^T = \mathbf{B}^T \mathbf{A}^T$.

Suppose that $\mathbf{A}$ is an $n \times 1$ vector with elements $a_1, a_2, \ldots, a_n$. Then the product $\mathbf{A}^T \mathbf{A}$ is well defined, and it is the $1 \times 1$ matrix given by

$$
\mathbf{A}^T \mathbf{A} = a_1^2 + a_2^2 + \cdots + a_n^2 = \sum a_i^2 \tag{2A.1}
$$

That is, it is the sum of squares of the elements of the vector $\mathbf{A}$. The square root of this quantity is called the *norm* or the *length* of the vector $\mathbf{A}$. For example, let $\mathbf{Y}$ be the observed $n \times 1$ data vector, $\hat{\mathbf{Y}}$ the $n \times 1$ vector of fitted values. Then the residual vector is given by $\hat{\mathbf{e}} = \mathbf{Y} - \hat{\mathbf{Y}}$, and the residual sum of squares is simply $\hat{\mathbf{e}}^T \hat{\mathbf{e}} = (\mathbf{Y} - \hat{\mathbf{Y}})^T (\mathbf{Y} - \hat{\mathbf{Y}})$.

In the context of this book, $\mathbf{X}$ is the $n \times p'$ matrix giving the values of the predictors. The $i$th row of $\mathbf{X}$ is denoted by $\mathbf{x}_i^T$. In models with an intercept $\mathbf{x}_i$ is $(p+1) \times 1$, with a 1 in the first place. An important matrix obtained from $\mathbf{X}$ is $\mathbf{X}^T \mathbf{X}$, a $p' \times p'$ symmetric matrix of uncorrected sums of squares and cross products.

**Partitioned matrix.**  Occasionally, notation for a part of a matrix is useful. The $n \times p'$ matrix $\mathbf{X}$ can be partitioned by columns into matrices $\mathbf{X}_1$ and $\mathbf{X}_2$ where

$$
\mathbf{X} = (\mathbf{X}_1 \quad \mathbf{X}_2) \tag{2A.2}
$$

and $\mathbf{X}_1$ is the first $q$ columns of $\mathbf{X}$ and $\mathbf{X}_2$ is the last $p' - q$ columns of $\mathbf{X}$. The matrix product $\mathbf{X}^T \mathbf{X}$ is

$$\mathbf{X}^T\mathbf{X} = (\mathbf{X}_1 \quad \mathbf{X}_2)^T(\mathbf{X}_1 \quad \mathbf{X}_2)$$

$$= \begin{pmatrix} \mathbf{X}_1^T\mathbf{X}_1 & \mathbf{X}_1^T\mathbf{X}_2 \\ \mathbf{X}_2^T\mathbf{X}_1 & \mathbf{X}_2^T\mathbf{X}_2 \end{pmatrix}$$

In some applications, it is convenient to think of $\mathbf{X}$ and $\mathbf{Y}$ as a single array of numbers, say $\mathbf{Z} = (\mathbf{X} \quad \mathbf{Y})$. Then $\mathbf{Z}^T\mathbf{Z}$ is just a $(p'+1) \times (p'+1)$ matrix

$$\mathbf{Z}^T\mathbf{Z} = \begin{pmatrix} \mathbf{X}^T\mathbf{X} & \mathbf{X}^T\mathbf{Y} \\ \mathbf{Y}^T\mathbf{X} & \mathbf{Y}^T\mathbf{Y} \end{pmatrix}$$

giving the uncorrected sums of squares and cross products for both $\mathbf{X}$ and $\mathbf{Y}$.

**Inverse of a matrix.**   Suppose we have a $k \times k$ matrix $\mathbf{C}$. If we can find another $k \times k$ matrix, say $\mathbf{D}$, such that $\mathbf{CD} = \mathbf{I}_k$, then we shall say that $\mathbf{C}$ has an inverse, usually written as $\mathbf{C}^{-1}$, or $\mathbf{C}^{-1} = \mathbf{D}$. If the inverse exists, it is unique.

The inverse is easy to compute only in special cases. The easiest is the identity matrix $\mathbf{I}$, which is its own inverse. If $\mathbf{C}$ is a diagonal matrix, say,

$$\mathbf{C} = \begin{pmatrix} 3 & 0 & 0 & 0 \\ 0 & -1 & 0 & 0 \\ 0 & 0 & 4 & 0 \\ 0 & 0 & 0 & 1 \end{pmatrix}$$

then the inverse of $\mathbf{C}$ is the diagonal matrix

$$\mathbf{C}^{-1} = \begin{pmatrix} \frac{1}{3} & 0 & 0 & 0 \\ 0 & -1 & 0 & 0 \\ 0 & 0 & \frac{1}{4} & 0 \\ 0 & 0 & 0 & 1 \end{pmatrix}$$

It is easily verified that $\mathbf{C}^{-1}\mathbf{C} = \mathbf{CC}^{-1} = \mathbf{I}_4$. This will work for inverting any diagonal matrix, as long as none of the diagonal elements is zero. If there are zeros on the diagonal, then no inverse exists.

The most important class of matrices for which finding an inverse is easy are the orthogonal matrices. An $n \times n$ matrix $\mathbf{Q}$ is *orthogonal* if $\mathbf{Q}^T\mathbf{Q} = \mathbf{QQ}^T = \mathbf{I}_n$. Thus $\mathbf{Q}^{-1} = \mathbf{Q}^T$. For example, the matrix

$$\mathbf{Q} = \begin{pmatrix} \dfrac{1}{\sqrt{3}} & \dfrac{1}{\sqrt{2}} & \dfrac{1}{\sqrt{6}} \\ \dfrac{1}{\sqrt{3}} & 0 & -\dfrac{2}{\sqrt{6}} \\ \dfrac{1}{\sqrt{3}} & -\dfrac{1}{\sqrt{2}} & \dfrac{1}{\sqrt{6}} \end{pmatrix}$$

is orthogonal,

$$\mathbf{Q}^T = \mathbf{Q}^{-1} = \begin{pmatrix} \dfrac{1}{\sqrt{3}} & \dfrac{1}{\sqrt{3}} & \dfrac{1}{\sqrt{3}} \\ \dfrac{1}{\sqrt{2}} & 0 & -\dfrac{1}{\sqrt{2}} \\ \dfrac{1}{\sqrt{6}} & -\dfrac{2}{\sqrt{6}} & \dfrac{1}{\sqrt{6}} \end{pmatrix}$$

since $\mathbf{QQ}^T = \mathbf{I}$.

In most regression problems where matrix inversion is required, the best approach is to change the original problem into an equivalent one for which finding the inverse is either easy or not actually necessary. An example of this approach is outlined in Appendix 2A.3.

**Rank of a matrix.** Not all square matrices will have inverses. For real numbers only zero has no inverse; the inverse of any nonzero real number $k$ is $1/k$. If a square matrix has an inverse, we shall say that it is of *full rank* or *nonsingular*.

To illustrate matrix manipulations, we will substitute $\hat{\boldsymbol{\beta}} = (\mathbf{X}^T\mathbf{X})^{-1}\mathbf{X}^T\mathbf{Y}$ into $RSS = (\mathbf{Y} - \mathbf{X}\hat{\boldsymbol{\beta}})^T(\mathbf{Y} - \mathbf{X}\hat{\boldsymbol{\beta}})$ and simplify. First, performing the indicated multiplications, $RSS$ is

$$RSS = (\mathbf{Y} - \mathbf{X}\hat{\boldsymbol{\beta}})^T(\mathbf{Y} - \mathbf{X}\hat{\boldsymbol{\beta}})$$

$$= \mathbf{Y}^T\mathbf{Y} - \hat{\boldsymbol{\beta}}^T\mathbf{X}^T\mathbf{Y} - \mathbf{Y}^T\mathbf{X}\hat{\boldsymbol{\beta}} + \hat{\boldsymbol{\beta}}^T\mathbf{X}^T\mathbf{X}\hat{\boldsymbol{\beta}} \qquad (2A.3)$$

All terms on the right-hand side of (2A.3) are $1 \times 1$, so $(\mathbf{Y}^T\mathbf{X}\hat{\boldsymbol{\beta}}) = (\mathbf{Y}^T\mathbf{X}\hat{\boldsymbol{\beta}})^T = \hat{\boldsymbol{\beta}}^T\mathbf{X}^T\mathbf{Y}$ and

$$RSS = \mathbf{Y}^T\mathbf{Y} - 2\hat{\boldsymbol{\beta}}^T\mathbf{X}^T\mathbf{Y} + \hat{\boldsymbol{\beta}}^T\mathbf{X}^T\mathbf{X}\hat{\boldsymbol{\beta}} \qquad (2A.4)$$

Substituting $(\mathbf{X}^T\mathbf{X})^{-1}\mathbf{X}^T\mathbf{Y}$ for $\hat{\boldsymbol{\beta}}$ in the last term of (2A.4),

$$RSS = \mathbf{Y}^T\mathbf{Y} - 2\hat{\boldsymbol{\beta}}^T\mathbf{X}^T\mathbf{Y} + \hat{\boldsymbol{\beta}}^T(\mathbf{X}^T\mathbf{X})(\mathbf{X}^T\mathbf{X})^{-1}\mathbf{X}^T\mathbf{Y}$$

$$= \mathbf{Y}^T\mathbf{Y} - 2\hat{\boldsymbol{\beta}}^T\mathbf{X}^T\mathbf{Y} + \hat{\boldsymbol{\beta}}^T\mathbf{I}\mathbf{X}^T\mathbf{Y}$$

$$= \mathbf{Y}^T\mathbf{Y} - \hat{\boldsymbol{\beta}}^T\mathbf{X}^T\mathbf{Y} \qquad (2A.5)$$

since $\hat{\boldsymbol{\beta}}^T\mathbf{I}\mathbf{X}^T\mathbf{Y} = \hat{\boldsymbol{\beta}}^T\mathbf{X}^T\mathbf{Y}$. We can continue by substituting for $\hat{\boldsymbol{\beta}}^T$ in (2A.5) to get

$$RSS = \mathbf{Y}^T\mathbf{Y} - [(\mathbf{X}^T\mathbf{X})^{-1}\mathbf{X}^T\mathbf{Y}]^T\mathbf{X}^T\mathbf{Y}$$

$$= \mathbf{Y}^T\mathbf{Y} - \mathbf{Y}^T\mathbf{X}(\mathbf{X}^T\mathbf{X})^{-1}\mathbf{X}^T\mathbf{Y}$$

This last result requires recognizing that $(\mathbf{X}^T\mathbf{X})^{-T}$, the inverse of the transpose, equals $(\mathbf{X}^T\mathbf{X})^{-1}$ because $\mathbf{X}^T\mathbf{X}$ is symmetric. Similar manipulations will give the other forms of (2.22).

## 2A.2   Random vectors

A vector whose elements are random variables is called a *random vector*. In regression, the $n \times 1$ vector $\mathbf{e}$ of errors is a random vector. Other important random vectors in regression are the estimated parameter vector $\hat{\boldsymbol{\beta}}$, the vectors of observed values $\mathbf{Y}$ and of fitted values $\hat{\mathbf{Y}}$, and the residual vector $\hat{\mathbf{e}}$.

The mean or expected value of a random vector is the vector of means of the random variables in that vector. Thus, for example, the mean of $\mathbf{e}$ is a vector of all zeros. We write this as $E(\mathbf{e}) = \mathbf{0}$.

As in Appendix 1A.2, the mean is linear, which says that if $\mathbf{z}$ is a random vector of size $n \times 1$, and $\mathbf{C}$ is any $q \times n$ matrix, and $\mathbf{d}$ any $q \times 1$ fixed vector, then the mean of the random variable $\mathbf{Cz} + \mathbf{d}$ is $E(\mathbf{Cz} + \mathbf{d}) = \mathbf{C}E(\mathbf{z}) + \mathbf{d}$. We can apply this rule to find the mean of $\hat{\boldsymbol{\beta}}$,

$$E(\hat{\boldsymbol{\beta}}) = E[(\mathbf{X}^T\mathbf{X})^{-1}\mathbf{X}^T\mathbf{Y}] = E[(\mathbf{X}^T\mathbf{X})^{-1}\mathbf{X}^T(\mathbf{X}\boldsymbol{\beta} + \mathbf{e})]$$

since, according to the model, $\mathbf{Y} = \mathbf{X}\boldsymbol{\beta} + \mathbf{e}$. Then,

$$E(\hat{\boldsymbol{\beta}}) = (\mathbf{X}^T\mathbf{X})^{-1}\mathbf{X}^T\mathbf{X}\boldsymbol{\beta} + (\mathbf{X}^T\mathbf{X})^{-1}\mathbf{X}^T E(\mathbf{e}) \qquad (2A.6)$$

But, since $E(\mathbf{e}) = \mathbf{0}$,

$$E(\hat{\boldsymbol{\beta}}) = (\mathbf{X}^T\mathbf{X})^{-1}\mathbf{X}^T\mathbf{X}\boldsymbol{\beta} = \boldsymbol{\beta}$$

and $\hat{\boldsymbol{\beta}}$ is an unbiased estimate of $\boldsymbol{\beta}$. Incidentally, this shows exactly what the bias in the model would be if the model were not correct and if $E(\mathbf{e}) \neq \mathbf{0}$. Suppose, for example, that $E(\mathbf{e}) = \mathbf{Z}\boldsymbol{\gamma}$ for some $n \times q$ matrix $\mathbf{Z}$ of variables different from those in $\mathbf{X}$ although $\mathbf{Z}$ may be transformations or combinations of the columns of $\mathbf{X}$, and $\boldsymbol{\gamma}$ is a $q \times 1$ unknown parameter vector. Then

$$E(\hat{\boldsymbol{\beta}}) = \boldsymbol{\beta} + (\mathbf{X}^T\mathbf{X})^{-1}\mathbf{X}^T\mathbf{Z}\boldsymbol{\gamma}$$

and the bias in $\hat{\boldsymbol{\beta}}$ as an estimate of $\boldsymbol{\beta}$ is given by

$$\text{bias} = \boldsymbol{\beta} - E(\hat{\boldsymbol{\beta}}) = -(\mathbf{X}^T\mathbf{X})^{-1}\mathbf{X}^T\mathbf{Z}\boldsymbol{\gamma} \qquad (2A.7)$$

The matrix $(\mathbf{X}^T\mathbf{X})^{-1}\mathbf{X}^T\mathbf{Z}$ is called the *alias* matrix, as each element in $\hat{\boldsymbol{\beta}}$ is confused or aliased with the elements in $\boldsymbol{\gamma}$ in a way that is determined by the alias matrix. The bias will be zero if (1) the product $\mathbf{X}^T\mathbf{Z} = \mathbf{0}$, or (2) if $\boldsymbol{\gamma} = \mathbf{0}$. This situation will arise if we fit the model $\mathbf{Y} = \mathbf{X}\boldsymbol{\beta} + \mathbf{e}$, but the true model is actually $\mathbf{Y} = \mathbf{X}\boldsymbol{\beta} + \mathbf{Z}\boldsymbol{\gamma} + \mathbf{e}$. It is left as an exercise to show that, if the smaller model is true, and the larger one is fit, then the estimator of $\boldsymbol{\beta}$ is unbiased.

**Variance-covariance.** A random vector has a variance-covariance matrix associated with it. The diagonal entries of this matrix are the variances of the elements of the random vector, while the off-diagonal entries are the covariances between the elements; the $(i, j)$-th element of the variance covariance matrix is the covariance between the $i$th element of the random vector and the $j$th element of the random vector. We use the symbol var($\mathbf{z}$) to denote the variance-covariance matrix of the random vector $\mathbf{z}$.

The error vector $\mathbf{e}$ has been assumed to have elements with common variance and zero covariances. This is summarized as var($\mathbf{e}$) $= \sigma^2 \mathbf{I}_n$. A random vector with unequal variances but uncorrelated elements is given by a diagonal matrix,

$$\begin{pmatrix} \sigma_1^2 & & & 0 \\ & \sigma_2^2 & & \\ & & \ddots & \\ 0 & & & \sigma_n^2 \end{pmatrix}$$

The formula for var($\mathbf{Cz} + \mathbf{d}$) is given by

$$\text{var}(\mathbf{Cz} + \mathbf{d}) = \mathbf{C}[\text{var}(\mathbf{z})]\mathbf{C}^T$$

Applying this to

$$\hat{\boldsymbol{\beta}} = (\mathbf{X}^T\mathbf{X})^{-1}\mathbf{X}^T\mathbf{Y} = (\mathbf{X}^T\mathbf{X})^{-1}\mathbf{X}^T(\mathbf{X}\boldsymbol{\beta} + \mathbf{e})$$
$$= (\mathbf{X}^T\mathbf{X})^{-1}\mathbf{X}^T\mathbf{X}\boldsymbol{\beta} + (\mathbf{X}^T\mathbf{X})^{-1}\mathbf{X}^T\mathbf{e}$$

we see that the first term does not involve $\mathbf{e}$, and therefore corresponds to $\mathbf{d}$ in the formula. Associating $\mathbf{C}$ with $(\mathbf{X}^T\mathbf{X})^{-1}\mathbf{X}^T$,

$$\text{var}(\hat{\boldsymbol{\beta}}) = (\mathbf{X}^T\mathbf{X})^{-1}\mathbf{X}^T[\text{var}(\mathbf{e})][(\mathbf{X}^T\mathbf{X})^{-1}\mathbf{X}^T]^T$$
$$= (\mathbf{X}^T\mathbf{X})^{-1}\mathbf{X}^T(\sigma^2\mathbf{I})\mathbf{X}(\mathbf{X}^T\mathbf{X})^{-1}$$
$$= \sigma^2(\mathbf{X}^T\mathbf{X})^{-1}(\mathbf{X}^T\mathbf{X})(\mathbf{X}^T\mathbf{X})^{-1}$$
$$= \sigma^2(\mathbf{X}^T\mathbf{X})^{-1}$$

as given by (2.20).

Another important application of this result is to find the variance of the fitted value corresponding to the $p' \times 1$ vector $\mathbf{x}$. The fitted value is given by $\hat{y} = \mathbf{x}^T\hat{\boldsymbol{\beta}}$ and therefore

$$\text{var}(\hat{y}|\mathbf{x}) = \mathbf{x}^T[\text{var}(\hat{\boldsymbol{\beta}})]\mathbf{x} = \sigma^2\mathbf{x}^T(\mathbf{X}^T\mathbf{X})^{-1}\mathbf{x} \tag{2A.8}$$

For a prediction of an as yet unobserved future value $y_*$ at $\mathbf{x}_*$, the prediction is $\tilde{y}_* = \mathbf{x}_*^T\hat{\boldsymbol{\beta}}$, with variance

$$\text{var}(\tilde{y}_*|\mathbf{x}_*) = \sigma^2[1 + \mathbf{x}_*^T(\mathbf{X}^T\mathbf{X})^{-1}\mathbf{x}_*] \tag{2A.9}$$

## 2A.3  Least squares

Rather than repeat the derivation of the least squares estimators from Chapter 1, we find the least squares estimators in a way that gives an important computational method. The estimator $\hat{\boldsymbol{\beta}}$ is that value of $\boldsymbol{\beta}$ that minimizes the function

$$RSS(\boldsymbol{\beta}) = (\mathbf{Y} - \mathbf{X}\boldsymbol{\beta})^T(\mathbf{Y} - \mathbf{X}\boldsymbol{\beta}) \qquad (2A.10)$$

Our goal is to replace this minimization problem by one that can be solved trivially. Suppose we can find an $n \times p'$ matrix $\mathbf{Q}$ such that $\mathbf{Q}^T\mathbf{Q} = I_{p'}$, and a $p' \times p'$ upper triangular matrix $\mathbf{R}$ (all elements below the main diagonal are zero) such that

$$\mathbf{X} = \mathbf{QR} \qquad (2A.11)$$

We postpone discussion of the existence of these matrices and how to find them. Multiplying out (2A.10), and then substituting for X using (2A.11) gives,

$$RSS(\boldsymbol{\beta}) = \mathbf{Y}^T\mathbf{Y} - 2\mathbf{Y}^T\mathbf{X}\boldsymbol{\beta} + \boldsymbol{\beta}^T\mathbf{X}^T\mathbf{X}\boldsymbol{\beta}$$
$$= \mathbf{Y}^T\mathbf{Y} - 2\mathbf{Y}^T\mathbf{QR}\boldsymbol{\beta} + \boldsymbol{\beta}^T\mathbf{R}^T\mathbf{Q}^T\mathbf{QR}\boldsymbol{\beta}$$

Adding and subtracting $\mathbf{Y}^T\mathbf{QQ}^T\mathbf{Y}$ from the right-hand side of this equation will allow us to write $RSS(\boldsymbol{\beta})$ as a sum of two terms, only one of which includes $\boldsymbol{\beta}$:

$$RSS(\boldsymbol{\beta}) = \mathbf{Y}^T\mathbf{Y} - \mathbf{Y}^T\mathbf{QQ}^T\mathbf{Y} + (\mathbf{Y}^T\mathbf{QQ}^T\mathbf{Y} - 2\mathbf{Y}^T\mathbf{QR}\boldsymbol{\beta} + \boldsymbol{\beta}^T\mathbf{R}^T\mathbf{R}\boldsymbol{\beta})$$
$$= \mathbf{Y}^T(\mathbf{I} - \mathbf{QQ}^T)\mathbf{Y} + (\mathbf{Q}^T\mathbf{Y} - \mathbf{R}\boldsymbol{\beta})^T(\mathbf{Q}^T\mathbf{Y} - \mathbf{R}\boldsymbol{\beta})$$

$RSS(\boldsymbol{\beta})$ will be minimized by making the second term zero. This can be accomplished by setting

$$\mathbf{Q}^T\mathbf{Y} - \mathbf{R}\boldsymbol{\beta} = 0 \qquad (2A.12)$$

or,

$$\mathbf{R}\boldsymbol{\beta} = \mathbf{Q}^T\mathbf{Y}$$

and, as long as $\mathbf{R}$ has an inverse,

$$\hat{\boldsymbol{\beta}} = \mathbf{R}^{-1}\mathbf{Q}^T\mathbf{Y} \qquad (2A.13)$$

That (2A.13) is equivalent to (2.15) is left as an exercise.

This derivation of least squares estimators makes use of the $QR$ *factorization*, introduced by A. S. Householder (1958) and G. Golub (1965). An excellent source concerning this procedure as a basis of a computational method is Stewart (1974, Chapter 7). The existence and uniqueness up to changes in sign of $\mathbf{Q}$ and $\mathbf{R}$ for any $\mathbf{X}$ can be proved by finding an algorithm for computing them. High-quality Fortran subroutines for least squares computations based on this factorization are available in the Linpack package (Dongarra et al., 1979). The matrix $\mathbf{R}$ in the QR factorization is called the *Cholesky factor* of $\mathbf{X}^T\mathbf{X}$, and alternative computational methods can be based on $\mathbf{R}$, avoiding computation of $\mathbf{Q}$; see Stewart (1974).

## 5A.1 Relating regression equations

The diagnostic statistics examined in this book are practical because simple formulas can be derived to obtain various statistics when cases are deleted. Suppose that $\mathbf{X}$ is $n \times p'$, $\mathbf{Y}$ is $n \times 1$, and the matrix $(\mathbf{X}^T\mathbf{X})^{-1}$ has been computed. To compute $(\mathbf{X}_{(i)}^T\mathbf{X}_{(i)})^{-1}$ from $(\mathbf{X}^T\mathbf{X})^{-1}$, we use the following fundamental identity:

$$(\mathbf{X}_{(i)}^T\mathbf{X}_{(i)})^{-1} = (\mathbf{X}^T\mathbf{X})^{-1} + \frac{(\mathbf{X}^T\mathbf{X})^{-1}\mathbf{x}_i\mathbf{x}_i^T(\mathbf{X}^T\mathbf{X})^{-1}}{1 - h_{ii}} \tag{5A.1}$$

This remarkable formula was used by Gauss (1821); a history of it and many variations is given by Henderson and Searle (1981). It can be applied to give all the results that one would want relating the regressions with and without the $i$th case. For example, the distance from $\mathbf{x}_i$ to the center of the remaining $n-1$ cases is defined by $\mathbf{x}_i^T(\mathbf{X}_{(i)}^T\mathbf{X}_{(i)})^{-1}\mathbf{x}_i$. Using (5A.1),

$$\mathbf{x}_i^T(\mathbf{X}_{(i)}^T\mathbf{X}_{(i)})^{-1}\mathbf{x}_i = \mathbf{x}_i^T(\mathbf{X}^T\mathbf{X})^{-1}\mathbf{x}_i + \frac{\mathbf{x}_i^T(\mathbf{X}^T\mathbf{X})^{-1}\mathbf{x}_i\mathbf{x}_i^T(\mathbf{X}^T\mathbf{X})^{-1}\mathbf{x}_i}{1 - h_{ii}}$$

$$= h_{ii} + \frac{h_{ii}^2}{1 - h_{ii}} = \frac{h_{ii}}{1 - h_{ii}} \tag{5A.2}$$

Similar computations can be done to find any statistic for the regression deleting the $i$th case from the data. Estimate of $\boldsymbol{\beta}$:

$$\hat{\boldsymbol{\beta}}_{(i)} = \hat{\boldsymbol{\beta}} - \frac{(\mathbf{X}^T\mathbf{X})^{-1}\mathbf{x}_i\hat{e}_i}{1 - h_{ii}} \tag{5A.3}$$

Estimate of $\sigma^2$:

$$\hat{\sigma}_{(i)}^2 = \frac{1}{n - p' - 1}\hat{\sigma}^2(n - p' - r_i^2) \tag{5A.4}$$

## 8A.1 Derivation of $C_p$

A fixed subset model is specified by a partition of $\mathbf{X} = (\mathbf{X}_1 \ \mathbf{X}_2)$ so that the subset model is

$$Y = X_1 \boldsymbol{\beta}_1 + e$$

define further notation. Let the $i$th fitted value for the subset model be given by $\hat{y}_i$, and let $u_{ii}$ be the corresponding diagonal element of $U = X_1(X_1^T X_1)^{-1} X_1^T$, with $\text{var}(\hat{y}_i) = \sigma^2 u_{ii}$. Similarly, for the full model, as usual let $H = X(X^T X)^{-1} X^T$ with diagonals $h_{ii}$, and let $\hat{Y}_i$ be the $i$th fitted value; this is new notation used only in this section to distinguish the two sets of fitted values. As defined in the text,

$$J_p = \frac{1}{\sigma^2} \sum_{i=1}^{n} \text{mse}(\hat{y}_i) \qquad (8A.1)$$

where

$$\text{mse}(\hat{y}_i) = \text{var}(\hat{y}_i) + (\text{bias})^2 = \sigma^2 u_{ii} + [E(\hat{y}_i) - E(y_i)]^2$$

Since the full model is assumed to be unbiased, $E(y_i) = E(\hat{Y}_i)$, and hence $[E(\hat{y}_i) - E(y_i)]^2 = [E(\hat{y}_i) - E(\hat{Y}_i)]^2$. One can show (Weisberg, 1981) that

$$[E(\hat{y}_i) - E(\hat{Y}_i)]^2 = E(\hat{y}_i - \hat{Y}_i)^2 - \sigma^2(h_{ii} - u_{ii})$$

This result is not obvious and is most easily proved by transforming to a problem in which $X_1$ and $X_2$ are orthogonal. Substituting into the formula for the mse, replacing $E(\hat{y}_i - \hat{Y}_i)^2$ by its observed value, and replacing $\sigma^2$ by its estimate $\hat{\sigma}^2$ from the full model,

$$\widehat{\text{mse}}(\hat{y}_i) = (\hat{y}_i - \hat{Y}_i)^2 + \hat{\sigma}^2[u_{ii} - (h_{ii} - u_{ii})]$$

and hence

$$C_p = \frac{1}{\hat{\sigma}^2} \sum_{i=1}^{n} \widehat{\text{mse}}(\hat{y}_i)$$

$$= \sum_{i=1}^{n} \left[ \frac{(\hat{y}_i - \hat{Y}_i)^2}{\hat{\sigma}^2} + u_{ii} - (h_{ii} - u_{ii}) \right] \qquad (8A.2)$$

There are several interesting features of this last formula. Since $(\hat{y}_i - \hat{Y}_i)^2 = [(y_i - \hat{y}_i) - (y_i - \hat{Y}_i)]^2 =$ squared change in the $i$th residual from the full and subset models, we see that each term in the sum has three parts: the part due to the residual, a penalty $u_{ii}$, the potential of the $i$th row of $X_1$, and the change in potential, $h_{ii} - u_{ii}$. The term in square brackets in (8A.2) is called $C_{pi}$ because $\sum C_{pi} = C_p$. A plot of $C_{pi}$ versus the $u_{ii}$ provides a diagnostic check on the $C_p$ statistic. For good models, nearly all points should fall near the line $C_{pi} = u_{ii}$.

To get the usual form for $C_p$, we need three results (left for proof to the interested reader):

1. $\sum (\hat{y}_i - \hat{Y}_i)^2 =$ additional sum of squares for regression on $\mathbf{X}_2$ after $\mathbf{X}_1$.
2. $\sum u_{ii} = p$.
3. $\sum h_{ii} = k'$.

Substituting these into (8A.2) gives (8.19).

# TABLES

## Table A    Student's *t*-distribution

The tabled values are two-tailed values $t(\alpha; \nu)$, such that

$$\text{prob}\{|t_\nu \text{ variate}| > t(\alpha; \nu)\} = \alpha$$

The entries in the table were computed on a CDC Cyber 172 computer at the University of Minnesota using IMSL subroutine MDSTI

|         |         |         | $\alpha$ |         |         |
| ------- | ------- | ------- | -------- | ------- | ------- |
| $\nu$   | 0.200   | 0.100   | 0.050    | 0.010   | 0.001   |
| 1       | 3.08    | 6.31    | 12.71    | 63.66   | 636.62  |
| 2       | 1.89    | 2.92    | 4.30     | 9.92    | 31.60   |
| 3       | 1.64    | 2.35    | 3.18     | 5.84    | 12.92   |
| 4       | 1.53    | 2.13    | 2.78     | 4.60    | 8.61    |
| 5       | 1.48    | 2.02    | 2.57     | 4.03    | 6.87    |
| 6       | 1.44    | 1.94    | 2.45     | 3.71    | 5.96    |
| 7       | 1.41    | 1.89    | 2.36     | 3.50    | 5.41    |
| 8       | 1.40    | 1.86    | 2.31     | 3.36    | 5.04    |
| 9       | 1.38    | 1.83    | 2.26     | 3.25    | 4.78    |
| 10      | 1.37    | 1.81    | 2.23     | 3.17    | 4.59    |
| 11      | 1.36    | 1.80    | 2.20     | 3.11    | 4.44    |
| 12      | 1.36    | 1.78    | 2.18     | 3.05    | 4.32    |
| 13      | 1.35    | 1.77    | 2.16     | 3.01    | 4.22    |
| 14      | 1.35    | 1.76    | 2.14     | 2.98    | 4.14    |
| 15      | 1.34    | 1.75    | 2.13     | 2.95    | 4.07    |
| 16      | 1.34    | 1.75    | 2.12     | 2.92    | 4.01    |
| 17      | 1.33    | 1.74    | 2.11     | 2.90    | 3.97    |
| 18      | 1.33    | 1.73    | 2.10     | 2.88    | 3.92    |

## Table A (continued)

| | | | $\alpha$ | | |
|---|---|---|---|---|---|
| $v$ | 0.200 | 0.100 | 0.050 | 0.010 | 0.001 |
| 19 | 1.33 | 1.73 | 2.09 | 2.86 | 3.88 |
| 20 | 1.33 | 1.72 | 2.09 | 2.85 | 3.85 |
| 21 | 1.32 | 1.72 | 2.08 | 2.83 | 3.82 |
| 22 | 1.32 | 1.72 | 2.07 | 2.82 | 3.79 |
| 23 | 1.32 | 1.71 | 2.07 | 2.81 | 3.77 |
| 24 | 1.32 | 1.71 | 2.06 | 2.80 | 3.75 |
| 25 | 1.32 | 1.71 | 2.06 | 2.79 | 3.73 |
| 26 | 1.31 | 1.71 | 2.06 | 2.78 | 3.71 |
| 27 | 1.31 | 1.70 | 2.05 | 2.77 | 3.69 |
| 28 | 1.31 | 1.70 | 2.05 | 2.76 | 3.67 |
| 29 | 1.31 | 1.70 | 2.05 | 2.76 | 3.66 |
| 30 | 1.31 | 1.70 | 2.04 | 2.75 | 3.65 |
| 31 | 1.31 | 1.70 | 2.04 | 2.74 | 3.63 |
| 32 | 1.31 | 1.69 | 2.04 | 2.74 | 3.62 |
| 33 | 1.31 | 1.69 | 2.03 | 2.73 | 3.61 |
| 34 | 1.31 | 1.69 | 2.03 | 2.73 | 3.60 |
| 35 | 1.31 | 1.69 | 2.03 | 2.72 | 3.59 |
| 36 | 1.31 | 1.69 | 2.03 | 2.72 | 3.58 |
| 37 | 1.30 | 1.69 | 2.03 | 2.72 | 3.57 |
| 38 | 1.30 | 1.69 | 2.02 | 2.71 | 3.57 |
| 39 | 1.30 | 1.68 | 2.02 | 2.71 | 3.56 |
| 40 | 1.30 | 1.68 | 2.02 | 2.70 | 3.55 |
| 41 | 1.30 | 1.68 | 2.02 | 2.70 | 3.54 |
| 42 | 1.30 | 1.68 | 2.02 | 2.70 | 3.54 |
| 43 | 1.30 | 1.68 | 2.02 | 2.70 | 3.53 |
| 44 | 1.30 | 1.68 | 2.02 | 2.69 | 3.53 |
| 45 | 1.30 | 1.68 | 2.01 | 2.69 | 3.52 |
| 46 | 1.30 | 1.68 | 2.01 | 2.69 | 3.51 |
| 47 | 1.30 | 1.68 | 2.01 | 2.68 | 3.51 |
| 48 | 1.30 | 1.68 | 2.01 | 2.68 | 3.51 |
| 49 | 1.30 | 1.68 | 2.01 | 2.68 | 3.50 |
| 50 | 1.30 | 1.68 | 2.01 | 2.68 | 3.50 |
| 60 | 1.30 | 1.67 | 2.00 | 2.66 | 3.46 |
| 70 | 1.29 | 1.67 | 1.99 | 2.65 | 3.44 |
| 80 | 1.29 | 1.66 | 1.99 | 2.64 | 3.42 |
| 90 | 1.29 | 1.66 | 1.99 | 2.63 | 3.40 |
| 100 | 1.29 | 1.66 | 1.98 | 2.63 | 3.39 |
| 120 | 1.29 | 1.66 | 1.98 | 2.62 | 3.37 |
| $\infty$ | 1.28 | 1.64 | 1.96 | 2.58 | 3.29 |

## Table B F-distribution[a]

The tabled values are $F(\alpha, v_1, v_2)$ such that $\text{prob}\{F(v_1, v_2) \text{ variate} \geqslant F(\alpha, v_1, v_2)\} = \alpha$.

$$\alpha = 0.05$$

Degrees of freedom for numerator

| $v_2$ \ $v_1$ | 1 | 2 | 3 | 4 | 5 | 6 | 7 | 8 | 9 | 10 | 12 | 15 | 20 | 24 | 30 | 40 | 60 | 120 | ∞ |
|---|---|---|---|---|---|---|---|---|---|---|---|---|---|---|---|---|---|---|---|
| 1 | 161.4 | 199.5 | 215.7 | 224.6 | 230.2 | 234.0 | 236.8 | 238.9 | 240.5 | 241.9 | 243.9 | 245.9 | 248.0 | 249.0 | 250.1 | 251.1 | 252.2 | 253.3 | 254.3 |
| 2 | 18.51 | 19.00 | 19.16 | 19.25 | 19.30 | 19.33 | 19.35 | 19.37 | 19.38 | 19.40 | 19.41 | 19.43 | 19.45 | 19.45 | 19.46 | 19.47 | 19.48 | 19.49 | 19.50 |
| 3 | 10.13 | 9.55 | 9.28 | 9.12 | 9.01 | 8.94 | 8.89 | 8.85 | 8.81 | 8.79 | 8.74 | 8.70 | 8.66 | 8.64 | 8.62 | 8.59 | 8.57 | 8.55 | 8.53 |
| 4 | 7.71 | 6.94 | 6.59 | 6.39 | 6.26 | 6.16 | 6.09 | 6.04 | 6.00 | 5.96 | 5.91 | 5.86 | 5.80 | 5.77 | 5.75 | 5.72 | 5.69 | 5.66 | 5.63 |
| 5 | 6.61 | 5.79 | 5.41 | 5.19 | 5.05 | 4.95 | 4.88 | 4.82 | 4.77 | 4.74 | 4.68 | 4.62 | 4.56 | 4.53 | 4.50 | 4.46 | 4.43 | 4.40 | 4.36 |
| 6 | 5.99 | 5.14 | 4.76 | 4.53 | 4.39 | 4.28 | 4.21 | 4.15 | 4.10 | 4.06 | 4.00 | 3.94 | 3.87 | 3.84 | 3.81 | 3.77 | 3.74 | 3.70 | 3.67 |
| 7 | 5.59 | 4.74 | 4.35 | 4.12 | 3.97 | 3.87 | 3.79 | 3.73 | 3.68 | 3.64 | 3.57 | 3.51 | 3.44 | 3.41 | 3.38 | 3.34 | 3.30 | 3.27 | 3.23 |
| 8 | 5.32 | 4.46 | 4.07 | 3.84 | 3.69 | 3.58 | 3.50 | 3.44 | 3.39 | 3.35 | 3.28 | 3.22 | 3.15 | 3.12 | 3.08 | 3.04 | 3.01 | 2.97 | 2.93 |
| 9 | 5.12 | 4.26 | 3.86 | 3.63 | 3.48 | 3.37 | 3.29 | 3.23 | 3.18 | 3.14 | 3.07 | 3.01 | 2.94 | 2.90 | 2.86 | 2.83 | 2.79 | 2.75 | 2.71 |
| 10 | 4.96 | 4.10 | 3.71 | 3.48 | 3.33 | 3.22 | 3.14 | 3.07 | 3.02 | 2.98 | 2.91 | 2.85 | 2.77 | 2.74 | 2.70 | 2.66 | 2.62 | 2.58 | 2.54 |
| 11 | 4.84 | 3.98 | 3.59 | 3.36 | 3.20 | 3.09 | 3.01 | 2.95 | 2.90 | 2.85 | 2.79 | 2.72 | 2.65 | 2.61 | 2.57 | 2.53 | 2.49 | 2.45 | 2.40 |
| 12 | 4.75 | 3.89 | 3.49 | 3.26 | 3.11 | 3.00 | 2.91 | 2.85 | 2.80 | 2.75 | 2.69 | 2.62 | 2.54 | 2.51 | 2.47 | 2.43 | 2.38 | 2.34 | 2.30 |
| 13 | 4.67 | 3.81 | 3.41 | 3.18 | 3.03 | 2.92 | 2.83 | 2.77 | 2.71 | 2.67 | 2.60 | 2.53 | 2.46 | 2.42 | 2.38 | 2.34 | 2.30 | 2.25 | 2.21 |
| 14 | 4.60 | 3.74 | 3.34 | 3.11 | 2.96 | 2.85 | 2.76 | 2.70 | 2.65 | 2.60 | 2.53 | 2.46 | 2.39 | 2.35 | 2.31 | 2.27 | 2.22 | 2.18 | 2.13 |
| 15 | 4.54 | 3.68 | 3.29 | 3.06 | 2.90 | 2.79 | 2.71 | 2.64 | 2.59 | 2.54 | 2.48 | 2.40 | 2.33 | 2.29 | 2.25 | 2.20 | 2.16 | 2.11 | 2.07 |
| 16 | 4.49 | 3.63 | 3.24 | 3.01 | 2.85 | 2.74 | 2.66 | 2.59 | 2.54 | 2.49 | 2.42 | 2.35 | 2.28 | 2.24 | 2.19 | 2.15 | 2.11 | 2.06 | 2.01 |
| 17 | 4.45 | 3.59 | 3.20 | 2.96 | 2.81 | 2.70 | 2.61 | 2.55 | 2.49 | 2.45 | 2.38 | 2.31 | 2.23 | 2.19 | 2.15 | 2.10 | 2.06 | 2.01 | 1.96 |
| 18 | 4.41 | 3.55 | 3.16 | 2.93 | 2.77 | 2.66 | 2.58 | 2.51 | 2.46 | 2.41 | 2.34 | 2.27 | 2.19 | 2.15 | 2.11 | 2.06 | 2.02 | 1.97 | 1.92 |
| 19 | 4.38 | 3.52 | 3.13 | 2.90 | 2.74 | 2.63 | 2.54 | 2.48 | 2.42 | 2.38 | 2.31 | 2.23 | 2.16 | 2.11 | 2.07 | 2.03 | 1.98 | 1.93 | 1.88 |
| 20 | 4.35 | 3.49 | 3.10 | 2.87 | 2.71 | 2.60 | 2.51 | 2.45 | 2.39 | 2.35 | 2.28 | 2.20 | 2.12 | 2.08 | 2.04 | 1.99 | 1.95 | 1.90 | 1.84 |
| 21 | 4.32 | 3.47 | 3.07 | 2.84 | 2.68 | 2.57 | 2.49 | 2.42 | 2.37 | 2.32 | 2.25 | 2.18 | 2.10 | 2.05 | 2.01 | 1.96 | 1.92 | 1.87 | 1.81 |
| 22 | 4.30 | 3.44 | 3.05 | 2.82 | 2.66 | 2.55 | 2.46 | 2.40 | 2.34 | 2.30 | 2.23 | 2.15 | 2.07 | 2.03 | 1.98 | 1.94 | 1.89 | 1.84 | 1.78 |
| 23 | 4.28 | 3.42 | 3.03 | 2.80 | 2.64 | 2.53 | 2.44 | 2.37 | 2.32 | 2.27 | 2.20 | 2.13 | 2.05 | 2.01 | 1.96 | 1.91 | 1.86 | 1.81 | 1.76 |
| 24 | 4.26 | 3.40 | 3.01 | 2.78 | 2.62 | 2.51 | 2.42 | 2.36 | 2.30 | 2.25 | 2.18 | 2.11 | 2.03 | 1.98 | 1.94 | 1.89 | 1.84 | 1.79 | 1.73 |
| 25 | 4.24 | 3.39 | 2.99 | 2.76 | 2.60 | 2.49 | 2.40 | 2.34 | 2.28 | 2.24 | 2.16 | 2.09 | 2.01 | 1.96 | 1.92 | 1.87 | 1.82 | 1.77 | 1.71 |
| 26 | 4.23 | 3.37 | 2.98 | 2.74 | 2.59 | 2.47 | 2.39 | 2.32 | 2.27 | 2.22 | 2.15 | 2.07 | 1.99 | 1.95 | 1.90 | 1.85 | 1.80 | 1.75 | 1.69 |
| 27 | 4.21 | 3.35 | 2.96 | 2.73 | 2.57 | 2.46 | 2.37 | 2.31 | 2.25 | 2.20 | 2.13 | 2.06 | 1.97 | 1.93 | 1.88 | 1.84 | 1.79 | 1.73 | 1.67 |
| 28 | 4.20 | 3.34 | 2.95 | 2.71 | 2.56 | 2.45 | 2.36 | 2.29 | 2.24 | 2.19 | 2.12 | 2.04 | 1.96 | 1.91 | 1.87 | 1.82 | 1.77 | 1.71 | 1.65 |
| 29 | 4.18 | 3.33 | 2.93 | 2.70 | 2.55 | 2.43 | 2.35 | 2.28 | 2.22 | 2.18 | 2.10 | 2.03 | 1.94 | 1.90 | 1.85 | 1.81 | 1.75 | 1.70 | 1.64 |
| 30 | 4.17 | 3.32 | 2.92 | 2.69 | 2.53 | 2.42 | 2.33 | 2.27 | 2.21 | 2.16 | 2.09 | 2.01 | 1.93 | 1.89 | 1.84 | 1.79 | 1.74 | 1.68 | 1.62 |
| 40 | 4.08 | 3.23 | 2.84 | 2.61 | 2.45 | 2.34 | 2.25 | 2.18 | 2.12 | 2.08 | 2.00 | 1.92 | 1.84 | 1.79 | 1.74 | 1.69 | 1.64 | 1.58 | 1.51 |
| 60 | 4.00 | 3.15 | 2.76 | 2.53 | 2.37 | 2.25 | 2.17 | 2.10 | 2.04 | 1.99 | 1.92 | 1.84 | 1.75 | 1.70 | 1.65 | 1.59 | 1.53 | 1.47 | 1.39 |
| 120 | 3.92 | 3.07 | 2.68 | 2.45 | 2.29 | 2.17 | 2.09 | 2.02 | 1.96 | 1.91 | 1.83 | 1.75 | 1.66 | 1.61 | 1.55 | 1.50 | 1.43 | 1.35 | 1.25 |
| ∞ | 3.84 | 3.00 | 2.60 | 2.37 | 2.21 | 2.10 | 2.01 | 1.94 | 1.88 | 1.83 | 1.75 | 1.67 | 1.57 | 1.52 | 1.46 | 1.39 | 1.32 | 1.22 | 1.00 |

Degrees of Freedom for Denominator

$$\alpha = 0.01$$

Degrees of freedom for numerator

| $v_2$ \ $v_1$ | 1 | 2 | 3 | 4 | 5 | 6 | 7 | 8 | 9 | 10 | 12 | 15 | 20 | 24 | 30 | 40 | 60 | 120 | ∞ |
|---|---|---|---|---|---|---|---|---|---|---|---|---|---|---|---|---|---|---|---|
| 1 | 4052 | 4999.5 | 5403 | 5625 | 5764 | 5859 | 5928 | 5982 | 6022 | 6056 | 6106 | 6157 | 6209 | 6235 | 6261 | 6287 | 6313 | 6339 | 6366 |
| 2 | 98.50 | 99.00 | 99.17 | 99.25 | 99.30 | 99.33 | 99.36 | 99.37 | 99.39 | 99.40 | 99.42 | 99.43 | 99.45 | 99.46 | 99.47 | 99.47 | 99.48 | 99.49 | 99.50 |
| 3 | 34.12 | 30.82 | 29.46 | 28.71 | 28.24 | 27.91 | 27.67 | 27.49 | 27.35 | 27.23 | 27.05 | 26.87 | 26.69 | 26.60 | 26.50 | 26.41 | 26.32 | 26.22 | 26.13 |
| 4 | 21.20 | 18.00 | 16.69 | 15.98 | 15.52 | 15.21 | 14.98 | 14.80 | 14.66 | 14.55 | 14.37 | 14.20 | 14.02 | 13.93 | 13.84 | 13.75 | 13.65 | 13.56 | 13.46 |
| 5 | 16.26 | 13.27 | 12.06 | 11.39 | 10.97 | 10.67 | 10.46 | 10.29 | 10.16 | 10.05 | 9.89 | 9.72 | 9.55 | 9.47 | 9.38 | 9.29 | 9.20 | 9.11 | 9.02 |
| 6 | 13.75 | 10.92 | 9.78 | 9.15 | 8.75 | 8.47 | 8.26 | 8.10 | 7.98 | 7.87 | 7.72 | 7.56 | 7.40 | 7.31 | 7.23 | 7.14 | 7.06 | 6.97 | 6.88 |
| 7 | 12.25 | 9.55 | 8.45 | 7.85 | 7.46 | 7.19 | 6.99 | 6.84 | 6.72 | 6.62 | 6.47 | 6.31 | 6.16 | 6.07 | 5.99 | 5.91 | 5.82 | 5.74 | 5.65 |
| 8 | 11.26 | 8.65 | 7.59 | 7.01 | 6.63 | 6.37 | 6.18 | 6.03 | 5.91 | 5.81 | 5.67 | 5.52 | 5.36 | 5.28 | 5.20 | 5.12 | 5.03 | 4.95 | 4.86 |
| 9 | 10.56 | 8.02 | 6.99 | 6.42 | 6.06 | 5.80 | 5.61 | 5.47 | 5.35 | 5.26 | 5.11 | 4.96 | 4.81 | 4.73 | 4.65 | 4.57 | 4.48 | 4.40 | 4.31 |
| 10 | 10.04 | 7.56 | 6.55 | 5.99 | 5.64 | 5.39 | 5.20 | 5.06 | 4.94 | 4.85 | 4.71 | 4.56 | 4.41 | 4.33 | 4.25 | 4.17 | 4.08 | 4.00 | 3.91 |
| 11 | 9.65 | 7.21 | 6.22 | 5.67 | 5.32 | 5.07 | 4.89 | 4.74 | 4.63 | 4.54 | 4.40 | 4.25 | 4.10 | 4.02 | 3.94 | 3.86 | 3.78 | 3.69 | 3.60 |
| 12 | 9.33 | 6.93 | 5.95 | 5.41 | 5.06 | 4.82 | 4.64 | 4.50 | 4.39 | 4.30 | 4.16 | 4.01 | 3.86 | 3.78 | 3.70 | 3.62 | 3.54 | 3.45 | 3.36 |
| 13 | 9.07 | 6.70 | 5.74 | 5.21 | 4.86 | 4.62 | 4.44 | 4.30 | 4.19 | 4.10 | 3.96 | 3.82 | 3.66 | 3.59 | 3.51 | 3.43 | 3.34 | 3.25 | 3.17 |
| 14 | 8.86 | 6.51 | 5.56 | 5.04 | 4.69 | 4.46 | 4.28 | 4.14 | 4.03 | 3.94 | 3.80 | 3.66 | 3.51 | 3.43 | 3.35 | 3.27 | 3.18 | 3.09 | 3.00 |
| 15 | 8.68 | 6.36 | 5.42 | 4.89 | 4.56 | 4.32 | 4.14 | 4.00 | 3.89 | 3.80 | 3.67 | 3.52 | 3.37 | 3.29 | 3.21 | 3.13 | 3.05 | 2.96 | 2.87 |
| 16 | 8.53 | 6.23 | 5.29 | 4.77 | 4.44 | 4.20 | 4.03 | 3.89 | 3.78 | 3.69 | 3.55 | 3.41 | 3.26 | 3.18 | 3.10 | 3.02 | 2.93 | 2.84 | 2.75 |
| 17 | 8.40 | 6.11 | 5.18 | 4.67 | 4.34 | 4.10 | 3.93 | 3.79 | 3.68 | 3.59 | 3.46 | 3.31 | 3.16 | 3.08 | 3.00 | 2.92 | 2.83 | 2.75 | 2.65 |
| 18 | 8.29 | 6.01 | 5.09 | 4.58 | 4.25 | 4.01 | 3.84 | 3.71 | 3.60 | 3.51 | 3.37 | 3.23 | 3.08 | 3.00 | 2.92 | 2.84 | 2.75 | 2.66 | 2.57 |
| 19 | 8.18 | 5.93 | 5.01 | 4.50 | 4.17 | 3.94 | 3.77 | 3.63 | 3.52 | 3.43 | 3.30 | 3.15 | 3.00 | 2.92 | 2.84 | 2.76 | 2.67 | 2.58 | 2.49 |
| 20 | 8.10 | 5.85 | 4.94 | 4.43 | 4.10 | 3.87 | 3.70 | 3.56 | 3.46 | 3.37 | 3.23 | 3.09 | 2.94 | 2.86 | 2.78 | 2.69 | 2.61 | 2.52 | 2.42 |
| 21 | 8.02 | 5.78 | 4.87 | 4.37 | 4.04 | 3.81 | 3.64 | 3.51 | 3.40 | 3.31 | 3.17 | 3.03 | 2.88 | 2.80 | 2.72 | 2.64 | 2.55 | 2.46 | 2.36 |
| 22 | 7.95 | 5.72 | 4.82 | 4.31 | 3.99 | 3.76 | 3.59 | 3.45 | 3.35 | 3.26 | 3.12 | 2.98 | 2.83 | 2.75 | 2.67 | 2.58 | 2.50 | 2.40 | 2.31 |
| 23 | 7.88 | 5.66 | 4.76 | 4.26 | 3.94 | 3.71 | 3.54 | 3.41 | 3.30 | 3.21 | 3.07 | 2.93 | 2.78 | 2.70 | 2.62 | 2.54 | 2.45 | 2.35 | 2.26 |
| 24 | 7.82 | 5.61 | 4.72 | 4.22 | 3.90 | 3.67 | 3.50 | 3.36 | 3.26 | 3.17 | 3.03 | 2.89 | 2.74 | 2.66 | 2.58 | 2.49 | 2.40 | 2.31 | 2.21 |
| 25 | 7.77 | 5.57 | 4.68 | 4.18 | 3.85 | 3.63 | 3.46 | 3.32 | 3.22 | 3.13 | 2.99 | 2.85 | 2.70 | 2.62 | 2.54 | 2.45 | 2.36 | 2.27 | 2.17 |
| 26 | 7.72 | 5.53 | 4.64 | 4.14 | 3.82 | 3.59 | 3.42 | 3.29 | 3.18 | 3.09 | 2.96 | 2.81 | 2.66 | 2.58 | 2.50 | 2.42 | 2.33 | 2.23 | 2.13 |
| 27 | 7.68 | 5.49 | 4.60 | 4.11 | 3.78 | 3.56 | 3.39 | 3.26 | 3.15 | 3.06 | 2.93 | 2.78 | 2.63 | 2.55 | 2.47 | 2.38 | 2.29 | 2.20 | 2.10 |
| 28 | 7.64 | 5.45 | 4.57 | 4.07 | 3.75 | 3.53 | 3.36 | 3.23 | 3.12 | 3.03 | 2.90 | 2.75 | 2.60 | 2.52 | 2.44 | 2.35 | 2.26 | 2.17 | 2.06 |
| 29 | 7.60 | 5.42 | 4.54 | 4.04 | 3.73 | 3.50 | 3.33 | 3.20 | 3.09 | 3.00 | 2.87 | 2.73 | 2.57 | 2.49 | 2.41 | 2.33 | 2.23 | 2.14 | 2.03 |
| 30 | 7.56 | 5.39 | 4.51 | 4.02 | 3.70 | 3.47 | 3.30 | 3.17 | 3.07 | 2.98 | 2.84 | 2.70 | 2.55 | 2.47 | 2.39 | 2.30 | 2.21 | 2.11 | 2.01 |
| 40 | 7.31 | 5.18 | 4.31 | 3.83 | 3.51 | 3.29 | 3.12 | 2.99 | 2.89 | 2.80 | 2.66 | 2.52 | 2.37 | 2.29 | 2.20 | 2.11 | 2.02 | 1.92 | 1.80 |
| 60 | 7.08 | 4.98 | 4.13 | 3.65 | 3.34 | 3.12 | 2.95 | 2.82 | 2.72 | 2.63 | 2.50 | 2.35 | 2.20 | 2.12 | 2.03 | 1.94 | 1.84 | 1.73 | 1.60 |
| 120 | 6.85 | 4.79 | 3.95 | 3.48 | 3.17 | 2.96 | 2.79 | 2.66 | 2.56 | 2.47 | 2.34 | 2.19 | 2.03 | 1.95 | 1.86 | 1.76 | 1.66 | 1.53 | 1.38 |
| ∞ | 6.63 | 4.61 | 3.78 | 3.32 | 3.02 | 2.80 | 2.64 | 2.51 | 2.41 | 2.32 | 2.18 | 2.04 | 1.88 | 1.79 | 1.70 | 1.59 | 1.47 | 1.32 | 1.00 |

Degrees of Freedom for Denominator

[a]Taken from Draper and Smith (1966) and reproduced with permission from E. S. Pearson and H. O. Hartley (1966), *Biometrika Tables for Statisticans*, Vol. 1, 3rd ed., London: Cambridge University.

## Table C Percentage points of the chi-squared distribution

Tabled values are $\chi^2(\alpha; n)$ such that

$$\text{prob}\{\chi^2(n) \text{ variate} \geqslant \chi^2(\alpha; n)\} = \alpha$$

The table entries were computed using subroutine MDCHI from the IMSL (1977) library at the University of Minnesota.

|        |        |        | $\alpha$ |        |         |
|--------|--------|--------|----------|--------|---------|
| d.f.   | 0.20   | 0.10   | 0.05     | 0.01   | 0.001   |
| 1      | 1.64   | 2.71   | 3.84     | 6.64   | 10.81   |
| 2      | 3.22   | 4.60   | 5.99     | 9.22   | 13.69   |
| 3      | 4.64   | 6.25   | 7.82     | 11.32  | 16.29   |
| 4      | 5.99   | 7.78   | 9.49     | 13.28  | 18.43   |
| 5      | 7.29   | 9.24   | 11.07    | 15.09  | 20.75   |
| 6      | 8.56   | 10.65  | 12.60    | 16.81  | 22.68   |
| 7      | 9.80   | 12.02  | 14.07    | 18.47  | 24.53   |
| 8      | 11.03  | 13.36  | 15.51    | 20.08  | 26.32   |
| 9      | 12.24  | 14.69  | 16.93    | 21.65  | 28.06   |
| 10     | 13.44  | 15.99  | 18.31    | 23.19  | 29.76   |
| 11     | 14.63  | 17.28  | 19.68    | 24.75  | 31.43   |
| 12     | 15.81  | 18.55  | 21.03    | 26.25  | 33.07   |
| 13     | 16.99  | 19.81  | 22.37    | 27.72  | 34.68   |
| 14     | 18.15  | 21.07  | 23.69    | 29.17  | 36.27   |
| 15     | 19.31  | 22.31  | 25.00    | 30.61  | 37.84   |
| 16     | 20.47  | 23.55  | 26.30    | 32.03  | 39.39   |
| 17     | 21.62  | 24.77  | 27.59    | 33.44  | 40.93   |
| 18     | 22.76  | 25.99  | 28.88    | 34.83  | 42.44   |
| 19     | 23.90  | 27.21  | 30.15    | 36.22  | 43.95   |
| 20     | 25.04  | 28.42  | 31.42    | 37.59  | 45.44   |
| 21     | 26.17  | 29.62  | 32.68    | 38.96  | 46.92   |
| 22     | 27.30  | 30.82  | 33.93    | 40.31  | 48.39   |
| 23     | 28.43  | 32.01  | 35.18    | 41.66  | 49.85   |
| 24     | 29.56  | 33.20  | 36.42    | 43.00  | 51.29   |
| 25     | 30.68  | 34.38  | 37.66    | 44.34  | 52.73   |
| 26     | 31.80  | 35.57  | 38.89    | 45.66  | 54.16   |
| 27     | 32.91  | 36.74  | 40.12    | 46.99  | 55.58   |
| 28     | 34.03  | 37.92  | 41.34    | 48.30  | 57.00   |
| 29     | 35.14  | 39.09  | 42.56    | 49.61  | 58.41   |
| 30     | 36.25  | 40.26  | 43.78    | 50.91  | 59.81   |
| 40     | 47.26  | 51.80  | 55.75    | 63.71  | 73.49   |
| 50     | 58.16  | 63.16  | 67.50    | 76.17  | 86.74   |
| 60     | 68.97  | 74.39  | 79.08    | 88.40  | 99.68   |
| 70     | 79.71  | 85.52  | 90.53    | 100.44 | 112.38  |
| 80     | 90.40  | 96.57  | 101.88   | 112.34 | 124.90  |
| 90     | 101.05 | 107.56 | 113.14   | 124.13 | 137.27  |
| 100    | 111.66 | 118.49 | 124.34   | 135.82 | 149.50  |

## Table D Rankits for $n \leqslant 20^a$

Values of the rankits (or expected values of normal order statistics) not shown are found by symmetry. For example, the fifteenth largest rankit for a sample of size $n = 17$ is equal to the negative of the third rankit for $n = 17$, or 1.03.

| | | | | | $n$ | | | | | |
|---|---|---|---|---|---|---|---|---|---|---|
| $i$ | 1 | 2 | 3 | 4 | 5 | 6 | 7 | 8 | 9 | 10 |
| 1 | 0 | −0.56 | −0.85 | −1.03 | −1.16 | −1.27 | −1.35 | −1.42 | −1.49 | −1.54 |
| 2 | | 0.56 | 0.00 | −0.30 | −0.50 | −0.64 | −0.76 | −0.85 | −0.93 | −1.00 |
| 3 | | | 0.85 | 0.30 | 0.00 | −0.20 | −0.35 | −0.47 | −0.57 | −0.66 |
| 4 | | | | 1.03 | 0.50 | 0.20 | 0.00 | −0.15 | −0.27 | −0.38 |
| 5 | | | | | 1.16 | 0.64 | 0.35 | 0.15 | 0.00 | −0.12 |
| 6 | | | | | | 1.27 | 0.76 | 0.47 | 0.27 | 0.12 |

| | | | | | $n$ | | | | | |
|---|---|---|---|---|---|---|---|---|---|---|
| $i$ | 11 | 12 | 13 | 14 | 15 | 16 | 17 | 18 | 19 | 20 |
| 1 | −1.59 | −1.63 | −1.67 | −1.70 | −1.74 | −1.77 | −1.79 | −1.82 | −1.84 | −1.87 |
| 2 | −1.06 | −1.12 | −1.16 | −1.21 | −1.25 | −1.28 | −1.32 | −1.35 | −1.38 | −1.41 |
| 3 | −0.73 | −0.79 | −0.85 | −0.90 | −0.95 | −0.99 | −1.03 | −1.07 | −1.10 | −1.13 |
| 4 | −0.46 | −0.54 | −0.60 | −0.66 | −0.71 | −0.76 | −0.81 | −0.85 | −0.89 | −0.92 |
| 5 | −0.22 | −0.31 | −0.39 | −0.46 | −0.52 | −0.57 | −0.62 | −0.66 | −0.71 | −0.75 |
| 6 | 0.00 | −0.10 | −0.19 | −0.27 | −0.34 | −0.40 | −0.45 | −0.50 | −0.55 | −0.59 |
| 7 | 0.22 | 0.10 | 0.00 | −0.09 | −0.17 | −0.23 | −0.30 | −0.35 | −0.40 | −0.45 |
| 8 | 0.46 | 0.31 | 0.19 | 0.09 | 0.00 | −0.08 | −0.15 | −0.21 | −0.26 | −0.31 |
| 9 | 0.73 | 0.54 | 0.39 | 0.27 | 0.17 | 0.08 | 0.00 | −0.07 | −0.13 | −0.19 |
| 10 | 1.06 | 0.79 | 0.60 | 0.46 | 0.34 | 0.23 | 0.15 | 0.07 | 0.00 | −0.06 |

[a]Abridged with permission from Table 28 of E. S. Pearson and H. O. Hartley (1966), *Biometrika Tables for Statisticians*, Vol. 1, 3rd ed. London: Cambridge University.

## Table E  Critical values for the outlier test

The values in the table in row $n$ and column $p'$ are $t(\alpha/n; n-p'-1)$ for the choices of $\alpha=0.01$ and $0.05$. The layout of the table was suggested by Christopher Bingham. The table was computed on a CDC 6400 computer at the University of Minnesota using IMSI subroutine MDSTI.

| $n$ \ $p'$ | Critical values for outlier test, $\alpha=.05$ | | | | | | | | | | | | | | | | | |
|---|---|---|---|---|---|---|---|---|---|---|---|---|---|---|---|---|---|---|
| | 1 | 2 | 3 | 4 | 5 | 6 | 7 | 8 | 9 | 10 | 11 | 12 | 13 | 14 | 15 | 20 | 25 | 30 |
| 6 | 4.85 | 6.23 | 10.89 | 76.39 | | | | | | | | | | | | | | |
| 7 | 4.38 | 5.07 | 6.58 | 11.77 | 89.12 | | | | | | | | | | | | | |
| 8 | 4.12 | 4.53 | 5.26 | 6.90 | 12.59 | 101.9 | | | | | | | | | | | | |
| 9 | 3.95 | 4.22 | 4.66 | 5.44 | 7.18 | 13.36 | 114.6 | | | | | | | | | | | |
| 10 | 3.83 | 4.03 | 4.32 | 4.77 | 5.60 | 7.45 | 14.09 | 127.3 | | | | | | | | | | |
| 11 | 3.75 | 3.90 | 4.10 | 4.40 | 4.88 | 5.75 | 7.70 | 14.78 | 140.1 | | | | | | | | | |
| 12 | 3.69 | 3.81 | 3.96 | 4.17 | 4.49 | 4.98 | 5.89 | 7.94 | 15.44 | 152.8 | | | | | | | | |
| 13 | 3.65 | 3.74 | 3.86 | 4.02 | 4.24 | 4.56 | 5.08 | 6.02 | 8.16 | 16.08 | 165.5 | | | | | | | |
| 14 | 3.61 | 3.69 | 3.79 | 3.91 | 4.07 | 4.30 | 4.63 | 5.16 | 6.14 | 8.37 | 16.69 | 178.2 | | | | | | |
| 15 | 3.58 | 3.65 | 3.73 | 3.83 | 3.95 | 4.12 | 4.36 | 4.70 | 5.25 | 6.25 | 8.58 | 17.28 | 191.0 | | | | | |
| 16 | 3.56 | 3.62 | 3.68 | 3.77 | 3.87 | 4.00 | 4.17 | 4.41 | 4.76 | 5.33 | 6.36 | 8.77 | 17.85 | 203.7 | | | | |
| 17 | 3.54 | 3.59 | 3.65 | 3.72 | 3.80 | 3.90 | 4.04 | 4.21 | 4.46 | 4.82 | 5.40 | 6.47 | 8.95 | 18.40 | 216.4 | | | |
| 18 | 3.53 | 3.57 | 3.62 | 3.68 | 3.75 | 3.83 | 3.94 | 4.08 | 4.26 | 4.51 | 4.88 | 5.47 | 6.57 | 9.13 | 18.93 | | | |
| 19 | 3.52 | 3.56 | 3.60 | 3.65 | 3.71 | 3.78 | 3.86 | 3.97 | 4.11 | 4.30 | 4.55 | 4.93 | 5.54 | 6.67 | 9.30 | | | |
| 20 | 3.51 | 3.54 | 3.58 | 3.62 | 3.67 | 3.73 | 3.81 | 3.89 | 4.00 | 4.15 | 4.33 | 4.59 | 4.98 | 5.60 | 6.76 | | | |
| 21 | 3.50 | 3.53 | 3.57 | 3.60 | 3.65 | 3.70 | 3.76 | 3.83 | 3.92 | 4.03 | 4.18 | 4.37 | 4.64 | 5.03 | 5.67 | | | |
| 22 | 3.50 | 3.52 | 3.55 | 3.59 | 3.63 | 3.67 | 3.72 | 3.78 | 3.86 | 3.95 | 4.06 | 4.21 | 4.40 | 4.68 | 5.08 | 280.1 | | |
| 23 | 3.49 | 3.52 | 3.54 | 3.57 | 3.61 | 3.65 | 3.69 | 3.75 | 3.81 | 3.88 | 3.98 | 4.09 | 4.24 | 4.44 | 4.71 | 21.41 | | |
| 24 | 3.49 | 3.51 | 3.53 | 3.56 | 3.59 | 3.63 | 3.67 | 3.71 | 3.77 | 3.83 | 3.91 | 4.00 | 4.12 | 4.27 | 4.47 | 10.07 | | |
| 25 | 3.48 | 3.50 | 3.53 | 3.55 | 3.58 | 3.61 | 3.65 | 3.69 | 3.73 | 3.79 | 3.85 | 3.93 | 4.02 | 4.14 | 4.30 | 7.17 | | |
| 26 | 3.48 | 3.50 | 3.52 | 3.54 | 3.57 | 3.60 | 3.63 | 3.66 | 3.70 | 3.75 | 3.81 | 3.87 | 3.95 | 4.05 | 4.17 | 5.95 | | |

|     | | | | | | | | | | | | | | | | | | |
| --- | --- | --- | --- | --- | --- | --- | --- | --- | --- | --- | --- | --- | --- | --- | --- | --- | --- | --- |
| 27 | 3.48 | 3.50 | 3.52 | 3.54 | 3.56 | 3.58 | 3.61 | 3.65 | 3.68 | 3.72 | 3.77 | 3.83 | 3.89 | 3.97 | 4.07 | 5.29 | 343.8 | 407.4 |
| 28 | 3.48 | 3.50 | 3.51 | 3.53 | 3.55 | 3.58 | 3.60 | 3.63 | 3.66 | 3.70 | 3.74 | 3.79 | 3.84 | 3.91 | 3.99 | 4.88 | 23.63 | 25.66 |
| 29 | 3.48 | 3.49 | 3.51 | 3.53 | 3.55 | 3.57 | 3.59 | 3.62 | 3.64 | 3.68 | 3.71 | 3.76 | 3.81 | 3.86 | 3.93 | 4.61 | 10.74 | 11.34 |
| 30 | 3.48 | 3.49 | 3.51 | 3.52 | 3.54 | 3.56 | 3.58 | 3.60 | 3.63 | 3.66 | 3.69 | 3.73 | 3.77 | 3.82 | 3.88 | 4.42 | 7.53 | 7.84 |
| 31 | 3.48 | 3.49 | 3.50 | 3.52 | 3.54 | 3.55 | 3.57 | 3.59 | 3.62 | 3.64 | 3.67 | 3.71 | 3.74 | 3.79 | 3.84 | 4.28 | 6.18 | 6.39 |
| 32 | 3.48 | 3.49 | 3.50 | 3.52 | 3.53 | 3.55 | 3.57 | 3.59 | 3.61 | 3.63 | 3.66 | 3.69 | 3.72 | 3.76 | 3.80 | 4.17 | 5.47 | 5.62 |
| 33 | 3.48 | 3.49 | 3.50 | 3.52 | 3.53 | 3.54 | 3.56 | 3.58 | 3.60 | 3.62 | 3.64 | 3.67 | 3.70 | 3.74 | 3.77 | 4.08 | 5.03 | 5.16 |
| 34 | 3.48 | 3.49 | 3.50 | 3.51 | 3.53 | 3.54 | 3.56 | 3.57 | 3.59 | 3.61 | 3.63 | 3.66 | 3.68 | 3.71 | 3.75 | 4.01 | 4.74 | 4.84 |
| 35 | 3.48 | 3.49 | 3.50 | 3.51 | 3.52 | 3.54 | 3.55 | 3.57 | 3.58 | 3.60 | 3.62 | 3.64 | 3.67 | 3.70 | 3.73 | 3.96 | 4.53 | 4.62 |
| 36 | 3.48 | 3.49 | 3.50 | 3.51 | 3.52 | 3.54 | 3.55 | 3.56 | 3.58 | 3.60 | 3.61 | 3.63 | 3.65 | 3.68 | 3.71 | 3.91 | 4.37 | |
| 37 | 3.48 | 3.49 | 3.50 | 3.51 | 3.52 | 3.53 | 3.55 | 3.56 | 3.57 | 3.59 | 3.61 | 3.62 | 3.64 | 3.67 | 3.69 | 3.87 | 4.26 | |
| 38 | 3.49 | 3.49 | 3.50 | 3.51 | 3.52 | 3.53 | 3.54 | 3.56 | 3.57 | 3.58 | 3.60 | 3.62 | 3.64 | 3.66 | 3.68 | 3.84 | 4.16 | |
| 39 | 3.49 | 3.49 | 3.50 | 3.51 | 3.52 | 3.53 | 3.54 | 3.55 | 3.57 | 3.58 | 3.59 | 3.61 | 3.62 | 3.65 | 3.67 | 3.81 | 4.09 | |
| 40 | 3.49 | 3.49 | 3.50 | 3.51 | 3.52 | 3.53 | 3.54 | 3.54 | 3.56 | 3.58 | | | | 3.64 | 3.66 | 3.79 | 4.03 | |
| 50 | 3.51 | 3.51 | 3.51 | 3.52 | 3.53 | 3.54 | 3.54 | 3.55 | 3.55 | 3.56 | 3.57 | 3.57 | 3.58 | 3.59 | 3.60 | 3.66 | 3.75 | 3.88 |
| 60 | 3.53 | 3.53 | 3.53 | 3.54 | 3.54 | 3.56 | 3.55 | 3.56 | 3.56 | 3.56 | 3.57 | 3.57 | 3.57 | 3.58 | 3.59 | 3.62 | 3.67 | 3.73 |
| 70 | 3.55 | 3.55 | 3.55 | 3.55 | 3.56 | 3.58 | 3.56 | 3.58 | 3.57 | 3.57 | 3.58 | 3.58 | 3.58 | 3.59 | 3.59 | 3.61 | 3.64 | 3.67 |
| 80 | 3.57 | 3.57 | 3.57 | 3.57 | 3.57 | 3.59 | 3.58 | 3.60 | 3.58 | 3.58 | 3.59 | 3.59 | 3.59 | 3.60 | 3.60 | 3.61 | 3.63 | 3.66 |
| 90 | 3.58 | 3.59 | 3.59 | 3.59 | 3.59 | 3.61 | 3.59 | 3.61 | 3.60 | 3.60 | 3.60 | 3.60 | 3.60 | 3.61 | 3.61 | 3.62 | 3.63 | 3.65 |
| 100 | 3.60 | 3.60 | 3.60 | 3.60 | 3.61 | | 3.61 | | 3.61 | 3.61 | 3.61 | 3.62 | 3.62 | 3.62 | 3.62 | 3.63 | 3.64 | 3.65 |
| 200 | 3.73 | 3.73 | 3.73 | 3.73 | 3.73 | 3.73 | 3.73 | 3.73 | 3.73 | 3.73 | 3.73 | 3.73 | 3.73 | 3.73 | 3.74 | 3.74 | 3.74 | 3.74 |
| 300 | 3.81 | 3.81 | 3.81 | 3.81 | 3.81 | 3.81 | 3.81 | 3.81 | 3.81 | 3.81 | 3.82 | 3.82 | 3.82 | 3.82 | 3.82 | 3.82 | 3.82 | 3.82 |
| 400 | 3.87 | 3.87 | 3.87 | 3.87 | 3.87 | 3.87 | 3.87 | 3.88 | 3.88 | 3.88 | 3.88 | 3.88 | 3.88 | 3.88 | 3.88 | 3.88 | 3.88 | 3.88 |
| 500 | 3.92 | 3.92 | 3.92 | 3.92 | 3.92 | 3.92 | 3.92 | 3.92 | 3.92 | 3.92 | 3.92 | 3.92 | 3.92 | 3.92 | 3.92 | 3.92 | 3.92 | 3.92 |

**Table E** (continued)

Critical values for outlier test, $\alpha = .01$

| $n$ \ $p'$ | 1 | 2 | 3 | 4 | 5 | 6 | 7 | 8 | 9 | 10 | 11 | 12 | 13 | 14 | 15 | 20 | 25 | 30 |
|---|---|---|---|---|---|---|---|---|---|---|---|---|---|---|---|---|---|---|
| 6 | 7.53 | 10.87 | 24.46 | 382.0 | | | | | | | | | | | | | | |
| 7 | 6.35 | 7.84 | 11.45 | 26.43 | 445.6 | | | | | | | | | | | | | |
| 8 | 5.71 | 6.54 | 8.12 | 11.98 | 28.26 | 509.3 | | | | | | | | | | | | |
| 9 | 5.31 | 5.84 | 6.71 | 8.38 | 12.47 | 29.97 | 573.0 | | | | | | | | | | | |
| 10 | 5.04 | 5.41 | 5.96 | 6.97 | 8.61 | 12.92 | 31.60 | 636.6 | | | | | | | | | | |
| 11 | 4.85 | 5.12 | 5.50 | 6.07 | 7.01 | 8.83 | 13.35 | 33.14 | 700.3 | | | | | | | | | |
| 12 | 4.71 | 4.91 | 5.19 | 5.58 | 6.17 | 7.15 | 9.03 | 13.75 | 34.62 | 763.9 | | | | | | | | |
| 13 | 4.60 | 4.76 | 4.97 | 5.25 | 5.66 | 6.26 | 7.27 | 9.22 | 14.12 | 36.03 | 827.6 | | | | | | | |
| 14 | 4.51 | 4.64 | 4.81 | 5.02 | 5.32 | 5.73 | 6.35 | 7.39 | 9.40 | 14.48 | 37.40 | 891.3 | | | | | | |
| 15 | 4.44 | 4.55 | 4.68 | 4.85 | 5.08 | 5.37 | 5.80 | 6.43 | 7.50 | 9.57 | 14.82 | 38.71 | 954.9 | | | | | |
| 16 | 4.38 | 4.48 | 4.59 | 4.72 | 4.90 | 5.12 | 5.43 | 5.86 | 6.51 | 7.60 | 9.73 | 15.15 | 39.98 | | | | | |
| 17 | 4.34 | 4.41 | 4.51 | 4.62 | 4.76 | 4.94 | 5.17 | 5.48 | 5.92 | 6.59 | 7.70 | 9.88 | 15.46 | 41.21 | | | | |
| 18 | 4.30 | 4.36 | 4.44 | 4.54 | 4.66 | 4.80 | 4.98 | 5.21 | 5.53 | 5.98 | 6.66 | 7.80 | 10.03 | 15.76 | 42.41 | | | |
| 19 | 4.26 | 4.32 | 4.39 | 4.47 | 4.57 | 4.69 | 4.83 | 5.01 | 5.25 | 5.57 | 6.03 | 6.72 | 7.89 | 10.17 | 16.05 | | | |
| 20 | 4.23 | 4.29 | 4.35 | 4.42 | 4.50 | 4.60 | 4.72 | 4.86 | 5.05 | 5.29 | 5.62 | 6.08 | 6.79 | 7.98 | 10.31 | | | |
| 21 | 4.21 | 4.26 | 4.31 | 4.37 | 4.44 | 4.52 | 4.62 | 4.74 | 4.89 | 5.08 | 5.33 | 5.66 | 6.13 | 6.85 | 8.06 | | | |
| 22 | 4.19 | 4.23 | 4.28 | 4.33 | 4.39 | 4.46 | 4.55 | 4.65 | 4.77 | 4.92 | 5.11 | 5.36 | 5.70 | 6.18 | 6.91 | | | |
| 23 | 4.17 | 4.21 | 4.25 | 4.30 | 4.35 | 4.41 | 4.49 | 4.57 | 4.67 | 4.80 | 4.95 | 5.14 | 5.40 | 5.74 | 6.22 | 47.94 | | |
| 24 | 4.15 | 4.19 | 4.22 | 4.27 | 4.32 | 4.37 | 4.43 | 4.51 | 4.59 | 4.70 | 4.82 | 4.98 | 5.17 | 5.43 | 5.78 | 17.36 | | |
| 25 | 4.14 | 4.17 | 4.20 | 4.24 | 4.28 | 4.33 | 4.39 | 4.45 | 4.53 | 4.62 | 4.72 | 4.85 | 5.00 | 5.20 | 5.46 | 10.92 | | |
| 26 | 4.12 | 4.15 | 4.18 | 4.22 | 4.26 | 4.30 | 4.35 | 4.41 | 4.47 | 4.55 | 4.64 | 4.74 | 4.87 | 5.03 | 5.23 | 8.43 | | |

| | | | | | | | | | | | | | | | | | | |
|---|---|---|---|---|---|---|---|---|---|---|---|---|---|---|---|---|---|---|
| 27 | 4.11 | 4.14 | 4.17 | 4.20 | 4.24 | 4.27 | 4.32 | 4.37 | 4.43 | 4.49 | 4.57 | 4.66 | 4.76 | 4.89 | 5.05 | 7.17 | | |
| 28 | 4.10 | 4.13 | 4.15 | 4.18 | 4.21 | 4.25 | 4.29 | 4.33 | 4.38 | 4.44 | 4.51 | 4.59 | 4.68 | 4.78 | 4.91 | 6.43 | 52.90 | |
| 29 | 4.09 | 4.12 | 4.14 | 4.17 | 4.20 | 4.23 | 4.26 | 4.30 | 4.35 | 4.40 | 4.46 | 4.53 | 4.60 | 4.69 | 4.80 | 5.94 | 18.50 | |
| 30 | 4.09 | 4.11 | 4.13 | 4.15 | 4.18 | 4.21 | 4.24 | 4.28 | 4.32 | 4.36 | 4.42 | 4.47 | 4.54 | 4.62 | 4.71 | 5.60 | 11.44 | |
| 31 | 4.08 | 4.10 | 4.12 | 4.34 | 4.14 | 4.19 | 4.22 | 4.26 | 4.29 | 4.33 | 4.38 | 4.43 | 4.49 | 4.56 | 4.64 | 5.35 | 8.75 | |
| 32 | 4.07 | 4.09 | 4.11 | 4.13 | 4.15 | 4.18 | 4.21 | 4.24 | 4.27 | 4.31 | 4.35 | 4.39 | 4.45 | 4.50 | 4.57 | 5.16 | 7.40 | |
| 33 | 4.07 | 4.08 | 4.10 | 4.12 | 4.14 | 4.17 | 4.19 | 4.22 | 4.25 | 4.28 | 4.32 | 4.36 | 4.41 | 4.46 | 4.52 | 5.01 | 6.60 | 57.43 |
| 34 | 4.06 | 4.08 | 4.09 | 4.11 | 4.13 | 4.15 | 4.18 | 4.20 | 4.23 | 4.26 | 4.29 | 4.33 | 4.37 | 4.42 | 4.47 | 4.89 | 6.09 | 19.51 |
| 35 | 4.06 | 4.07 | 4.09 | 4.11 | 4.12 | 4.14 | 4.16 | 4.19 | 4.21 | 4.24 | 4.27 | 4.31 | 4.34 | 4.39 | 4.43 | 4.79 | 5.72 | 11.90 |
| 36 | 4.05 | 4.07 | 4.08 | 4.10 | 4.12 | 4.13 | 4.15 | 4.18 | 4.20 | 4.22 | 4.25 | 4.28 | 4.32 | 4.36 | 4.40 | 4.71 | 5.46 | 9.03 |
| 37 | 4.05 | 4.06 | 4.08 | 4.09 | 4.11 | 4.13 | 4.14 | 4.16 | 4.19 | 4.21 | 4.24 | 4.26 | 4.29 | 4.33 | 4.37 | 4.64 | 5.26 | 7.60 |
| 38 | 4.05 | 4.06 | 4.07 | 4.09 | 4.10 | 4.12 | 4.13 | 4.15 | 4.17 | 4.20 | 4.22 | 4.25 | 4.27 | 4.31 | 4.34 | 4.59 | 5.10 | 6.76 |
| 39 | 4.04 | 4.06 | 4.07 | 4.08 | 4.10 | 4.11 | 4.13 | 4.14 | 4.16 | 4.18 | 4.21 | 4.23 | 4.26 | 4.28 | 4.32 | 4.54 | 4.97 | 6.21 |
| 40 | 4.04 | 4.05 | 4.06 | 4.08 | 4.09 | 4.10 | 4.12 | 4.14 | 4.15 | 4.17 | 4.19 | 4.22 | 4.24 | 4.27 | 4.29 | 4.49 | 4.87 | 5.83 |
| 50 | 4.03 | 4.03 | 4.04 | 4.05 | 4.06 | 4.07 | 4.07 | 4.08 | 4.09 | 4.10 | 4.12 | 4.13 | 4.14 | 4.15 | 4.17 | 4.25 | 4.38 | 4.59 |
| 60 | 4.03 | 4.03 | 4.04 | 4.04 | 4.05 | 4.05 | 4.06 | 4.06 | 4.07 | 4.08 | 4.08 | 4.09 | 4.10 | 4.11 | 4.12 | 4.17 | 4.23 | 4.32 |
| 70 | 4.03 | 4.03 | 4.04 | 4.04 | 4.05 | 4.05 | 4.05 | 4.06 | 4.06 | 4.07 | 4.07 | 4.08 | 4.08 | 4.09 | 4.09 | 4.13 | 4.17 | 4.22 |
| 80 | 4.04 | 4.04 | 4.04 | 4.05 | 4.05 | 4.05 | 4.06 | 4.06 | 4.06 | 4.07 | 4.07 | 4.07 | 4.08 | 4.08 | 4.09 | 4.11 | 4.13 | 4.17 |
| 90 | 4.05 | 4.05 | 4.05 | 4.05 | 4.06 | 4.06 | 4.06 | 4.06 | 4.07 | 4.07 | 4.07 | 4.07 | 4.08 | 4.08 | 4.08 | 4.10 | 4.12 | 4.14 |
| 100 | 4.06 | 4.06 | 4.06 | 4.06 | 4.06 | 4.07 | 4.07 | 4.07 | 4.07 | 4.07 | 4.08 | 4.08 | 4.08 | 4.08 | 4.09 | 4.10 | 4.11 | 4.13 |
| 200 | 4.15 | 4.15 | 4.15 | 4.15 | 4.15 | 4.15 | 4.15 | 4.15 | 4.15 | 4.15 | 4.15 | 4.15 | 4.15 | 4.15 | 4.15 | 4.16 | 4.16 | 4.16 |
| 300 | 4.21 | 4.21 | 4.21 | 4.21 | 4.21 | 4.21 | 4.22 | 4.22 | 4.22 | 4.22 | 4.22 | 4.22 | 4.22 | 4.22 | 4.22 | 4.22 | 4.22 | 4.22 |
| 400 | 4.26 | 4.27 | 4.27 | 4.27 | 4.27 | 4.27 | 4.27 | 4.27 | 4.27 | 4.27 | 4.27 | 4.27 | 4.27 | 4.27 | 4.27 | 4.27 | 4.27 | 4.27 |
| 500 | 4.31 | 4.31 | 4.31 | 4.31 | 4.31 | 4.31 | 4.31 | 4.31 | 4.31 | 4.31 | 4.31 | 4.31 | 4.31 | 4.31 | 4.31 | 4.31 | 4.31 | 4.31 |

# REFERENCES

Afifi, A. A. and R. M. Elashoff (1966). "Missing values in multivariate statistics I. Review of the literature." *J. Am. Statist. Assoc.*, **61**, 595–604.

Aitchison, J. and I. R. Dunsmore (1975). *Statistical Prediction Analysis*. New York: Cambridge University.

Aitken, A. C. (1934). "Note on selection from a multivariate normal population." *Proc. Edinburgh Math. Soc.*, **4**, 106–110.

Allen, D. M. (1974). "The relationship between variable selection and prediction." *Technometrics*, **16**, 125–127.

Allison, T. and D. V. Cicchetti (1976). "Sleep in mammals: Ecological and constiutional correlates." *Science*, **194**, 732–734.

Anderson, T. W. (1976). "Estimation of linear functional relationships: Approximate distributions and connections with simultaneous equations in econometrics (with discussion)." *J. Roy. Statist. Soc. Ser. B*, **38**, 1–36.

Andrews, D. F. (1974). "A robust method for multiple linear regression." *Technometrics*, **16**, 523–531.

Andrews, D. F., P. Bickel, F. Hampel, P. Huber, W. Rogers, and J. W. Tukey (1972). *Robust Estimates of Location*. Princeton: Princeton University Press.

Anscombe, F. J. (1973). "Graphs in statistical analysis." *Am. Statist.* **27**, 17–21.

Atkinson, A. C. (1973). "Testing transformations to normality." *J. Roy. Statist. Soc. Ser. B*, **35**, 473–479.

Atkinson, A. C. (1981). "Robustness, transformations and two graphical displays for outlying and influential observations in regression." *Biometrika*, **68**, 13–20.

Atkinson, A. C. (1982). "Regression diagnostics, transformations and constructed variables (with discussion)." *J. Roy. Statist. Soc. Ser. B*, **44**, 1–35.

Atkinson, A. C. (1983). "Diagnostics, regression analysis and shifted power transformations." *Technometrics*, **25**, 23–34.

**307**

Baes, C. F. and H. H. Kellogg (1953). "Effect of dissolved sulphur on the surface tension of liquid copper." *J. Metals*, **5**, 643–648.

Baker, R. J. and J. A. Nelder (1978). "The GLIM System, Release 3, Generalized Linear Interactive Modeling." Oxford: Numerical Algorithms Group.

Barnett, V. and T. Lewis (1978). *Outliers in Statistical Data*. Chichester: Wiley.

Bates, D. and D. Watts (1980). "Relative curvature measures of nonlinearity (with discussion)." *J. Roy. Statist. Soc. Ser. B*, **22**, 41–88.

Beale, E. M. L. (1960). "Confidence regions in nonlinear estimation (with discussion)." *J. Roy. Statist. Soc. Ser. B*, **22**, 41–88.

Beale, E. M. L., M. G. Kendall, and D. W. Mann (1967). "The discarding of variables in multivariate analysis." *Biometrika*, **54**, 537–566.

Beale, E. M. L. and R. J. Little (1975). "Missing values in multivarite analysis." *J. Roy. Statist. Soc. Ser. B*, **37**, 129–145.

Beaton, A. E. (1964). "The use of special matrix operators in statistical calculations." Unpublished Ph.D. dissertation, Harvard University.

Beaton, A. E., D. B. Rubin, and J. L. Barone (1976). "The acceptability of regression solutions: Another look at computational stability." *J. Am. Statist. Assoc.*, **71**, 158–168.

Beckman, R. and R. D. Cook (1983). "Outlier . . . s." *Technometrics*, **25**, 119–149.

Belsley, D. A. (1984). "Demeaning condition diagnostics through centering." *Am. Statist.*, **38**, 73–77.

Belsley, D. A., E. Kuh, and R. E. Welsch (1980). *Regression Diagnostics*. New York: Wiley.

Berger, J. (1975). "Minimax estimation of location vectors for a wide variety of densities." *Ann. Statist.*, **3**, 1318–1328.

Berk, K. (1977). "Tolerance and condition in regression computations." *J. Am. Statist. Assoc.*, **72**, 863–866.

Berkson, J. (1950). "Are there two regressions?" *J. Am. Statist. Assoc.*, **45**, 164–180.

Bickel, P. and K. Doksum (1981). "An analysis of transformations revisited." *J. Am. Statist. Assoc.*, **76**, 296–311.

Bingham, C. and K. Larntz (1977). "Comment on 'A simulation study of alternatives to ordinary least squares'." *J. Am. Statist. Assoc.*, **72**, 97–102.

Bland, J. (1978). "A comparison of certain aspects of ontogeny in the long and short shoots of McIntosh apple during one annual growth cycle." Unpublished Ph.D. dissertation, University of Minnesota, St. Paul, MN.

Blom, G. (1958). *Statistical Estimates and Transformed Beta Variates*. New York: Wiley.

Box, G. E. P. (1953). "Non-normality and tests on variances." *Biometrika*, **40**, 318–335.

Box, G. E. P. (1980). "Sampling and Bayes' inference in scientific modelling and robustness (with discussion)." *J. Roy. Statist. Soc. Ser. A*, **143**, 383–430.

Box, G. E. P. and D. R. Cox (1964). "An analysis of transformations (with discussion)." *J. Roy. Statist. Soc. Ser. B*, **26**, 211–246.

Box, G. E. P., J. S. Hunter, and W. G. Hunter (1978). *Statistics for Experimenters.* New York: Wiley.

Box, G. E. P. and P. W. Tidwell (1962). "Transformations of the independent variables." *Technometrics*, **4**, 531–550.

Box, G. E. P. and K. B. Wilson (1951). "On the experimental attainment of optimal conditions (with discussion)." *J. Roy. Statist. Soc. Ser. B*, **13**, 1–45.

Box, M. J. (1971). "Bias in non-linear estimation (with discussion)." *J. Roy. Statist. Soc. Ser. B*, **33**, 171–201.

Buck, S. F. (1960). "A method of estimating missing values in multivariate data suitable for use with an electronic computer." *J. Roy. Statist. Soc. Ser. B*, **22**, 302–306.

Burt, C. (1966). "The genetic determination of differences in intelligence: A study of monozygotic twins readed together and apart." *Br. J. Psych.*, **57**, 137–53.

Butler, R. W. (1984). "The significance attained by the best fitting regressor variable." *J. Am. Statist. Assoc.*, **79**, 341–48.

Carroll, R. (1980). "A robust method for testing transformations to achieve approximate normality." *J. Roy. Statist. Soc. Ser. B*, **42**, 71–78.

Carroll, R. J. (1982). "Adapting for heteroscedasticity in linear models." *Ann. Statist.*, **4**, 1224–1233.

Chambers, J. M., W. S. Cleveland, B. Kleiner, and P. Tukey (1983). *Graphical Methods for Data Analysis.* Belmont, CA: Wadsworth.

Chan, T., G. Golub, and R. LeVeque (1983). "Algorithms for computing the sample variance: Analysis and recommendations." *Am. Statist.*, **37**, 242–47.

Chen, C. F. (1983). "Score tests for regression models." *J. Am. Statist. Assoc.*, **78**, 158–161.

Clapham, A. W. (1934). *English Romanesque Architecture After the Conquest.* Oxford: Clarendon Press.

Cook, R. D. (1977). "Detection of influential observations in linear regression." *Technometrics*, **19**, 15–18.

Cook, R. D. (1979). "Influential observations in linear regression." *J. Am. Statist. Assoc.*, **74**, 169–174.

Cook, R. D. (1984). "Comment on Belsley (1984)." *Am. Statist.*, **38**, 78–79.

Cook, R. D. and J. O. Jacobsen (1978). "Analysis of 1977 West Hudson Bay snow goose surveys." Unpublished report, Canadian Wildlife Service.

Cook, R. D. and P. Prescott (1981). "Approximate significance levels for detecting outliers in linear regression." *Technometrics*, **23**, 59–64.

Cook, R. D. and Wang, P. C. (1983). "Transformations and influential cases in regression." *Technometrics*, **25**, 337–344.

Cook, R. D. and S. Weisberg (1980). "Characterizations of an empirical influence function for detecting influential cases in regression." *Technometrics*, **22**, 495–508.

Cook, R. D. and S. Weisberg (1982a). *Residuals and Influence in Regression.* London: Chapman Hall.

Cook, R. D. and S. Weisberg (1982b). "Criticism in regression," in Leinhardt, S. (ed.), *Sociological Methodology*. San Francisco: Jossey-Bass, Chapter 8.

Cook, R. D. and S. Weisberg (1983). "Diagnostics for heteroscedasticity in regression." *Biometrika*, **70**, 1–10.

Copas, J. B. (1983). "Regression, prediction and shrinkage (with discussion)." *J. Roy. Statist. Soc. Ser. B*, **45**, 311–354.

Cox, D. R. (1958). *The Planning of Experiments*. New York: Wiley.

Cox, D. R. (1970). *The Analysis of Binary Data*. London: Chapman Hall.

Cox, D. R. (1977). "Nonlinear models, residuals and transformations." *Math. Operationsforsch. Statist. Ser. Statist.*, **8**, 3–22.

Cox, D. R. and D. Oakes (1984). *Analysis of Survival Data*. London: Chapman Hall.

Daniel, C. (1976). *Applications of Statistics to Industrial Experiments*. New York: Wiley.

Daniel, C. and F. Wood (1980). *Fitting Equations to Data*, 2nd ed. New York: Wiley.

Dalziel, C. F., J. B. Lagen, and J. L. Thurston (1941). "Electric Shocks." *Trans. IEEE*, **60**, 1073–1079.

Davies, R. B. and B. Hutton (1975). "The effects of errors in the independent variables in linear regression." *Biometrika*, **62**, 383-391.

Dempster, A. P. (1973). "Alternatives to least squares in multiple regression," in Kale, D. G. and R. P. Gupta (eds.), *Multivariate Statistical Inference*. Amsterdam: North-Holland.

Dempster, A. P., N. M. Laird, and D. B. Rubin (1977). "Maximum likelihood from incomplete data via the EM algorithm (with discussion)." *J. Roy. Statist. Soc. Ser. B*, **39**, 1–38.

Dempster, A. P., M. Schatzoff, and N. Wermuth (1977). "A simulation study of alternatives to ordinary least squares (with discussion)." *J. Am. Statist. Assoc.*, **72**, 77–104.

Devlin, S. J. R. Gnanadesikan, and J. Kettenring (1975). "Robust estimation and outlier detection with correlation coefficients." *Biometrika*, **62**, 531–546.

Dixon, W. J. (ed.) (1983). *BMDP Biomedical Computer Programs*. Berkeley: University of California.

Dongarra, J., J. P. Bunch, C. B. Moler, and G. W. Stewart (1979). *The LINPACK Users Guide*. Philadelphia: Society for Industrial and Applied Mathematics.

Draper, N. and W. G. Hunter (1969). "Transformations: Some examples revisited." *Technometrics*, **11**, 23–40.

Draper, N. and H. Smith (1966). *Applied Regression Analysis*. New York: Wiley.

Draper, N. R. and R. C. Van Nostrand (1978). "Ridge regression: Is it worthwhile?" Technical Report No. 501, Dept. of Statistics, University of Wisconsin.

Draper, N. R. and R. C. Van Nostrand (1979). "Ridge regression and James–Stein estimation: Review and comments." *Technometrics*, **21**, 451–465.

Durbin, J. and G. S. Watson (1950). "Testing for serial correlation in least squares regression I." *Biometrika*, **58**, 1–19.

Durbin, J. and G. S. Watson (1951). "Testing for serial correlation in least squares regression II."*Biometrika*, **38**, 159–178.

Durbin, J. and G. S. Watson (1971). "Testing for serial correlation in least squares regression III." *Biometrika*, **58**, 1–19.

Efron, B. and C. Morris (1973). "Stein's estimation rule and its competitors—An empirical Bayes approach." *J. Am. Statist. Assoc.*, **68**, 117–130.

Efron, B. and C. Morris (1975). "Data analysis using Stein's estimator and its generalizations." *J. Am. Statist. Assoc.*, **70**, 311–319.

Ehrenberg, A. S. C. (1981). "The problem of numeracy." *Am. Statist.*, **35**, 67–71.

Ehrenberg, A. S. C. (1982). "How good is best?" *J. Roy. Statist. Soc. Ser. A*, **145**, 364–366.

Ezekiel, M. and F. A. Fox (1959). *Methods of Correlation and Regression Analysis*. New York: Wiley.

Fienberg, S. E. (1977). *The Analysis of Cross Classified Categorical Data*, 2nd ed. Cambridge: MIT Press.

Finkelstein, M. O. (1980). The judicial reception of multiple regression studies in race and sex discrimination cases." *Columbia Law Rev.*, **80**, 734–757.

Forbes, J. D. (1857). "Further experiments and remarks on the measurement of heights by the boiling point of water." *Trans. R. Soc. Edinburgh*, **21**, 135–143.

Freedman, D. (1983). "A note on screening regression equations." *Am. Statist.*, **37**, 152–155.

Freedman, D. and S. Peters (1984). "Bootstrapping a regression equation: Some empirical results." *J. Am. Statist. Assoc.*, **79**, 97–106.

Freeman, M. F. and J. W. Tukey (1950). "Transformations related to the angular and the square root." *Ann. Math. Statist.*, **21**, 607–611.

Furnival, G. and R. Wilson (1974). "Regression by leaps and bounds." *Technometrics*, **16**, 499–511.

Garside, M. J. (1971). "Some computational procedures for the best subset problem." *Appl. Statist.*, **20**, 8–15.

Gauss, C. F. (1821, collected works 1873). "Theoria combinationis observationum erroribus minimis obnoxiae," in Werke 4, Section 35, Göttingen.

Geisser, S. (1980). "A predictivistic primer," in Zellner, A. (ed.), *Bayesian Analysis in Econometrics and Statistics*. Amsterdam: North-Holland.

Geisser, S. and W. F. Eddy (1979). "A predictive approach to model selection." *J. Am. Statist. Assoc.*, **74**, 153–160.

Gentry, A. H. and J. Lopez-Parodi (1980). "Deforestation and increased flooding in the upper Amazon." *Science*, **210**, 1354–1356.

Gnanadesikan, R. (1977). *Methods for Statistical Analysis of Multivariate Data*. New York: Wiley.

Goldstein, M. and A. F. M. Smith (1974). "Ridge type estimators for regression analysis." *J. Roy. Statist. Soc. Ser. B*, **36**, 284–301.

Golub, G. H. (1965). "Numerical methods for solving linear least squares problems." *Numerical Mathematics*, **7**, 206–216.

Goodnight, J. H. (1979). "A tutorial on the SWEEP operator." *Am. Statist.*, **33**, 149–158.

Gould, S. J. (1966). "Allometry and size in ontogeny and phylogeny." *Biol. Rev.*, **41**, 587–640.

Gould, S. J. (1973). "The shape of things to come." *Syst. Zool.*, **22**, 401–404.

Gray, J. and R. Ling (1984). "K-clustering as a detection tool for influential subsets in regression." *Technometrics*, **26**, 304–318.

Graybill, F. A. (1969). *Introduction to Matrices with Statistical Applications.* Belmont, CA: Wadsworth.

Hald, A. (1960). *Statistical Theory with Engineering Applications.* New York: Wiley.

Hartley, H. O. and R. R. Hocking (1971). "The analysis of incomplete data." *Biometrics*, **27**, 783–808.

Hawkins, D. M. (1980). *Identification of Outliers.* London: Chapman Hall.

Hawkins, D. M., D. Bradu, and G. Kass (1984). "Location of several outliers in multiple regression using elemental sets." *Technometrics*, **26**, 197–208.

Henderson, H. V. and S. R. Searle (1981). "On deriving the inverse of a sum of matrices." *SIAM Rev.*, **23**, 53–60.

Hernandez, F. and R. A. Johnson (1980). "The large sample behavior of transformations to normality." *J. Am. Statist. Assoc.*, **75**, 855–861.

Hinkley, D. V. and G. Runger (1984). "The analysis of transformed data (with discussion)." *J. Am. Statist. Assoc.*, **79**, 302–319.

Hoaglin, D. C. and R. Welsch (1970). "The hat matrix in regression and ANOVA." *Am. Statist.*, **32**, 17–22.

Hocking, R. R. and R. N. Leslie (1967). "Selection of the best subset in regression analysis." *Technometrics*, **2**, 531–540.

Hocking, R. R. and O. J. Pendleton (1982). "The regression dilemma." *Commun. Statist. Ser. A*, **12**, 497–527.

Hodges, S. D. and P. G. Moore (1972). "Data uncertainties and least squares regression." *Appl. Statist.*, **21**, 185–195.

Hoerl, A. E. and R. W. Kennard (1970a). "Ridge regression: biased estimation for nonorthogonal problems." *Technometrics*, **12**, 55–67.

Hoerl, A. E. and R. W. Kennard (1970b). "Ridge regression: Applications to nonorthogonal problems." *Technometrics*, **12**, 69–82.

Holt, D. and A. J. Scott (1981). "Regression analysis using survey data." *The Statistician*, **30**, 169–178.

Householder, A. S. (1958). "The approximate solution of matrix problems." *J. Assoc. Comput. Mach.*, **5**, 204–243.

Huber, P. J. (1981). *Robust Statistics.* New York: Wiley.

IMSL (1979). *The IMSL Library.* Houston: International Mathematics and Statistics Library.

James, W. and C. Stein (1961). "Estimation with quadratic loss," in *Proceedings of the*

*Fourth Berkeley Symposium on Mathematical Statistics and Probability*, Vol. 1. Berkeley: University of California.

Jenson, R. (1977). "Evinrude's computerized quality control productivity." *Qual. Prog.*, **10**, 12–16.

Jevons, W. S. (1868). "On the condition of the gold coinage of the United Kingdom, with reference to the question of international currency." *J. [Roy.] Statist. Soc.*, **31**, 426–464.

John, P. W. M. (1971). *Statistical Design and Analysis of Experiments*. New York: Macmillan.

Johnson, M. P. and P. H. Raven (1973). "Species number and endemism: The Galápagos Archipelago revisited." *Science*, **179**, 893–895.

Kalbfleisch, J. D. and R. L. Prentice (1980). *The Statistical Analysis of Failure Time Data*. New York: Wiley.

Kennard, R. W. (1971). "A note on the $C_p$ statistic." *Technometrics*, **13**, 899–900.

Kennedy, W. and T. Bancroft (1971). "Model building for prediction in regression based on repeated significance tests," *Ann. Math. Statist.*, **42**, 1273–1284.

Kennedy, W. and J. Gentle (1980). *Statistical Computing*. New York: Marcel Dekker.

Kerr, R. A. (1982). "Test fails to confirm cloud seeding effect." *Science*, **217**, 234–236.

Krasker, W. S. and R. E. Welsch (1982). "Efficient bounded influence regression estimation." *J. Am. Statist. Assoc.*, **77**, 595–604.

LaMotte, L. R. and R. R. Hocking (1970). "Computational efficiency in the selection of variables." *Technometrics*, **12**, 83–93.

Land, C. E. (1974). "Confidence interval estimation for means after data transformations to normality." *J. Am. Statist. Assoc.*, **69**, 795–802 (correction, *ibid.*, **71**, 255).

Landwehr, J., D. Pregibon, and A. Shoemaker (1984). "Graphical methods for assessing logistic regression models (with discussion)." *J. Am. Statist. Assoc.*, **79**, 61–83.

Larson, W. A. and S. A. McCleary (1972). "The use of partial residual plots in regression analysis." *Technometrics*, **14**, 781–790.

Lawley, D. N. (1943). "A note on Karl Pearson's selection formulae." *Proc. R. Soc. Edinburgh*, **62**, 28–30.

Lindgren, B. L. (1976). *Statistical Theory*, 3rd ed. New York: Macmillan.

Little, R. J. A. (1979). "Maximum likelihood inference for multiple regression with missing values: A simulation study." *J. Roy. Statist. Soc. Ser. B*, **41**, 76–87.

Longley, J. W. (1967). "An appraisal of least squares programs for the electronic computer from the point of view of the user." *J. Am. Statist. Assoc.*, **62**, 819–841.

Louis, T. A. (1982). "Finding the observed information matrix when using the EM algorithm." *J. Roy. Statist. Soc. Ser. B*, **44**, 226–233.

Madansky, A. (1959). "The fitting of straight lines when both variables are subject to error." *J. Am. Statist. Assoc.* **54**, 173–206.

Mallows, C. L. (1973). "Some comments on $C_p$." *Technometrics*, **15**, 661–676.

Mansfield, E. R., J. T. Webster, and R. F. Gunst (1977). "An analytic variable selection technique for principal component regression." *Appl. Statist.*, **26**, 34–40.

Mantel, N. (1970). "Why stepdown procedures in variable selection?" *Technometrics*, **12**, 621–625.

Marler, G. D. (1969). *The Story of Old Faithful*. W. Yellowstone, Wyoming: Yellowstone Library and Museum Association.

Marquardt, D. W. (1970). "Generalized inverses, ridge regression and biased linear estimation." *Technometrics*, **12**, 591–612.

Marquardt, D. W. and R. Snee (1975). "Ridge regression in practice." *Am. Statist.*, **12**, 3–19.

McCullagh, P. and J. Nelder (1983). *Generalized Linear Models*. London: Chapman and Hall.

Miller, R. (1981). *Simultaneous Inference*, 2nd ed. New York: Springer.

Moore, J. A. (1975). "Total Biochemical Oxygen Demand of Animal Manures." Unpublished Ph.D. dissertation, University of Minnesota.

Morgan, J. A. and J. F. Tartar (1972). "Calculation of residual sum of squares for all possible regressions." *Technometrics*, **14**, 317–325.

Mosteller, F. and J. W. Tukey (1977). *Data Analysis and Linear Regression*. Reading, MA: Addison-Wesley.

Myers, R. H. (1971). *Response Surface Methodology*. Boston: Allyn and Bacon.

Narula, S. C. and J. W. Wellington (1977). "Prediction, linear regression and minimum sum of relative errors." *Technometrics*, **19**, 185–190.

Nelder, J. A. and R. W. M. Wedderburn (1972). "Generalized Linear Models." *J. Roy. Statist. Soc. Ser. A*, **135**, 370–384.

Noll, S. L., P. E. Waibel, R. D. Cook, and J. A. Witmer (1984). "Biopotency of methionine sources for young turkeys." *Poult. Sci.* **63**, 2458–2470.

Obenchain, R. L. (1975). "Ridge analysis following a preliminary test of the shrunken hypothesis." *Technometrics*, **17**, 431–445.

Orchard, T. and M. A. Woodbury (1972). "A missing information principle: Theory and applications," in *Proceedings of the Sixth Berkeley Symposium on Mathematical Statistics and Probability*, Vol. 1. Berkeley: University of California.

Pearson, K. (1930). *Life and Letters and Labours of Francis Galton*, Vol. IIIa. Cambridge, UK: Cambridge University Press.

Pereira, B. (1977). "Discriminating among separate models: A bibliography." *Int. Statist. Rev.*, **45**, 163–172.

Picard, R. R. and R. D. Cook (1984). "Cross-validation of regression models." *J. Am. Statist. Assoc.*, **79**, 575–583.

Pregibon, D. (1980). "Goodness of link tests for generalized linear models." *Appl. Statist.*, **29**, 15–24.

Pregibon, D. (1981). "Logistic regression diagnostics." *Ann. Statist.*, **9**, 705–724.

Ralston, A. (1960). *Mathematical Methods for Digital Computers*. New York: Wiley.

Ratkowsky, D. A. (1983). *Nonlinear regression modeling*. New York: Marcel Dekker.

Rencher, A. C. and F. C. Pun (1980). "Inflation of $R^2$ in best subset regression." *Technometrics*, **22**, 49–53.

Renshaw, E. (1958). "Scientific appraisal." *Natl. Tax J.*, **11**, 314–322.

Rolph, J. E. (1976). "Choosing shrinkage estimators for regression problems." *Commun. Statist. Ser. A*, **5**, 789–802.

Royston, J. P. (1982a). "An extension of Shapiro and Wilk's $W$ test for normality to large samples." *Appl. Statist.*, **31**, 115–124.

Royston, J. P. (1982b). "Expected normal order statistics (exact and approximate), Algorithm AS177." *Appl. Statist.*, **31**, 161–168.

Royston, J. P. (1982c). "The $W$ test for normality, Algorithm AS181." *Appl. Statist.*, **31**, 176–180.

Rubin, D. B. (1974). "Characterizing the estimation of parameters in incomplete data problems." *J. Am. Statist. Assoc.*, **69**, 467–474.

Rubin, D. B. (1976). "Inference and missing data." *Biometrika*, **63**, 581–592.

Rubin, D. B. (1980). "Using empirical Bayes' techniques in the Law School validity studies." *J. Am. Statist. Assoc.*, **67**, 801–827.

Ryan, T., B. Joiner and B. Ryan (1985). *Minitab Student Handbook*, 2nd ed. Belmont, CA: Duxbury.

Sandberg, J. S., M. J. Basso, and B. A. Okin (1978). "Winter rain and summer ozone: A predictive relationship." *Science*, **200**, 1051–1054.

Saw, J. G. (1966). "A conservative test for concurrence of several regression lines and related problems." *Biometrika*, **53**, 272–275.

Schatzoff, M., S. Fienberg, and R. Tsao (1968). "Efficient calculations of all possible regressions." *Technometrics*, **10**, 769–779.

Scheffé, H. (1959). *The Analysis of Variance*. New York: Wiley.

Sclove, S. (1968). "Improved estimators for coefficients in linear regression." *J. Am. Statist. Assoc.*, **63**, 596–606.

Sclove, S. (1972). "($Y$ vs $X$) or (log $Y$ vs $X$)?" *Technometrics*, **14**, 391–403.

Searle, S. R. (1971). *Linear Models*. New York: Wiley.

Searle, S. R. (1982). *Matrix Algebra Useful for Statistics*. New York: Wiley.

Seber, G. A. F. (1977). *Linear Regression Analysis*. New York: Wiley.

Shapiro, S. S. and M. B. Wilk (1965). "An analysis of variance test for normality (complete samples)." *Biometrika*, **52**, 591–611.

Smith, B. T., J. M. Boyle, J. J. Dongarra, B. S. Garbow, Y. Ikebe, V. C. Klema, and C. B. Moler (1976). *Matrix Eigensystem Routines—EISPACK Guide*, 2nd ed. Lecture Notes in Computer Science, No. 6. New York: Springer.

Smith, G. and F. Campbell (1980). "A critique of some ridge regression methods (with discussion)." *J. Am. Statist. Assoc.*, **75**, 74–103.

Snee, R. D. (1977). "Validation of regression models. Methods and examples." *Technometrics*, **19**, 415–428.

Sprent, P. (1972). "The mathematics of size and shape." *Biometrics*, **28**, 23–37.

Stein, C. (1956). "Inadmissibility of the usual estimator for the mean of a multivariate normal distribution," in *Third Berkeley Symposium on Probability and Statistics*, Vol. 1. Berkeley: University of California.

Stewart, G. W. (1974). *Introduction to Matrix Computations.* New York: Academic.

Stone, M. (1974). "Cross-validatory choice and assessment of statistical predictions (with discussion)." *J. Roy. Statist. Soc. Ser. B*, **36**, 111–147.

Strawderman, W. (1978). "Minimax adaptive generalized ridge regression estimators." *J. Am. Statist. Assoc.*, **73**, 623–627.

Thisted, R. (1978a). "On generalized ridge regression." Technical Report No. 57, Dept. of Statistics, University of Chicago.

Thisted, R. (1978b). "Multicollinearity, information and ridge regression," Technical Report No. 58, Dept. of Statistics, University of Chicago.

Titterington, D. M. (1978). "Estimation of correlation coefficients by elliptica trimming." *Appl. Statist.*, **27**, 227–234.

Tuddenham, R. D. and M. M. Snyder (1954). "Physical growth of California boys and girls from birth to age 18." *Calif. Publ. Child Develop.*, **1**, 183–364.

Weisberg, H., E. Beier, H. Brody, R. Patton, K. Raychaudhari, H. Takeda, R. Thern, and R. Van Berg (1978). "$s$-dependence of proton fragmentation by hadrons. II. Incident laboratory momenta 30–250 GeV/$c$." *Phys. Rev. D*, **17**, 2875–2887.

Weisberg, S. (1981). "A statistic for allocating $C_p$ to individual cases." *Technometrics*, **23**, 27–31.

Weisberg, S. (1983). "Principles for regression diagnostics and influence analysis, discussion of a paper by R. R. Hocking." *Technometrics*, **25**, 240–244.

Weisberg, S. and C. Bingham (1975). "An analysis of variance test for normality suitable for machine calculation." *Technometrics*, **17**, 133.

West, D. H. D. (1979). "Updating mean and variance estimates: An improved method." *Commun. ACM*, **22**, 532–535.

Wilk, M. B. and R. Gnanadesikan (1968). "Probability plotting methods for the analysis of data." *Biometrika*, **55**, 1–17.

Williams, E. (1959). *Regression Analysis.* New York: Wiley.

Wilm, H. G. (1950). "Statistical control in hydrologic forecasting." *Res. Notes*, **61**, Pacific Northwest Forest Range Experiment Station, Oregon.

Wood, F. S. (1973). "The use of individual effects and residuals in fitting equations to data." *Technometrics*, **15**, 677–695.

Woodley, W. L., J. Simpson, R. Biondini, and J. Berkeley (1977). "Rainfall results 1970–75: Florida area cumulus experiment." *Science*, **195**, 735–742.

# AUTHOR INDEX

# SUBJECT INDEX

Added variable plot, 37–40, 52–54, 129, 143
Additive effects, 171
Adjusted $R^2$, 217
Amazon River, 31
Analysis of covariance, 184
Analysis of variance, 15–19, 48–52
Asymptote, 261
Atkinson's score test, 151–152

Backward elimination, 211, 212, 213, 214
Bayes method:
    James-Stein estimates, 258
    prediction, 238
Bayes rules, 256
Biased regression, 253–259
Binomial distribution, 268–273
BMDP programs, 184, 218, 239, 250, 258, 264, 266
Bonferroni inequality, 116, 117, 213
    $t$ values, 146, 302–305
Box and Cox procedure, 147–152, 159, 160, 162, 167
Box and Tidwell procedure, 152–156

Case statistics, *see* Diagnostic statistics
Causal relationships, 56, 65, 219, 230
Censored data, 244
Centering, 165, 282–283
Central composite design, 167
Cholesky decomposition, 48, 293
Coefficient of determination, 19, 49
Collinearity, 196–203, 211, 251
Comparison of regression lines, 179–185
Computations, 32, 47–48, 59–62, 104–105, 282–283, 292–293
    all possible regressions, 218
    cross validation and PRESS, 239

leaps and bounds, 218
polynomial regression, 164–165
Condition number, 200
Confidence:
    intervals, 20–21, 23
    regions, 97–99
Cook's distance, 118–125, 128, 252
Correlated errors, 159–160
Correlation, partial, 40–41
Correlation coefficient, 9, 278
Criticism, 128–129
Cross product terms, 165
Cross validation, 217, 229, 233, 235
    computations, 239

Dependent variable, *see* Response
Deviance, 270
Diagnostic methods, 128–159
    statistics, 106
Dummy variables, 169–185
Durbin-Watson statistic, 160

Eigenvalues and eigenvectors, 187
Eispack, 187
EM algorithm, 248
Envelope, for probability plots, 159–160
Error function, 272
Errors, 5
    comparison to residuals, 8
    distribution, 6
    fixed and random, 5–6, 276–277
    multiplicative, 141
Experiments, 70
Exponential family, 272
Extrapolation, 228, 235–237

Fitted values, 8, 22

**321**